D1131298

Problem Books in Mathematics

Edited by P. R. Halmos

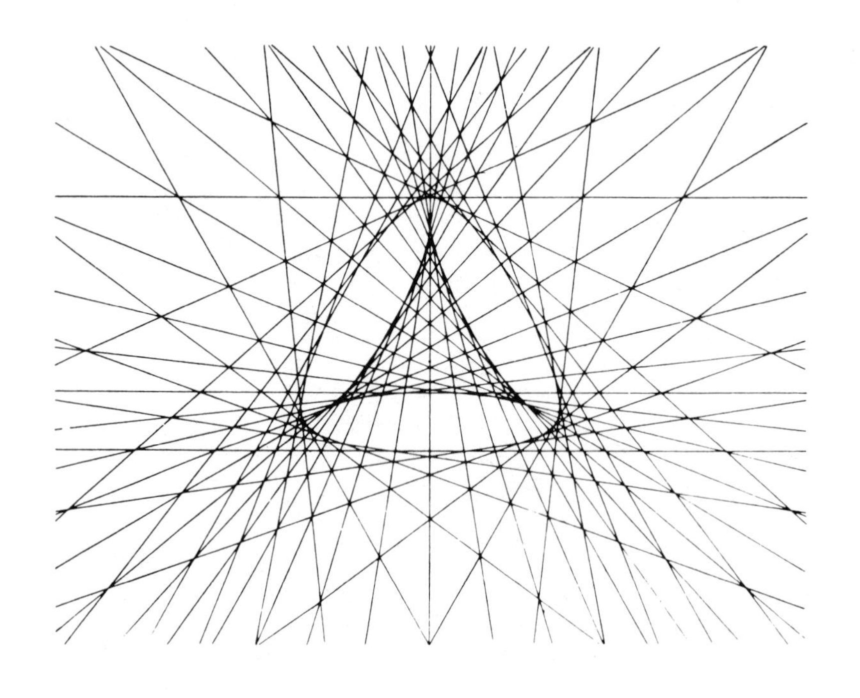

Marcel Berger
Pierre Pansu
Jean-Pic Berry
Xavier Saint-Raymond

Problems in Geometry

Translated by Silvio Levy

With 224 Illustrations

Springer-Verlag
New York Berlin Heidelberg Tokyo

Marcel Berger
U.E.R. de Mathematique
et Informatique
Université Paris VII
75251 Paris, Cedex 05
France

Pierre Pansu
Xavier Saint-Raymond
Centre National de la Recherche
Scientifique
Paris
France

Jean-Pic Berry
P.U.K.
Grenoble
France

Silvio Levy (*Translator*)
Mathematics Department
Princeton University
Princeton, NJ 08540
U.S.A.

Editor

Paul R. Halmos
Department of Mathematics
Indiana University
Bloomington, IN 47405
U.S.A.

AMS Subject Classifications: 51-01, 52-01, 53-01, 00A07

Library of Congress Cataloging in Publication Data
Géométrie, problèmes de géométrie commentés et rédigés.
English.
Problems in geometry.
(Problem books in mathematics)
Translation of: Géometrie, problèmes de géométrie
commentés et rédigés.
Includes index.
2. Geometry–Problems, exercises, etc. I. Berger,
Marcel, 1927- II. Title. III. Series.
QA445.G44513 1984 516′.0076 84-5495

Original French edition: Problèmes de Géométrie, Commentés et Rédigés, CEDIC, Paris, © 1982.

Media conversion by Science Typographers, Medford, New York.
Printed and bound by R.R. Donnelley & Sons, Harrisonburg, Virginia.
Printed in the United States of America.

9 8 7 6 5 4 3 2 1

ISBN 0-387-90971-0 Springer-Verlag New York Berlin Heidelberg Tokyo
ISBN 3-540-90971-0 Springer-Verlag Berlin Heidelberg New York Tokyo

Preface

The textbook *Geometry*, published in French by CEDIC/Fernand Nathan and in English by Springer-Verlag (scheduled for 1985) was very favorably received. Nevertheless, many readers found the text too concise and the exercises at the end of each chapter too difficult, and regretted the absence of any hints for the solution of the exercises.

This book is intended to respond, at least in part, to these needs. The length of the textbook (which will be referred to as [B] throughout this book) and the volume of the material covered in it preclude any thought of publishing an expanded version, but we considered that it might prove both profitable and amusing to some of our readers to have detailed solutions to some of the exercises in the textbook.

At the same time, we planned this book to be independent, at least to a certain extent, from the textbook; thus, we have provided summaries of each of its twenty chapters, condensing in a few pages and under the same titles the most important notions and results used in the solution of the problems. The statement of the selected problems follows each summary, and they are numbered in order, with a reference to the corresponding place in [B]. These references are not meant as indications for the solutions of the problems. In the body of each summary there are frequent references to [B], and these can be helpful in elaborating a point which is discussed too cursorily in this book.

Following the summaries we included a number of suggestions and hints for the solution of the problems; they may well be an intermediate step between your personal solution and ours!

The bulk of the book is dedicated to a fairly detailed solution of each problem, with references to both this book and the textbook. Following the practice in [B], we have made liberal use of illustrations throughout the text.

Finally, I would like to express my heartfelt thanks to Springer-Verlag, for including this work in their Problem Books in Mathematics series, and to Silvio Levy, for his excellent and speedy translation.

Marcel Berger

Contents

Chapter 1

Groups Operating on a Set: Nomenclature, Examples, Applications

1.A Operation of a Group on a Set ([B, 1.1])

Let X be a set; we *denote* by $\mathfrak{S}_X$ the set of all bijections (permutations) f: $X \to X$ from X into itself, and we endow $\mathfrak{S}_X$ with the law of composition given by $(f, g) \mapsto f \circ g$. Thus $\mathfrak{S}_X$ becomes a group, called the *permutation group* or *symmetric group* of X.

Now let G be a group and X a set; we say that G *operates* (or *acts*) *on* X *by* φ if φ is a homomorphism of G into $\mathfrak{S}_X$, i.e. if φ verifies the following axioms:

$$\varphi(g) \in \mathfrak{S}_X, \text{ all } g \in G;$$

$$\varphi(gh) = \varphi(g) \circ \varphi(h), \text{ all } g, h \in G$$

(the law of composition of G is written multiplicatively). We will generally use the *abbreviation* $g(x)$ instead of $\varphi(g)(x)$.

1.B Transitivity ([B, 1.4])

The operation φ of G on X is said to be *transitive* if for every pair (x, y) there is a g such that $g(x) = y$. If G operates transitively on X, we say also that G is a *homogeneous space* (*for* G).

1.C The Erlangen Program: Geometries

The situation where a group G operates transitively on a set X is a very frequent one in geometry. This in fact led Felix Klein, in his famous Erlangen program of 1872, to define a *geometry* as the study of the properties of a set which are invariant under its automorphism group (which is assumed to be transitive).

Note that this group can vary for the same set, thus giving rise to different geometries. For an affine space X, for instance, we can consider different groups that operate on it transitively: the group of translations, the group of dilatations (appropriate for the geometry of parallels), the group of all affine transformations (appropriate for barycenters), that of unimodular affine transformations. See 2.A, 2.C and problem 2.4.

If X is also Euclidean, we can in addition consider the group of proper motions (appropriate for metrical problems involving orientation) and the group of isometries (appropriate for metrical problems); see 9.A, 9.D.

The above notion of geometry is quite restrictive; cf. for instance convexity in chapters 11 and 12. Recently, mathematicians have started studying situations in which a group operates non-transitively but with orbits that are "big enough".

1.D Stabilizers ([B, 1.5])

For a group G acting on a set X the *stabilizer* or *isotropy subgroup* G_x of a point $x \in X$ is the subgroup of G defined by

$$G_x = \{ g \in G: g(x) = x \}.$$

If G is transitive, all stabilizers are essentially the same, being all conjugate to one another by some element of G.

Again if G acts transitively on X, for an arbitrary $x \in X$ the set X is isomorphic (as a set) to the set of left cosets G/G_x, i.e. the quotient of G by the equivalence relation "$g \sim h$ if and only if $g = hk$ for some $k \in G_x$."

This, at least in theory, allows one to study X algebraically. For instance, if G is finite and acts transitively on X, we have

$$\# X = \# G / \# G_x,$$

where $\#(\cdot)$ denotes the cardinality of a set.

The operation of G on X is said to be *simply transitive* if it is transitive and if for some (hence all) x in X the stabilizer G_x is trivial (i.e. contains just the identity).

1.E Orbits; the Class Formula ([B, 1.6])

If G is *not* transitive on X, it is natural to introduce the subsets of X where G is transitive, namely, the *orbits* of G: the orbit of $x \in X$ (under the action of G) is the set $O(x) = \{ g(x): g \in G \}$ of all the different images of x under elements of G.

In general, for a fixed group the nature of the orbits can vary with x; see problem 1.3.

The *class formula* says that if A is a set which contains exactly one element in each orbit of X under the action of G, and if both G and X are finite, the following holds:

$$\#X = \sum_{x \in A} (\#G/\#G_x).$$

1.F Regular Polyhedra ([B, 1.8])

One can show that there are exactly five types of regular polyhedra (which, however, derive from only three groups): the tetrahedron, the cube, the octahedron, the dodecahedron and the icosahedron. This is done by using the class formula and the fact that a rotation of three-dimensional Euclidean space (distinct from the identity) possesses exactly two fixed points on the unit sphere (see also 12.C).

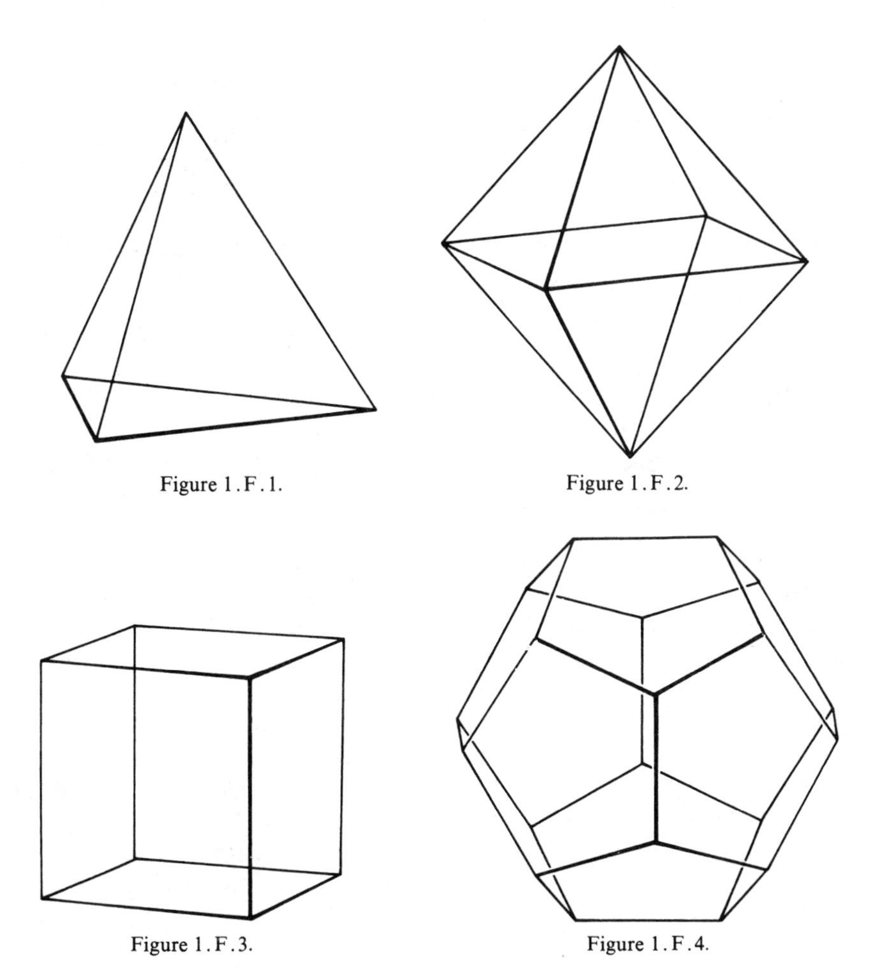

Figure 1.F.1.

Figure 1.F.2.

Figure 1.F.3.

Figure 1.F.4.

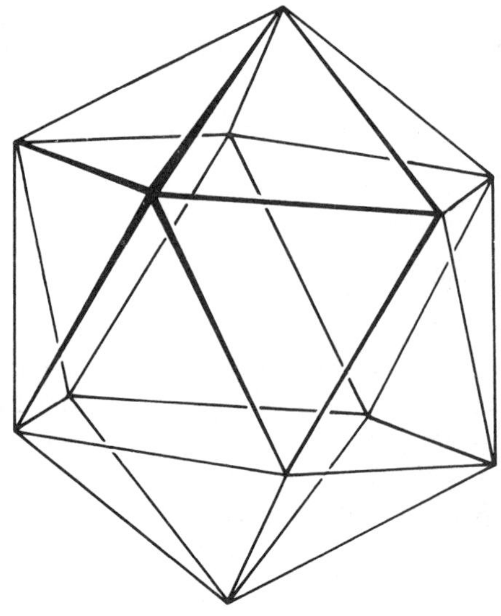

Figure 1.F.5.

1.G Plane Tilings and Crystallographic Groups ([B, 1.7])

We will now apply the general notions above to plane tilings; for more detailed explanations, refer to 1.H below.

Definition. A group G is called a *crystallographic group* (or *tesselation group*) for the Euclidean plane E if it can be obtained in the following way: G is a subgroup of the group $\mathrm{Is}^+(E)$ of proper motions (i.e. orientation-preserving isometries) of E, and there is a connected compact subset P of E, with non-empty interior $\mathring{P}$, such that the following two conditions are satisfied:

$$\bigcup_{g \in G} g(P) = E, \text{ and } g(\mathring{P}) \cap h(\mathring{P}) \neq \phi \text{ implies } g = h.$$

Observe that it is possible to *tile* the plane in such a way that there is no group that acts transitively on the tiles; see 1.H and problem 1.4. Notice also that there are many more types of tilings than there are groups G: see 1.H and [B, 1.7.7, 1.9.13].

We conclude that the fundamental step in the classification of tilings is finding all possible groups G, up to conjugation by an affine transformation of the plane. It can be shown ([B, 1.7]) that there are only *five* possible groups, corresponding to the five figures on page 5.

If we look for groups G not in $\mathrm{Is}^+(E)$ but, more generally, in $\mathrm{Is}(E)$ (i.e. if we allow the tiles to be *turned over*), we find another twelve additional groups (see figures in [B, 1.7.6]).

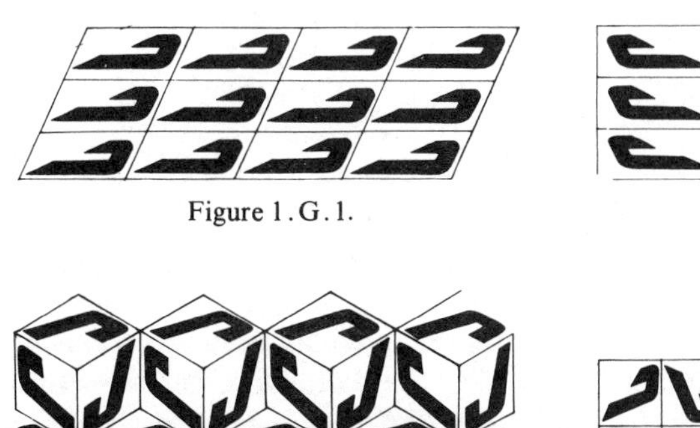

Figure 1.G.1.

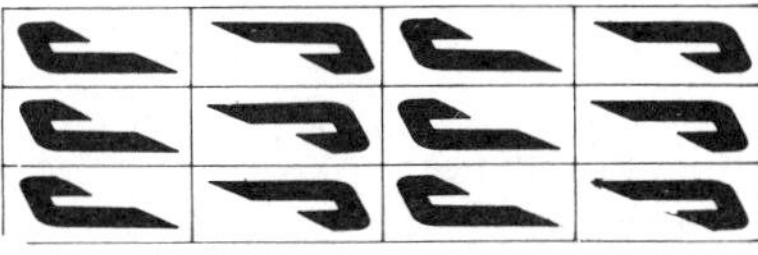

Figure 1.G.2.

Figure 1.G.3.

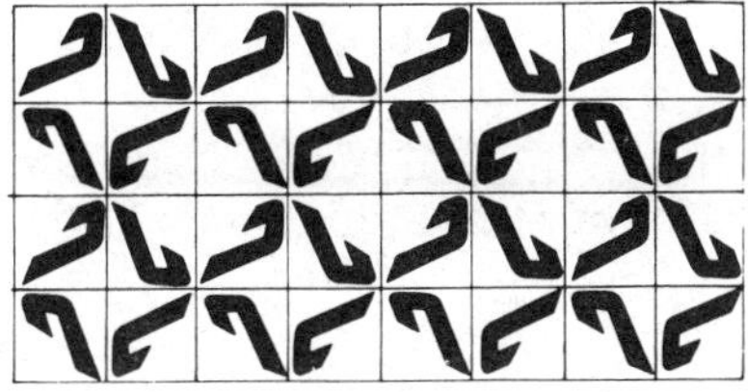

Figure 1.G.4.

Figure 1.G.5.

Taken from Y. Bossard: *Rosaces, frises et pavages*, CEDIC, 1977.

The proof of the classification consists initially in showing that the set of translations in G is a *lattice*, which means all the translations are given by the family of vectors $au + bv$, where $\{u, v\}$ is a fixed basis of the plane, and a and b range over the integers. Then we consider the elements which are not translations, i.e. rotations, whose structure is given in 9.C.

Finally, let us point out that the definition given above refers only to tilings obtained by repeating *a single tile*, but it is possible to work with a finite number of initial tiles (see for instance problem 1.4).

1.H More about Tilings; Exercises

The following remarks and illustrative exercises are intended to clarify the notion of a tiling, which is more complex than it might appear at first sight.

1.H.1

A tiling does not necessarily possess an underlying group; i.e., it is possible to fill the plane with sets $g(P)$, where P is a connected tile and g runs through a subset of the group of plane isometries, without this subset being a subgroup. In other words, the subgroup of isometries that leave invariant the full tiling (called the *invariance subgroup*) does not necessarily act transitively on the set of tiles. Non-periodic tilings (see problem 1.4) are a trivial example of this fact, in that the invariance subgroup is finite, so it had better not act transitively. But there are more "regular" examples: in the periodic tiling below, for instance, the invariance subgroup, though far from trivial, does not act transitively because P cannot be taken to P'.

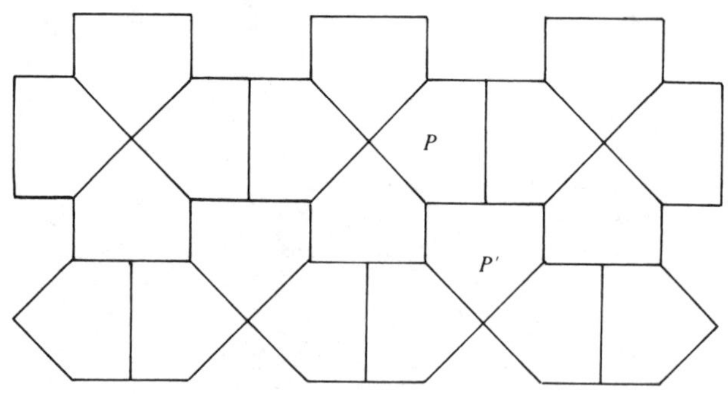

Figure 1.H.1.

1.H.2

From now on we will use the word *tiling* only when the invariance subgroup acts transitively. This means that any two tiles look the same, not only in shape but also in their relationship to their neighbors.

It is important to realize that in 1.G we studied the notion of a crystallographic group, not that of tilings in the visual sense. Here are some examples of tilings whose associated groups are the same, but which nevertheless appear essentially different:

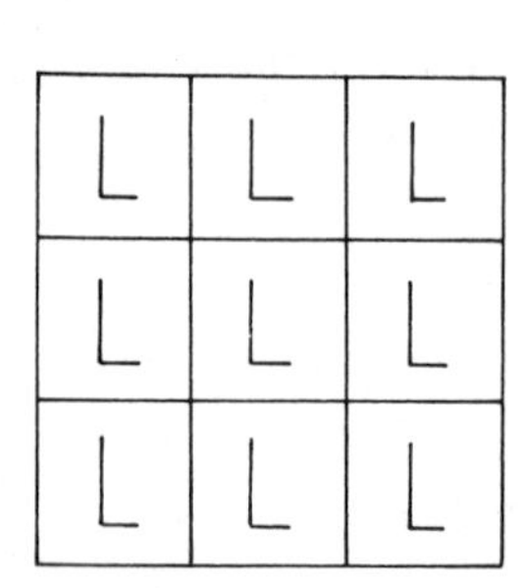

Figure 1.H.2.1.

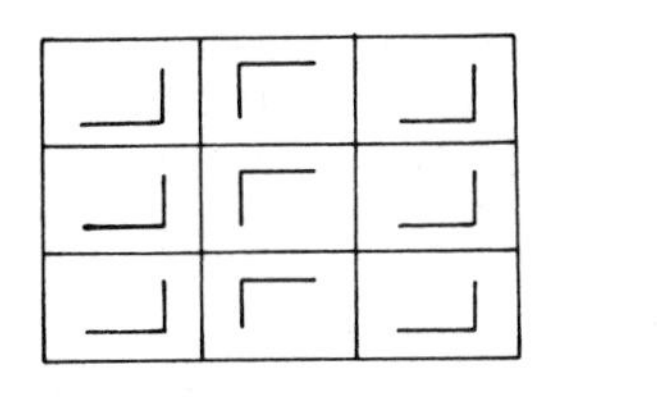

Figure 1.H.2.2.

The two figures 1.H.2.1 correspond to the group of figure 1.G.1, and figures 1.H.2.2 correspond to the group of figure 1.G.2.

1.H.3

If we want to study a tiling from the pictorial point of view, we can associate with it the following structures:

— the invariance group C, i.e. the group of isometries that leave the tiling globally invariant. This is one of the crystallographic groups.
— the subgroup $S(P)$ of C leaving a given tile P invariant. Notice that this group is the same (up to conjugacy) for every tile, so we can talk about the group $S(P)$ without specifying P. On the other hand, $S(P)$ is a subgroup of C, which limits the number of possibilities.
— the *degree* of each vertex (the degree or *valency* of a vertex is the number of tiles that share it). For an enumeration of the 11 possibilities for the degrees, see [B, 1.9.13].

The degrees and the group C give rise to classifications that are unrelated to each other. We have just seen two examples of tilings with the same group C but different degrees. To illustrate the converse, here is an example of two tilings with the same degree but different underlying groups:

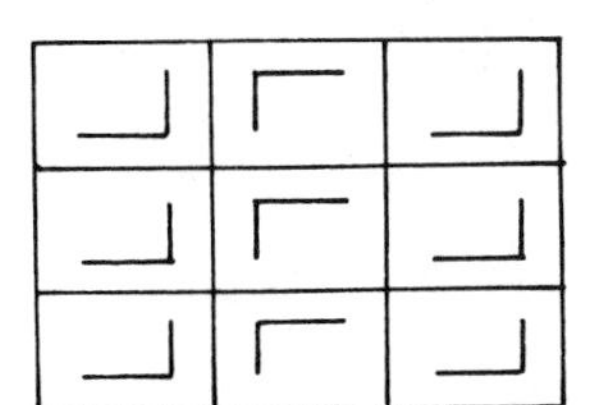
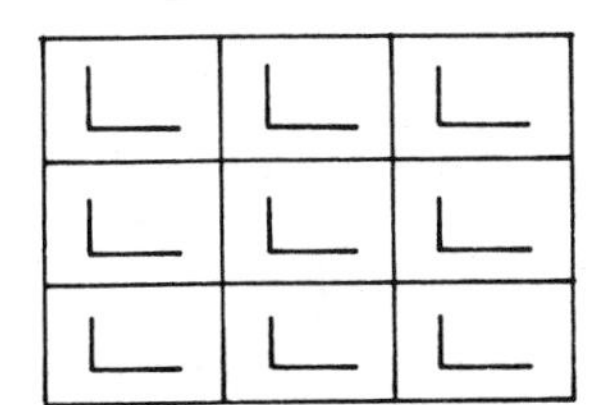

Figure 1.H.3.

So any classification of tilings must account independently for these two criteria. In fact, a more refined argument shows that even if C, $S(P)$ and the degrees are identical, two tilings can still display different patterns. We are thus led to define adjacency relations between the oriented boundary of a tile and that of its neighbors. This classification arises naturally, the more so when we verify that it is finer than a classification based on C, $S(P)$ or the degrees, or any combination of these three criteria; it divides all tilings into 93 patterns (see [B, 1.7.7.8]).

1.H.4 Remark on Tile Markings

Here and elsewhere in the book we have used markings, or drawings, on the tiles, to reveal their orientation. It is reasonable to ask if we could have done without them, by considering only the forms of the tiles. It is often possible to draw the tiles with an irregular shape that bars certain symmetries. In fact, from the 93 tiling patterns mentioned above, only 12 cannot be represented by means of unmarked tiles. So if one is interested only in the shape of the tiles, there are only 81 patterns left.

Exercise 1. Find representations by means of unmarked tilings for the five crystallographic groups that are contained in $\mathrm{Is}^+(E)$ (see 1.G).

1.H.5

Finally, we remark that the connection between the notion of a crystallographic group and that of a tiling is not canonical. We have defined the group by means of the plane tiling it generates when we make it act on a well-defined previously chosen tile; in this process, there is no reason why all the vertices should be isometric, or even why they should all have the same structure. (We say that two vertices are *isometric* if there is an isometry taking a neighborhood of one onto a neighborhood of the other. In the figure below, S_1 and S_2 are not isometric.)

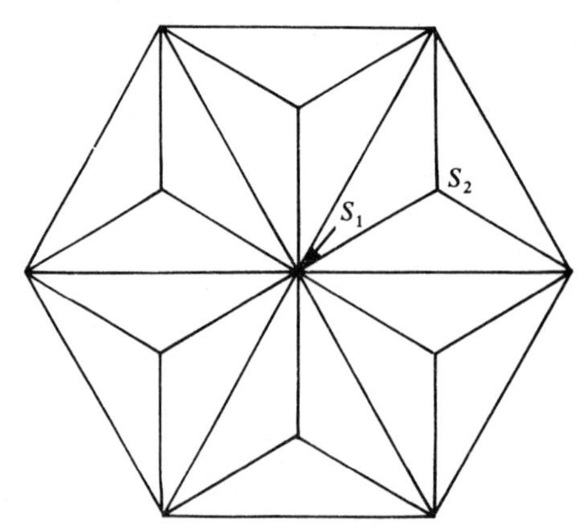

Figure 1.H.5.

We might equally well have the group act on a previously chosen *vertex*, thus obtaining a plane tiling that can contain different tiles. This kind of tiling is called *isogonal*, whereas the other kind is called *isohedral*. Using a classification based on an adjacency relation which is dual to that defined for isohedral tilings, we obtain again 93 types of isogonal tilings; of these, two are not representable by unmarked tilings (it is a bit more difficult, but still possible, to justify the marking of vertices in the same way as for tiles). Thus there are 91 (unmarked) isogonal tilings.

Note. Only tiles and vertices can give rise to classifications; the edges are always sided by the same number of tiles (namely, two), and bounded by the same number of vertices (again two).

Exercise 2. Find representations of the five crystallographic subgroups of $\mathrm{Is}^+(E)$ by means of unmarked isogonal tilings.

Problems

1.1 TRIANGLES AND QUADRILATERALS ([B, 1.9.14]). Does any triangle tile the plane? Any convex quadrilateral? Any quadrilateral?

1.2 FUNDAMENTAL DOMAIN OF A TILING ([B, 1.9.12]). Let $G \subset \mathrm{Is}(E)$ be a subgroup of the group of isometries of an affine Euclidean plane; we assume all orbits of G are discrete subgroups of E. For a fixed $a \in E$, show that G and the set P defined by

$$P = \{x \in E \colon \mathrm{d}(x, a) \leq \mathrm{d}(x, g(a)) \forall g \in G\}$$

verify the axioms of a crystallographic group.

1.3 THE FIVE ORIENTATION-PRESERVING CRYSTALLOGRAPHIC GROUPS ([B, 1.9.4]). For each of the five crystallographic groups containing only proper motions, find the following: the order of the stabilizers; the structure of the group; the different types of orbits; a presentation for the group (cf. [B, 1.8.7]).

1.4 THE ROBINSON NON-PERIODICAL TILING ([B, 1.9.16]). Show that with the six tiles below it is possible to tile the plane. Show also that any tiling with these tiles can never be periodic, i.e. its group of isometries cannot contain a non-trivial translation.

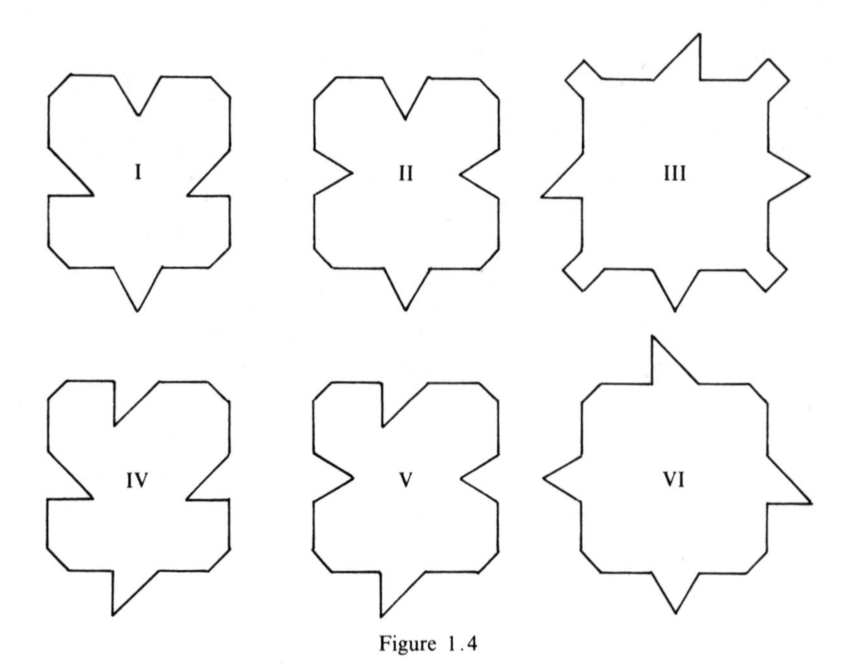

Figure 1.4

Chapter 2

Affine Spaces

All fields of scalars will be commutative.

2.A Affine Spaces; Affine Group ([B, 2.1, 2.3])

An affine space X is nothing more than a vector space under the action of the group generated by the linear automorphisms and the translations; this group is called the *affine group of* X and is *denoted* by $\mathrm{GA}(X)$. Its elements are the (affine) automorphisms of X, or again the affine maps of X into itself. The elements of X are called *points*.

This amounts to eliminating the privileged role previously played by the origin of X, and making all points of X equal. There are of course more sophisticated definitions of the notion of affine spaces; see [B, 2.1.1, 2.1.6].

The other side of the coin is that now the elements of X are not vectors anymore, but merely points, so they cannot be added or multiplied by scalars any longer! The calculations which are still possible to perform in X (we shall distinguish the affine space X from the vector space that gave rise to it, which we will denote by $\vec{X}$) are the following: for $x, y \in X$ we can take the *midpoint* $(x+y)/2$ of x, y in X (if the field has characteristic different from 2); and for $x, y \in X$ we can find the vector $y - x = \overrightarrow{xy}$ in $\vec{X}$. And if x, y, z are on the same line and $x \neq y$ we can calculate the ratio $\overrightarrow{xz}/\overrightarrow{xy}$, which is a scalar.

As a result, we can also perform calculations of the type $x = a + \lambda\vec{u}$, where a and x are in X, the vector $\vec{u}$ is in $\vec{X}$, and λ is a scalar.

Last but not least, we can fix some $a \in X$ and consider X as a vector space with origin a; this *vectorialization* of X at a will be denoted by X_a. A generalization of the notion of midpoint, which is still independent of any vectorialization, is the *barycenter*; we shall study it in chapter 3.

2.B Affine Maps ([B, 2.3])

Given two affine spaces X and Y (over the same field of scalars), we define an *affine map* from X into Y by means of their underlying vector spaces $\vec{X}$ and $\vec{Y}$. The set of such objects is *denoted* by $A(X; Y)$. For instance, if $Y = K$ is the field of scalars, an element of $A(X; K)$ will be called an *affine form* (which is essentially a linear form with a constant added).

For each affine map f there is a well-defined corresponding linear map $\vec{f}$ from $\vec{X}$ into $\vec{Y}$.

An (affine) *automorphism* of X is a bijective affine map; its inverse is necessarily affine. The *special* (or *unimodular*) affine group $\mathrm{SA}(X)$ consists of all affine automorphisms f of X such that the determinant of $\vec{f}$ equals 1. See problem 2.4 for the differential geometry of this situation.

The *translations* of X are the maps $x \mapsto x + \vec{\xi}$, where $\vec{\xi} \in \vec{X}$; the translation by $\vec{\xi}$ is *denoted* by $t_{\vec{\xi}}$.

2.C Homotheties and Dilatations ([B, 2.3.3])

Apart from translations, the simplest affine automorphisms are the homotheties. The *homothety of center $a \in X$ and ratio $\lambda \in K$, denoted by $H_{a,\lambda}$*, is defined by

$$x \mapsto a + \lambda \overrightarrow{ax}$$

Together with the translations, the set of homothethies forms a group, called the group of *dilatations*. The next section gives their geometrical characterization.

2.D Subspaces; Parallelism ([B, 2.4])

The *subspaces* (or, more precisely, *affine subspaces*) of an affine space X are the images under translations of vector subspaces of some vectorialization of X. The *direction* $\vec{Y}$ of the subspace Y is the vector subspace which gives rise to it. Two subspaces Y, Z are called *parallel* if $\vec{Y} = \vec{Z}$. One-dimensional (resp. two-dimensional) subspaces are called (straight) *lines* (resp. *planes*); we call Y a *hyperplane* of X if $\vec{Y}$ is a hyperplane (i.e. a codimension-1 subspace) of $\vec{X}$.

Dilatations can be characterized as follows: A bijection f is a dilatation if and only if for every line D the image $f(D)$ is a line parallel to D.

Calculation on subspaces is rendered possible by the fact that a hyperplane can be defined as the kernel of a non-constant affine form. For higher-codimension subspaces, we can take more than one affine form; but we can also, especially for lines, take parametric representations, such as $\{a+\lambda\vec{u}\colon \lambda\in K\}$ for the line passing through a and having direction $K\cdot\vec{u}$.

The smallest (affine) subspace of X which contains a subset A of X is called the subspace *spanned* by A, and is *denoted* by $\langle A\rangle$. In particular, if x and y are distinct points, $\langle x, y\rangle$ will denote the line passing through the two.

What are the involutive automorphisms of an affine space? (*Involutive* means that composition with itself gives the identity.) Notice first that if f belongs to $\mathrm{GA}(X)$ and $f^2=\mathrm{Id}_X$, then f has at least one fixed point a; just take $a=(x+f(x))/2$ for an arbitrary x (always assuming the characteristic is different from 2).

Considering now the vectorialization of X at a, we know that X_a is a direct sum $X_a=U\oplus V$, where f is the identity on U and minus the identity on V: $f=\mathrm{Id}_U-\mathrm{Id}_V$. Consequently, our involution will be the *affine reflection through Y and parallel to Z*, where Y and Z are two *complementary* subspaces (i.e. $\vec{Y}\oplus\vec{Z}=\vec{X}$). Observe also that $f(x)$ is uniquely determined by the two conditions $\overrightarrow{xf(x)}\in\vec{Z}$ and $(x+f(x))/2\in Y$.

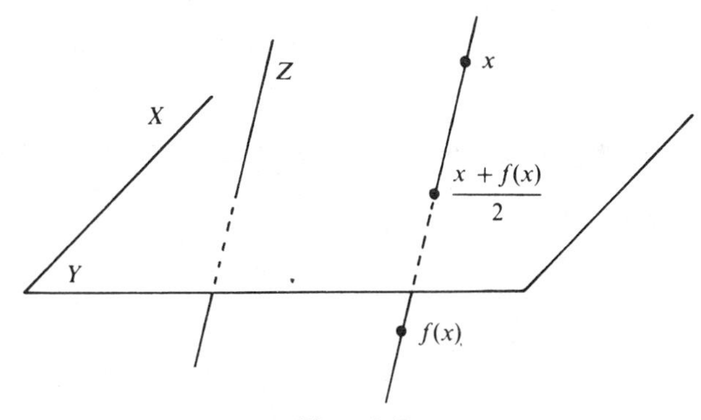

Figure 2.D.

2.E Independence; Affine Frames ([B, 2.2, 2.4])

If x, y, z are three points in X, is the subspace $\langle x, y, z\rangle$ a point, a line or a plane? By definition, points $\{x_i\}_{i=0,1,\ldots,k}$ of X are said to be (affinely) *independent* if the subspace $\langle x_0, x_1,\ldots, x_k\rangle$ which they span has dimension k. (The *dimension* of an affine subspace X is of course the dimension of $\vec{X}$.) If X has dimension n, a *simplex* in X is a set of $n+1$ independent points in X. We associate to each simplex $\{x_i\}_{i=0,1,\ldots,k}$ an *affine frame*, by taking x_0 as the origin of $\vec{X}$ and $\{\overrightarrow{x_0x_i}\}_{i=1,\ldots,k}$ as a basis of $\vec{X}$. The *coordinates* of a point

$x \in X$ are the scalars λ_i $(i=1,\ldots,n)$ such that

$$x = x_0 + \sum_i \lambda_i \overrightarrow{x_0 x_i} \quad \text{or} \quad \overrightarrow{x_0 x} = \sum_i \lambda_i \overrightarrow{x_0 x_i}.$$

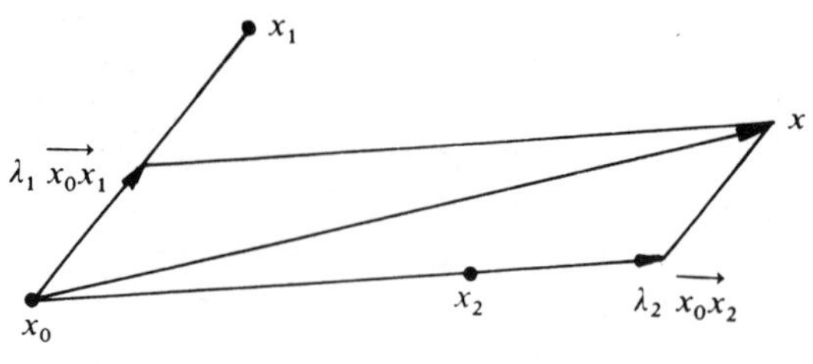

Figure 2.E.

An affine map is determined by its values on a simplex; in particular, for any two simplices in a finite-dimensional X, there is an automorphism taking one onto the other.

2.F The Fundamental Theorem of Affine Geometry ([B, 2.6])

This is the only delicate result following from the preceding notions. It concerns the set-theoretical bijections between two affine spaces X, X' (not taken *a priori* over the same field) that map lines into lines. Observe first that this condition is vacuous in dimension 1 and that, in the complex case, it is satisfied for maps which are not affine (for instance, $(z_1, z_2) \mapsto (\overline{z_1}, \overline{z_2})$). Our result says exactly that those two cases typify the only possible exceptions.

More precisely, if X and X' are defined over the fields K and K', a map f: $X \to X'$ is called *semi-affine* if the vectorialization $\vec{f}: X_a \to X'_{f(a)}$ of f is *semi-linear*, i.e. if there is field isomorphism σ: $K \to K'$ such that

$$\vec{f}(\lambda x + \mu y) = \sigma(\lambda)\vec{f}(x) + \sigma(\mu)\vec{f}(y)$$

for every $x, y \in X_a$ and every $\lambda, \mu \in K$.

The fundamental theorem of affine geometry says that if f: $X \to X'$ is a bijection that maps lines into lines, and if moreover X and X' have same finite dimension n with $n \geq 2$, then f is semi-affine.

2.G Finite-dimensional Real Affine Spaces ([B, 2.7])

From now on X will be a finite-dimensional affine space over the field of real numbers. These spaces (especially in dimension 1, 2 and 3) are those studied in classical geometry. They are especially rich and interesting as they relate to our physical world.

To begin with, X has a canonical topology, so we can talk about open sets, closed sets, compact sets and so on (see for instance problem 2.2). Differential calculus also applies in such spaces (see problem 2.4).

An important notion is that of a *half-space*; if Y is a hyperplane of X, its complement $X \backslash Y$ has exactly two connected components, which are called the *open half-spaces* determined by Y. Their closures are the corresponding *closed half-spaces*. This notion is fundamental in the study of convexity: see 11.B, 12.A.

Up to a positive scalar, a (finite-dimensional real) affine space possesses a canonical measure, called the *Lebesgue measure*. For example, if μ is such a measure, arising from the measure $\vec{\mu}$ on $\vec{X}$, we have the following definition for the *centroid* of a compact K in X with non-empty interior: If χ_K denotes the characteristic function of K and $\mu(K) = \int_X \chi_K \mu$ is the measure of K under μ, it can be shown that the vector integral

$$I_\mu(a) = \int_{x \in X} \chi_K \overrightarrow{ax}\, \vec{\mu}$$

is such that the point

$$a + (\mu(K))^{-1} I_\mu(a)$$

is independent of $a \in X$. This point is called the *centroid* of K; see problem 2.3.

A finite-dimensional vector space, and consequently a finite-dimensional affine space, can be *oriented*; this essentially amounts to choosing a basis and declaring it to be *positive*. Any other basis obtained from the first one by a linear automorphism of positive determinant is also positive. There are exactly two possible orientations, and neither is "canonical"; the choice is arbitrary.

Whether or not the real affine space X is oriented, we can always consider the maps $\vec{f} \in \mathrm{GL}(X)$ having positive determinant: $\det \vec{f} > 0$. We say that they preserve orientation (whichever one we choose) and *denote* by $\mathrm{GA}^+(X)$ the subgroup of $\mathrm{GA}(X)$ formed by such maps. Its complement is *denoted* by $\mathrm{GA}^-(X)$, and contains all maps f such that $\det \vec{f} < 0$.

Problems

2.1 THE THEOREMS OF CEVA AND MENELAUS ([B, 2.8.1, 2.8.2]). Let $\{a, b, c\}$ be a triangle in an affine plane, and let $a' \in \langle b, c \rangle$, $b' \in \langle c, a \rangle$, $c' \in \langle a\ b \rangle$ be three points on the sides of this triangle. Prove that the three lines $\langle a, a' \rangle$, $\langle b, b' \rangle$, $\langle c, c' \rangle$ are concurrent (or parallel) if and only if we have

$$\frac{\overrightarrow{a'b}}{\overrightarrow{a'c}} \frac{\overrightarrow{b'c}}{\overrightarrow{b'a}} \frac{\overrightarrow{c'a}}{\overrightarrow{c'b}} = -1 \text{ (Theorem of Ceva)}$$

With the same data, show that the three points a', b', c' are collinear if and only if

$$\frac{\overrightarrow{a'b}}{\overrightarrow{a'c}}\frac{\overrightarrow{b'c}}{\overrightarrow{b'a}}\frac{\overrightarrow{c'a}}{\overrightarrow{c'b}}=1 \text{ (Theorem of Menelaus).}$$

2.2 CONNECTEDNESS OF COMPLEMENTS OF SUBSPACES ([B, 2.8.8]). Show that if X is a finite-dimensional real affine space and Y is a subspace the complement $X\setminus Y$ is connected if $\dim Y \leq X-2$. Is $X\setminus Y$ simply connected? (See 18.A).

2.3 ASSOCIATIVE PROPERTIES OF CENTROIDS ([B, 2.8.11]). Let K be a compact subset of a finite-dimensional real affine space such that $\mathring{K} \neq \phi$; let H be a hyperplane of X and X', X'' its two closed half-spaces. We assume also that H is such that

$$\mathring{K}' \neq \phi \text{ and } \mathring{K}'' \neq \phi,$$

where $K'=K\cap X'$ and $K''=K\cap X''$. Show (cf. 3.A) that $\mathrm{cent}(K)$ is the weighted average of $\mathrm{cent}(K')$ and $\mathrm{cent}(K'')$ with weights $\mu(K')$ and $\mu(K'')$. or, in the terminology of 3.A, the barycenter of the family $\{(\mathrm{cent}(K'),\mu(K')),(\mathrm{cent}(K''),\mu(K''))\}$. Deduce that if in addition K is convex, then $\mathrm{cent}(K)\in\mathring{K}$.

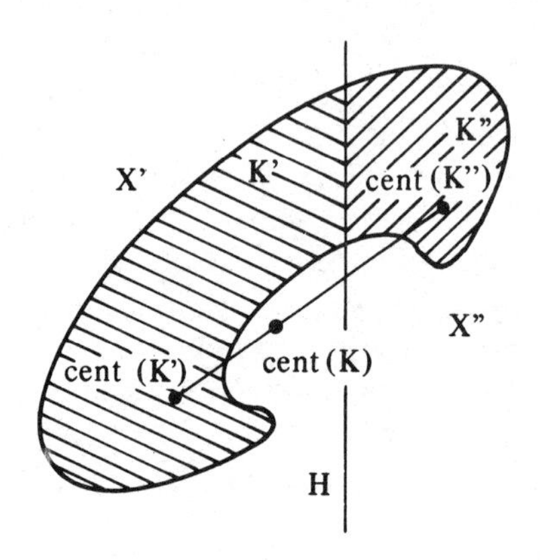

Figure 2.3.

2.4 EQUIAFFINE LENGTH AND CURVATURE ([B, 2.8.12]). Let X be a real affine plane, and *fix a basis for* $\vec{X}$. We can then define the determinant $\det(\vec{u},\vec{v})$ of any two vectors in $\vec{X}$ relative to this basis by writing the 2×2 matrix whose columns are $\vec{u},\vec{v}$ in this basis. Let $c\colon [a,b]\to X$ be a differentiable curve of class C^3 in X. The *equiaffine length* of c (in X, relative to this basis) is the real number

$$\int_a^b \left(\det(\vec{c}'(t),\vec{c}''(t))\right)^{1/3} dt.$$

Show that c has same equiaffine length as $f \circ c$ for any $f \in \mathrm{SA}(X)$, where we put $\mathrm{SA}(X) = \{f \in \mathrm{GA}(X) \colon |\det \vec{f}| = 1\}$ (see 2.B).

Show that we can reparametrize c by its equiaffine length if for every t we have $\det(\vec{c}'(t), \vec{c}''(t)) \neq 0$. The *equiaffine curvature* is the number $K = \det(\vec{c}'', \vec{c}''')$ when c is parametrized by its equiaffine length; show that this curvature, too, is invariant by $\mathrm{SA}(X)$. Find the equiaffine length and curvature for an ellipse, a parabola or a hyperbola in X (always fixing a basis). In the same way that a curve in the Euclidean plane is determined up to an isometry by giving the curvature as a function of the arclength (see for example M. P. do Carmo, *Differential Geometry of Curves and Surfaces*, Prentice-Hall 1976, p. 22, or M. Spivak, *Differential Geometry*, Publish or Perish 1970, vol. 2, p. 1-1), show that a curve in X is determined, up to an element of $\mathrm{SA}(X)$, by giving the equiaffine curvature as a function of its equiaffine arclength.

Chapter 3
Barycenters; the Universal Space

3.A Barycenters ([B, 3.4])

We have seen that in an affine space X we cannot add points or multiply them by scalars. However, there is an operation on points which generalizes the notion of midpoint $(x+y)/2$ (cf. 2.A): taking the barycenter.

If $\{(x_i, \lambda_i)\}_{i=1,\dots,n}$ is a family of points x_i in X, each together with a coefficient (or "mass") λ_i in the field of scalars K, and if $\Sigma_i\lambda_i = 1$, then the point

$$x = \sum_i \lambda_i x_i$$

is well-defined in X. Expressed in an arbitrary frame, its coordinates are defined by the linear combinations $\Sigma_i\lambda_i x_{ij}$ of the coordinates of the x_i. We say that x is the *barycenter of the* x_i with the masses λ_i. For instance, $\{(x,1/2), (y,1/2)\}$ gives us the midpoint of x and y.

If x and y are distinct points, the barycenter $g = \lambda x + (1-\lambda)y$ is located on the line $\langle x, y\rangle$ and is well defined by $\lambda = \overrightarrow{gy}/\overrightarrow{xy}$.

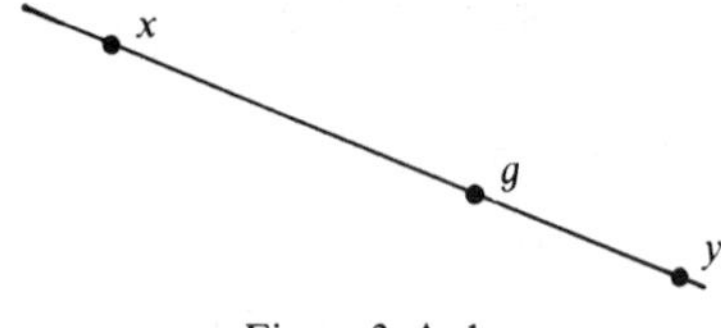

Figure 3.A.1.

More generally, if all the λ_i are equal to $1/n$ (so in particular the characteristic of K cannot divide n), the point $(x_1 + \cdots + x_n)/n$ is called the *centroid of the* x_i. This notion is obviously related to the center of mass in physics; see 2.G, problems 2.3, 3.2 and 12.3, or [B, 2.7.5.6].

The *barycentric subdivision* of a simplex (cf. 2.E or [B, 2.4.7]) is obtained by adding to the simplex its barycenter and the barycenters of all its subsimplices; see problem 3.3.

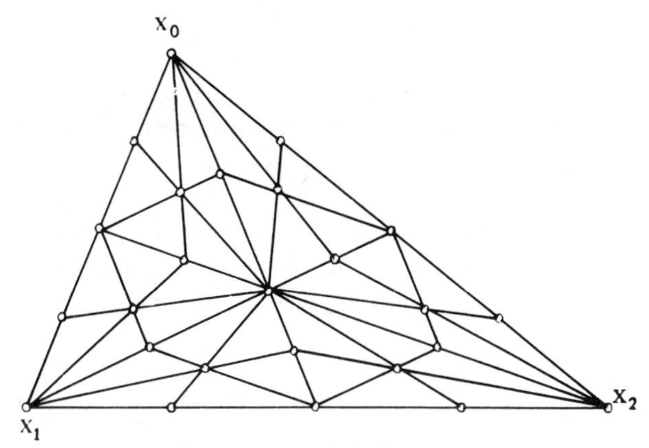

Figure 3.A.2.

A map is affine if and only if it preserves barycenters; the subspace spanned by $\{x_i\}_{i=1,\ldots,k}$ is exactly the set of all the barycenters of these points when the masses take on all possible values ([B, 3.5]).

In the more general case when $\Sigma_i\lambda_i \neq 1$ but nevertheless $\Sigma_i\lambda_i \neq 0$, it is still possible to define the barycenter of the family $\{(x_i, \lambda_i)\}_{i=1,\ldots,n}$ as

$$x = \frac{\Sigma_i\lambda_i x_i}{\Sigma_i\lambda_i} = \sum_i \left(\frac{\lambda_i}{\Sigma_j\lambda_j} \right) x_i .$$

In this case it is logical to ascribe to x the total mass $\Sigma_i\lambda_i$.

3.B Associativity of Barycenters ([B, 3.4.8])

The following is a simple result, but nevertheless rich in consequences. Consider the family $\{(x_i, \lambda_i)\} \cup \{(y_j, \mu_j)\}$, the union of two families, and assume

$$\sum_i \lambda_i \neq 0, \quad \sum_j \mu_j \neq 0, \quad \sum_i \lambda_i + \sum_j \mu_j \neq 0.$$

Then, denoting by g (resp. h) the barycenter of the family $\{(x_i, \lambda_i)\}$ (resp. $\{(y_j, \mu_j)\}$), we have that the barycenter of the union family is equal to the barycenter of the two points $\{(g, \Sigma_i\lambda_i), (h, \Sigma_j\lambda_j)\}$. (This is intuitive if you consider it from the physical point of view.)

It follows classically that: First, the three medians of a triangle intersect, and their intersection point is two thirds of the way from the vertices. Then, the midpoints of the sides of a quadrilateral form a parallelogram ([B, 3.4.10]).

3.C Barycentric Coordinates ([B, 3.6])

Let $\{x_i\}_{i=0,1,\ldots,n}$ be a simplex (cf. 2.E) in an affine space X of finite dimension n. Then every x in X can be uniquely written as $x=\Sigma_i\lambda_i x_i$ with $\Sigma_i\lambda_i=1$. The $n+1$ scalars which are thus uniquely determined are called the *barycentric coordinates of* x (in the simplex being considered).

In the case when $K=\mathbf{R}$, the set $\{\lambda x+(1-\lambda)y\colon \lambda\geq 0\}$ of barycenters of the pair (x, y) with positive mass is called the *segment* defined by x and y and is *denoted* by $[x, y]$. We will encounter this notion again, as well as the more general case of many positive masses, when we study convexity (11.A).

3.D A Universal Space ([B, 3.1, 3.2])

The preceding notions can be made more formal by the introduction of a *vector space* $\hat{X}$ attached to the affine space X; we define it as the union of points with a given non-zero mass (i.e. the product $X\times K^*$) and the vectors in $\vec{X}$ (cf. 2.A). In this space we can perform vector calculations; for instance, the elements of $\vec{X}$ correspond to the case $\Sigma_i\lambda_i=0$. The essential thing is that X is canonically embedded in $\hat{X}$ as the affine hyperplane formed by points with mass 1 (cf. 2.D). And the direction $\vec{X}$ of this hyperplane (in the sense of 2.D) is indeed the vector space which gives rise to X.

3.E Polynomials ([B, 3.3])

Starting from the classical notion of a homogeneous polynomial of degree k over a vector space, and using the vector space $\hat{X}$, we define for an affine space the notion of a *polynomial of degree less than or equal to* k over X. Conversely, it is possible to transform a polynomial of degree less than or equal to k over X into a polynomial of degree k over $\hat{X}$, by introducing a homogenizing variable (which is of course the mass). Moreover, a polynomial f over X possesses a *symbol* $\vec{f}$, which is a homogeneous polynomial over $\vec{X}$ and corresponds to the highest-degree term of f. In a less pedantic way, we can say that a polynomial of degree less than or equal to k is something which, in an arbitrary vectorialization of X, consists of a sum of homogeneous polynomials of degree $k, k-1,\ldots,1,0$.

Problems

3.1 ITERATED CENTERS OF MASS ([B, 3.7.16]). In a real affine space, consider p points $x_{1,1},\ldots,x_{1,p}(p \geq 2)$. For $i=1,2,\ldots,p$ denote by $x_{2,i}$ the center of mass of the $(x_{1,j})_{j \neq i}$. Then define by recurrence the center of mass $x_{k+1,i}$ of the $(x_{k,j})_{j \neq i}$, for all $k \geq 1$. Prove that every sequence $(x_{k,i})_{k \in \mathbf{N}}$ converges. What can you say about the limit of these sequences for different values of i?

3.2 CENTROIDS OF THE BOUNDARY OF A TRIANGLE AND OF A FOUR-SIDED PLATE ([B, 3.7.13 and 3.7.14]). Determine the centroid of the physical object consisting of three homogeneous pieces of wire of same linear density, lying on the three sides of a triangle. Give a geometrical construction for this point.

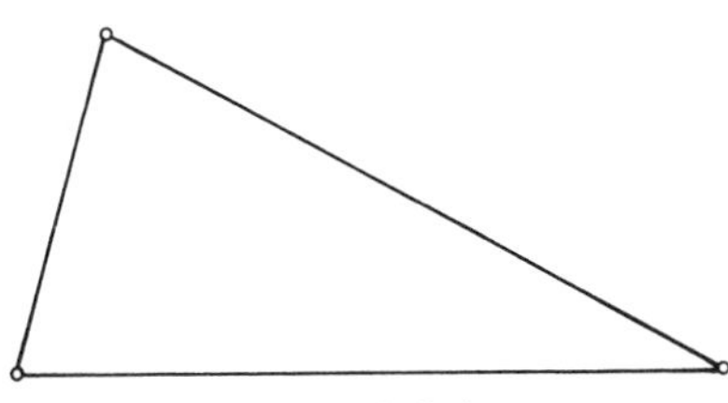

Figure 3.2.1.

Give a geometrical construction for the center of mass of a homogeneous plate in the shape of a quadrilateral. Compare this point with the centroid of the four vertices.

3.3 DIAMETERS OF THE BARYCENTRIC SUBDIVISIONS OF SIMPLICES ([B, 3.7.8]). Let Σ be a simplex in a Euclidean affine space of dimension n; its *diameter* d is the largest distance between any two of its points. Show that all simplices of the barycentric subdivision of Σ have diameter less than or equal to $nd/(n+1)$; deduce that when we iterate the process of barycentric subdivision, the diameter of all simplices tends towards zero.

3.4 DIRECT DEFINITION OF POLYNOMIAL MAPS ([B, 3.7.11]). Let X be an affine space and W a vector space. Find a direct definition (i.e. not using the universal space) for the space $\mathscr{P}_k(X; W)$ of polynomial maps of degree k from X into W. Show that your definition makes sense.

3.5 EXPLICIT CALCULATION OF THE POLAR FORM OF A POLYNOMIAL ([B, 3.7.15]). The polar form of a polynomial map f of degree k, from one vector space V into another vector space W, is the unique k-linear symmetric form φ with values in W and such that $f(v)=\varphi(v,\ldots,v)(K$ times).

Show that it is given by the formula

$$\varphi(v_1,\ldots,v_k)=\frac{1}{k!}\sum_{j=1}^{k}(-1)^{k-j}\sum_{1\leq i_1\leq\ldots\leq i_j\leq k}f(v_{i_1}+\cdots+v_{i_j}). \qquad (*)$$

3.6 THE EULER IDENTITY ([B, 3.7.12]). Recall that a real-valued function is said to be of class C^n if it can be differentiated n times, and its n-th derivative is continuous. Let X be a real vector space and $f\colon X\to\mathbf{R}$ a C^1 map such that $f(\lambda x)=\lambda^k f(x)$ for all $x\in X$ and $\lambda\in\mathbf{R}$. Show that the derivative f' of f satisfies the *Euler identity*:

$$f'(x)(x)=kf(x)\quad \forall x\in X.$$

Write and prove an analogous form for the p-th derivative of f, when f is of class C^p and again homogeneous of degree k. Deduce that if f is of class C^k and homogeneous of degree k, it is necessarily a polynomial.

Chapter 4

Projective Spaces

All fields considered here are commutative.

Affine spaces have the serious inconvenience that certain theorems have exceptional cases, for instance when lines become parallel. Projective spaces were created by Desargues in 1639 to remedy this situation. The affine space is completed with points at *infinity* which correspond to the directions of straight lines. Then one needs to know how to come back to the original affine space. This program is carried out here in two steps: Chapters 4 and 5.

4.A Definition ([B, 4.1])

A *projective space* is the space of lines (i.e. one-dimensional vector subspaces) of a vector space. If E is the vector space, we will *denote* by $P(E)$ the associated projective space, formed by the lines of E. Algebraically, $P(E)$ is the quotient of the complement $E \backslash 0$ of the origin in E by the equivalence relation "$x \sim y$ if and only if there is a non-zero scalar k such that $y = kx$". The *dimension* of $P(E)$ is one less than the dimension of E.

EXAMPLES. We *put* $P^n(K) = P(K^{n+1})$ and call this space the n-dimensional projective space over the field K. See for instance the case of finite fields in problems 4.1 and 4.5. An important case in geometry is when $K = \mathbf{R}$ or $\mathbf{C}$;

see 4.G. Finally, if E^* denotes the dual vector space to E, the projective space $P(E^*)$ is in one-to-one correspondence with the set of hyperplanes of E. In fact, a hyperplane is determined as the kernel of a non-zero linear form, defined up to a scalar.

4.B Subspaces, Intersections, Duality ([B,4.6])

The elements of $P(E)$ are called *points*. The (projective) subspaces of $P(E)$ are the images of (vector) subspaces of E. They are projective spaces in themselves; the dimension of a subspace is its dimension as a projective space. *Points* are zero-dimensional subspaces, *lines* are one-dimensional (images of planes in E), *planes* are two-dimensional. The hyperplanes of $P(E)$ are the images of hyperplanes of E; hence they too are in one-to-one correspondence with $P(E^*)$.

In projective geometry, as opposed to affine geometry, there are no exceptional cases for the intersection of subspaces. We always have, for any two subspaces V, W of $P(E)$:

$$\dim(\langle V \cup W \rangle) + \dim(V \cap W) = \dim V + \dim W.$$

We have introduced in this formula the *notation* $\langle A \rangle$ for the smallest subspace containing A (called the subspace *spanned* by A), where A is any subset of a projective space. Two distinct points x, y always span a line $\langle x, y \rangle$. We say that the points $\{m_i\}_{i=0,1,\ldots,k}$ are *projectively independent* if $\langle m_0, m_1, \ldots, m_k \rangle$ has dimension k.

What does a line of $P(E^*)$ represent in $P(E)$? It consists of the hyperplanes of $P(E)$ which contain a fixed codimension-2 subspace of $P(E)$ (i.e. the intersection of two different hyperplanes, or alternatively, if $P(E)$ is finite-dimensional, a subspace whose dimension is two less than that of $P(E)$). We call such a line in $P(E^*)$ a *pencil of hyperplanes*. If, for instance, $P(E)$ is a projective plane, a pencil of lines in $P(E)$ is the set of lines which go through a fixed point a in $P(E)$; as a line in $P(E^*)$ it is *denoted* by a^*.

4.C Homogeneous Coordinates; Charts ([B, 4.2])

The use of projective spaces is a trade-off: the statements have no special cases, but on the other hand calculation becomes more difficult. Let's suppose for now that we are in the finite-dimensional case. The first method of performing calculations is to cover $P(E)$ with (partial) charts: for instance, if $\{e_i\}_{i=0,1,\ldots,n}$ is a basis of E, and $p\colon E \to P(E)$ is the canonical projection (given by the equivalence relation in the definition), then the points in $P(E)$ of the form $p((1, x_1, \dots, x_n))$ make up exactly the complement of the hyperplane $P(H_0)$, where $H_0 = x_0^{-1}(0)$. So we can get $P(E)$ back by patching together the $n+1$ complements of the hyperplanes $P(H_i)$, where $H_i = x_i^{-1}(0)$. In order to perform calculations, we still need to know the relation between different charts; on the intersection of two complements associated with H_i and H_j, we have:

$$p((x_0, \dots, x_{i-1}, 1, x_{i+1}, \dots, x_n)) \mapsto$$

$$p\left(\left(\frac{x_0}{x_j}, \dots, \frac{x_{i-1}}{x_j}, \frac{1}{x_j}, \frac{x_{i+1}}{x_j}, \dots, \frac{x_{j-1}}{x_j}, 1, \frac{x_{j+1}}{x_j}, \dots, \frac{x_n}{x_j}\right)\right)$$

See problem 4.2. Another way to calculate is to somehow stay inside E, writing $m = p((x_0, x_1, \dots, x_n))$; the x_i are called the *homogeneous coordinates* of $m \in P(E)$ (associated with a basis of E). Observe, however, that they are only defined *up to a constant*. We have

$$p((kx_0, kx_1, \dots, kx_n)) = p((x_0, x_1, \dots, x_n))$$

for every $k \neq 0$.

4.D Projective Bases ([B, 4.4])

This notion is based on the following remark: Let $\{e_i\}, \{e_i'\}$ $(i = 1, \dots, n+1)$ be two bases of E such that we have both

$$p(e_i) = p(e_i'), \quad \forall\, i = 1, \dots, n+1,$$

and

$$p(e_i + \cdots + e_{n+1}) = p(e_i' + \cdots + e_{n+1}').$$

Then the two bases are necessarily proportional, i.e. there is some k such that $e_i' = ke_i$ for every i.

For an n-dimensional projective space $P(E)$, we will define a *projective base* as a set $\{m_i\}_{i=0,1,\dots,n+1}$ with $n+2$ points in $P(E)$, such that, for every $i = 0, 1, \dots, n+1$, the $n+1$ points $\{m_j\}_{j \neq i}$ are projectively independent. The preceding remark then implies that, up to a scalar, there is a unique basis $\{e_i\}_{i=0,1,\dots,n+1}$ of E (and consequently a system of homogeneous coordinates for $P(E)$) such that $p(e_i) = m_i$ $(i = 1, \dots, n+1)$, and $p(e_1 + \cdots + e_{n+1}) = m_0$.

For example, if $\dim(P(E)) = 2$, a projective base is composed of four points x, y, z, t such that the four triples $\{x, y, z\}$, $\{x, y, t\}$, $\{x, z, t\}$ and

$\{y, z, t\}$ are projectively independent. The homogeneous coordinates associated to the four points will be $(1,1,1)$, $(1,0,0)$, $(0,1,0)$ and $(0,0,1)$.

4.E Morphisms, Homography, Projective Group ([B, 4.5])

By definition, a *morphism* from $P(E)$ into $P(E')$ is a map $\underline{f}$ obtained from a linear map f from E into E' by taking quotients. It is important to notice that $\underline{f}$ is not defined on all of $P(E)$, only on the complement $P(E)\setminus P(f^{-1}(0))$ of the projective space $P(f^{-1}(0))$ defined by the kernel $f^{-1}(0)$ of f.

A *homography* from $P(E)$ into $P(E')$ is a morphism which comes from an isomorphism $E \to E'$. The homographies from $P(E)$ into itself form a group, called the *projective group* of $P(E)$, which is *denoted* by $\mathrm{GP}(E)$. This group is simply the quotient $GL(E)/K^*\cdot \mathrm{Id}_E$ of the linear group $\mathrm{GL}(E)$ of E by the (normal) subgroup of similarities $K^*\cdot \mathrm{Id}_E$.

A simple but very useful theorem is the "first fundamental theorem of projective geometry": Let $P(E)$, $P(E')$ be two projective spaces of same finite dimension, and $\{m_i\}$, $\{m_i'\}$ be projective frames in E and E' respectively. Then there exists one and only one homography g: $P(E) \to P(E')$ taking m_i to m_i' for every i. The "second fundamental theorem of projective geometry" is simply the projective version of 2.F (cf. [B, 5.4.8]).

In particular, $\mathrm{GP}(E)$ acts transitively (in fact, simply transitively; see 1.D) on the set of projective frames. Comparing this with the affine case, we conclude that projective groups are "bigger" than affine groups; for an affine group in dimension n we can only choose the images of $n+1$ points, whereas for a projective group we can choose $n+2$ points.

For example, if D and D' are two projective lines (whether in themselves or as subspaces of larger projective spaces), there is exactly one homography f: $D \to D'$ taking x, y, z into x', y', z' respectively, for any three distinct points x, y, z (resp. x', y', z') of D (resp. D').

To find out what we can do with four points on a line, see chapter 6. For five points in a plane, see problem 6.4.

4.F Perspectives ([B, 4.7])

Perspectives are defined by figure 4.F.1 below; we start from two hyperplanes H, H' in $P(E)$, and a point m which does not belong to either hyperplane. By 4.B, for every point x in H there is a unique line $\langle m, x\rangle$, and this line cuts H'

in a unique point $g(x)$. The perspective of center m from H to H' is given by $x \mapsto g(x)$; it is a homography.

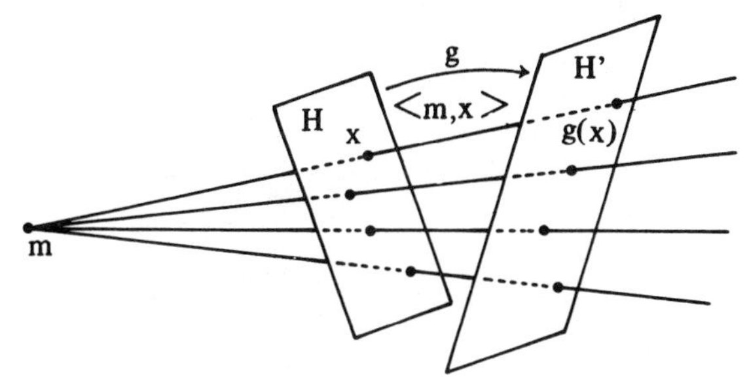

Figure 4.F.1.

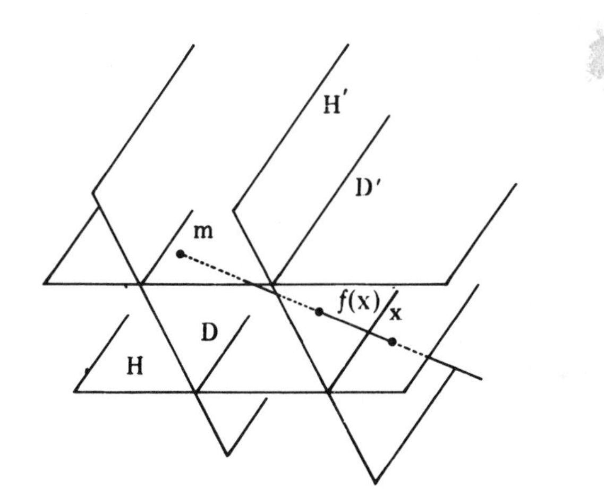

Figure 4.F.2.

When we return to affine spaces (see chapter 5), perspectives still exist, but they are not defined on the whole hyperplane H; we have to exclude one of its hyperplanes (figure 4.F.2).

4.G Topology ([B, 4.3])

If the field of scalars is **R** or **C** (and the dimension is finite), projective spaces have a *canonical topology* (the quotient topology of the underlying vector space). They are always compact. Besides the intersection properties not having exceptional cases, compactness is another advantage of projective over affine spaces.

Problems

4.1 A MODEL FOR $P^3(\mathbf{Z}_2)$ ([B, 4.9.9]). Find a model, in the usual three-dimensional space, for the configuration formed by the points, lines and planes of $P^3(\mathbf{Z}_2)$. Draw pictures.

4.2 ORIENTABILITY OF REAL PROJECTIVE SPACES (first method) ([B, 4.9.4]). For a real projective space of finite dimension n, find the sign of the Jacobian of the transition maps

$$\pi_j \circ \pi_i^{-1}: (v_{1,\dots,}v_n) \mapsto \left(\frac{v_1}{v_{j-1}},\dots,\frac{v_{i-1}}{v_{j-1}},\frac{1}{v_{j-1}},\frac{v_i}{v_{i-1}},\dots,\frac{v_{j-2}}{v_{j-1}},\frac{v_j}{v_{j-1}},\dots,\frac{v_n}{v_{j-1}}\right)$$

(defined on the overlap $\mathbf{R}^n \backslash v_{j-1}^{-1}(0)$).

4.3 ORIENTABILITY OF REAL PROJECTIVE SPACES (second method) ([B, 4.9.5]). Find out whether $P^n(\mathbf{R})$ is orientable by studying the connectedness of the projective group $\mathrm{GP}(P^n(\mathbf{R}))$.

4.4 HYPERPLANES AND DUALITY ([B, 4.9.10]). Let $\{H_i\}$ be a family of hyperplanes in the projective space $P(E)$ of finite dimension n; find a relation between $\dim(\cap_i H_i)$ in $P(E)$ and $\dim(\langle \cup_i H_i \rangle)$ in $P(E^*)$.

4.5 NUMBER OF POINTS AND OF SUBSPACES IN A PROJECTIVE SPACE OVER A FINITE FIELD ([B, 4.9.11]). Let K be a field with k elements, and $P(E)$ a projective space of dimension n over K. Show that the cardinality of the set of p-dimensional subspaces of $P(E)$ is equal to

$$\frac{(k^{n+1}-1)(k^{n+1}-k)\cdots(k^{n+1}-k^p)}{(k^{p+1}-1)(k^{p+1}-k)\cdots(k^{p+1}-k^p)}.$$

Show that the order of the projective group $\mathrm{GP}(E)$ is

$$(k^{n+1}-1)(k^{n+1}-k)\cdots(K^{n+1}-k^{n-1})K^n.$$

4.6 MÖBIUS TETRAHEDRA ([B, 4.9.12 and 5.5.3]). Construct, in a three-dimensional projective space, two tetrahedra $\{a,b,c,d\}$ and $\{a',b',c',d'\}$ such that each vertex of the first belongs to a face of the second and vice versa (i.e. $a \in \langle b',c',d'\rangle$ etc. and $a' \in \langle b,c,d\rangle$ etc.)

Construct such tetrahedra in a very simple way by sending as many points to infinity as possible (cf. 5.D).

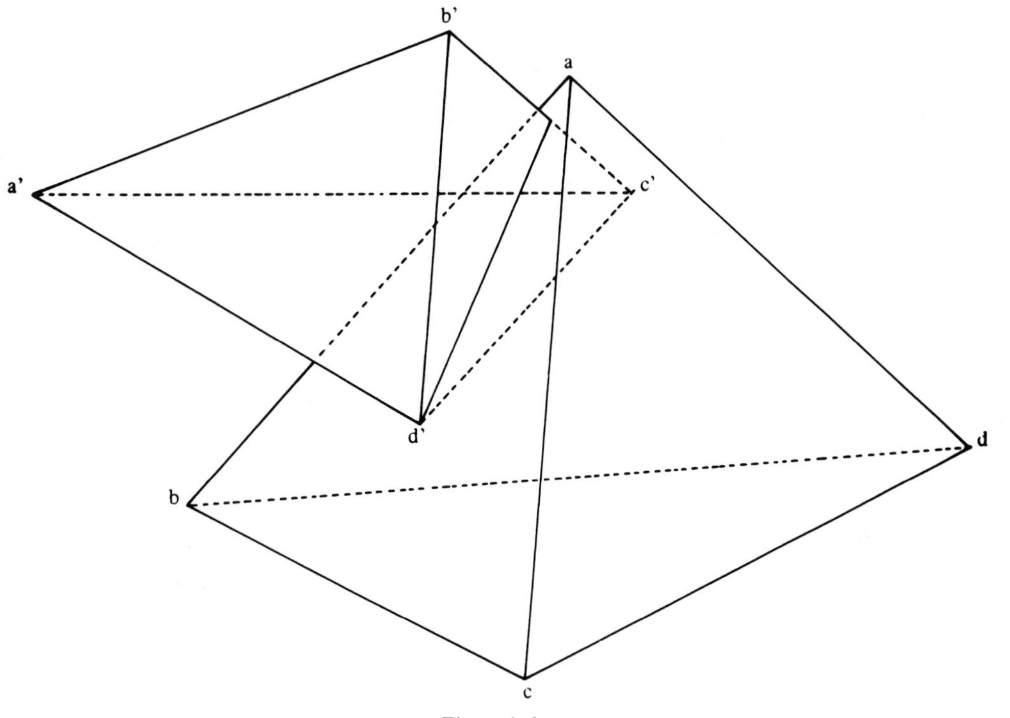

Figure 4.6.

Chapter 5
Affine-Projective Relationship: Applications

5.A The Projective Completion of an Affine Space ([B, 5.1])

We have shown, at the beginning of chapter 4, why it is necessary to go beyond the framework of affine spaces, adjoining to them points at infinity. This is now possible in the following way: we associate to the affine space X its universal space $\hat{X}$ (cf. 3.D), which is a vector space in which X is embedded as an affine hyperplane whose direction is a vector hyperplane of $\hat{X}$. Considering now the projectivization $\tilde{X} = P(\hat{X})$, we see that $\tilde{X}$ is the disjoint union of two sets: $P(X)$, which is canonically identified with X, and $P(\vec{X})$, which, being the space of lines in $\vec{X}$, is also the space of directions of lines in X. We denote it by $\infty_X = P(\vec{X})$. We write $\tilde{X} = X \cup \infty_X$, and say that ∞_X is the *hyperplane at infinity in* X.

A most important case is when $X = K$; then $\infty_X = \infty$ consists of a single point, the *point at infinity in* K, and $\tilde{K} = K \cup \infty$. This applies even for $K = \mathbf{R}$: the real line has two "ends", but $\tilde{\mathbf{R}}$ has only one point at infinity. Analogously, if $X = D$ is an affine line, ∞_D is a single point, the point at infinity of D.

5.B From Projective to Affine

Conversely, if we consider a hyperplane $P(H)$ in a projective space $P(E)$, the complement $P(E)\backslash P(H)$ has a natural affine structure, and the projective completion of the affine space thus obtained is of course $P(E)$.

The use of coordinates provides an approach which is more suitable for calculations, although less "natural": Take a point x expressed as $(x_1, \ldots, x_n)$ in terms of an affine frame for the n-dimensional affine space X; we embed X into the projective space $P^n(K)$ by means of the map

$$(x_1, \ldots, x_n) \mapsto p((1, x_1, \ldots, x_n)).$$

Conversely, we define a map $S \to \mathbf{R}^n \to X$ on the subset S of $P^n(K)$ where x_0 does not vanish:

$$p((x_0, x_1, \ldots, x_n)) \mapsto \left(\frac{x_1}{x_0}, \ldots, \frac{x_n}{x_0}\right)$$

5.C Correspondence between Subspaces ([B, 5.3])

If Y is an affine subspace of X, the completion $\tilde{Y}$ of Y is naturally embedded in $\tilde{X}$: we have $\tilde{Y} = Y \cup \infty_Y$, where in fact $\infty_Y = \infty_X \cap \tilde{Y}$ is a projective subspace of ∞_X (namely, $\infty_Y = P(\vec{Y})$). Two subspaces Y, Z are parallel if and only if $\infty_Y = \infty_Z$.

5.D Sending Points to Infinity and Back ([B, 5.4])

Using 5.A and 5.B we can apply the following, often very convenient, procedure in affine spaces: Consider, for example, a line D in an affine plane X; we start by taking the completion $\tilde{X}$ of X. Then the complement $X' = \tilde{X} \setminus \tilde{D}$ of $\tilde{D}$ is an affine space containing all of $X \setminus D$, but now the points of D are at infinity with respect to this subspace. For example, if E, F are two lines of X which intersect in a point of D, their images E' and F' in X' will be parallel.

Using this technique we can, to begin with, prove the second fundamental theorem of projective geometry (cf. 4.E, [B, 5.4.8]). Next we can prove the theorems of Pappus and Desargues by considering the affine cases with the greatest possible number of parallels (cf. [B, 5.4]). We give here the affine versions of the two (the projective versions can only be simpler):

THEOREM OF PAPPUS. *Let X be an affine plane, D and D' two distinct lines in X, and a, b, c, a', b', c' six distinct points such that $a, b, c \in D \setminus (D \cap D')$ and $a', b', c' \in D' \setminus (D \cap D')$. Then if all three points $\langle a, b' \rangle \cap \langle a', b \rangle$, $\langle b, c' \rangle \cap \langle b', c \rangle$, $\langle c, a' \rangle \cap \langle c', a \rangle$ exist, they lie on the same line.*

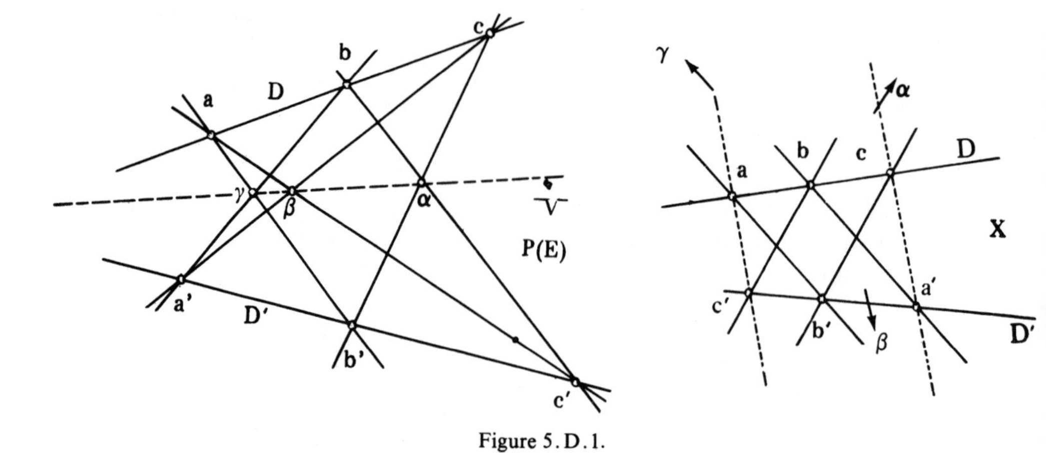

Figure 5.D.1.

THEOREM OF DESARGUES. *Let s, a, b, c, a', b', c' be points in an affine space such that s, a, a' (resp. s, b, b' and s, c, c') are aligned. Then if the three points $\langle a, b\rangle \cap \langle a', b'\rangle$, $\langle b, c\rangle \cap \langle b', c'\rangle$, $\langle c, a\rangle \cap \langle c', a'\rangle$ exist, they lie on the same line.*

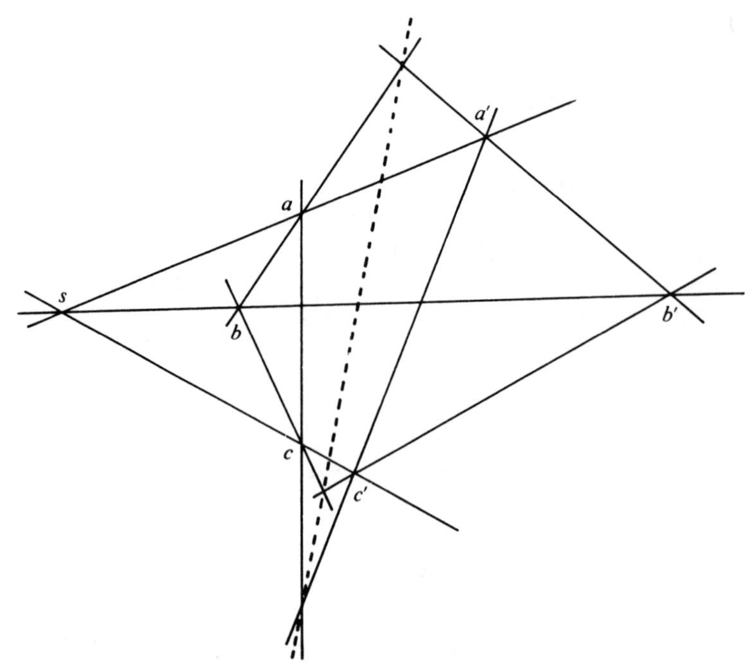

Figure 5.D.2.

Problems

5.1 THEOREM OF PAPPUS WHEN THERE ARE PARALLEL LINES ([B, 5.5.2]). Draw the figures for the theorem of Pappus (cf. 5.D) when there are points at infinity.

5.2 POINTS OUTSIDE YOUR DRAWING PAPER; RULER TOO SHORT ([B, 5.5.4 AND 5.5.5]). Suppose you are given a piece of paper with one point marked and segments of two lines which intersect outside the paper. Using only a straightedge (ruler), draw the line that joins the given point with the intersection point of the two lines.

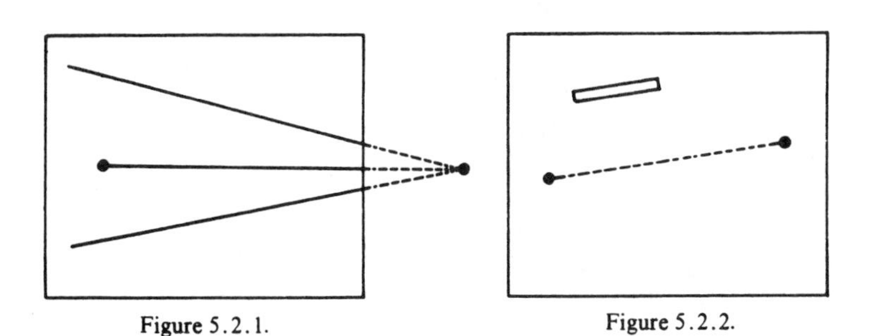

Figure 5.2.1. **Figure** 5.2.2.

Now suppose you are given two points, but your ruler is too short to connect them. Draw the line joining the points.

5.3 HEXAGONAL WEBS ([B, 5.5.8 AND 5.5.9]). We shall define a *web* in a real affine plane P as the following set of data: an open set A in P, and for each point a in A three distinct lines $d_i(a)(i=1,2,3)$ in P which go through a and which depend continuously on a. Show that for b on $d_1(a)$ close enough to a, we can define six points $(b_i)_{1,\dots,6}$ as follows:

$$b_1=d_3(b)\cap d_2(a),\quad b_2=d_1(b_1)\cap d_3(a),\quad b_3=d_2(b_2)\cap d_1(a),$$

$$b_4=d_3(b_3)\cap d_2(a),\quad b_5=d_1(b_4)\cap d_3(a),\quad b_6=d_2(b_5)\cap d_1(a).$$

A web is said to be *hexagonal* if $b_6=b$ for all sufficiently close a and b.

Let $(p_i)_{i=1,2,3}$ be three points of P, not on the same line, and let A be the complement of the three lines which connect each pair of points p_i. We define a web by setting $d_i(a)=\langle a,p_i\rangle$ for every $a\in A$. Show that this web is hexagonal.

More generally, consider a conic section C and a point p not situated on the conic. Assign to any point in the complement of $C \cup p$ the two tangents to C through x and the line xp. Show that the web thus obtained is hexagonal.

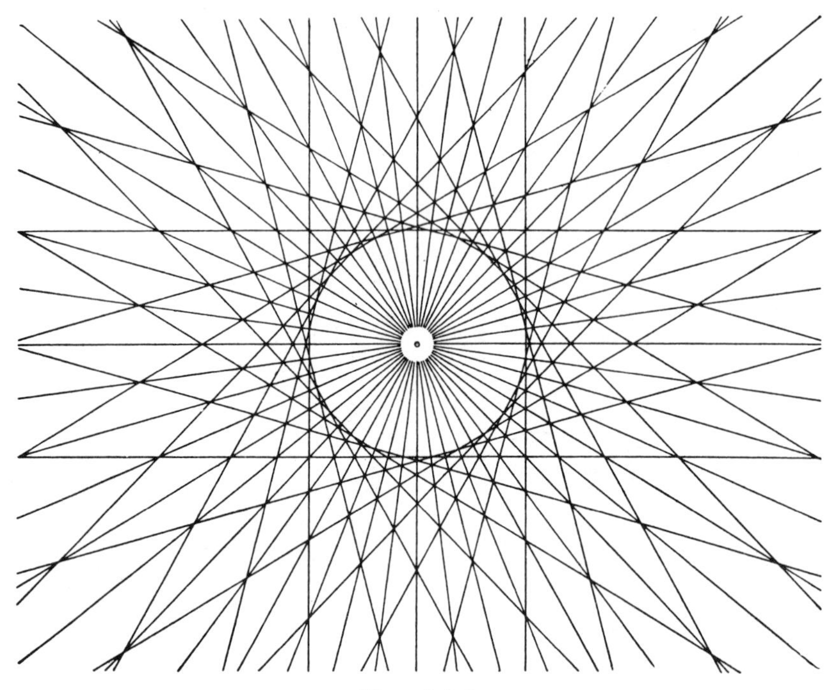

Figure 5.3.1.

Chapter 6

Projective Lines, Cross-Ratios, Homographies

6.A Cross-ratios ([B, 6.1, 6.2, 6.3])

To any four distinct points $(a_i)_{i=1,2,3,4}$ on a projective line (considered by itself or inside a projective space), we associate a scalar, *denoted* by $[a_i]=[a_1, a_2, a_3, a_4]$, and called the *cross-ratio* of these four points. For points on an affine line D, the cross-ratio is defined to be the same as on the completion $\tilde{D}=D\cup\infty_D$ (cf. 5.A).

The cross-ratio is a projective invariant, in the sense that if $(a_i)_{i=1,2,3,4}$ (resp. $(a_i')_{i=1,2,3,4}$) are four points on a line D (resp. D'), then the existence of a homography $f\colon D\to D'$ taking a_i to a_i', $i=1,2,3,4$, is equivalent to $[a_i]=[a_i']$. In particular, cross-ratios are invariant under homographies.

There are several equivalent ways of defining cross-ratios. The first one is by putting $[a,b,c,d]=f(d)$, where $f(d)$ is the element of K given by the unique homography $f\colon D\to\tilde{K}=K\cup\infty$ (cf. 5.A and 4.E) from our line into $\tilde{K}$ satisfying $f(a)=\infty$, $f(b)=0$ and $f(c)=1$.

Second way: For $D=D'\cup\infty_D$, where D' is an affine line, put

$$[a,b,c,d]=\frac{\overrightarrow{ca}/\overrightarrow{cb}}{\overrightarrow{da}/\overrightarrow{db}};\quad \text{in particular } [a,b,c,\infty_D]=\overrightarrow{ca}/\overrightarrow{cb}.$$

The following three relations describe the behavior of cross-ratios under permutations of the four points:

$$[a,b,c,d]=[b,a,c,d]^{-1}=[a,b,d,c]^{-1},[a,b,c,d]+[a,c,b,d]=1.$$

6.B Harmonic Division ([B, 6.4])

(We assume the field has characteristic different from 2.) A very important special case occurs when $[a, b, c, d] = -1$. Observe that then also $[b, a, c, d] = [a, b, d, c] = [c, d, a, b] = -1$; we say that the two pairs of points $\{a, b\}$ and $\{c, d\}$ are *in harmonic division*, and that a and b are *harmonic conjugates* with respect to c and d.

For instance, $[a, b, c, \infty_D] = -1$ is equivalent to $c = (a+b)/2$, i.e. c is the midpoint of a, b. Using this fact we can deduce the property of the *complete quadrilateral*: For any four points a, b, c, d on a plane (affine or projective), we have $[a, c, \gamma, \delta] = -1$ in the figure below:

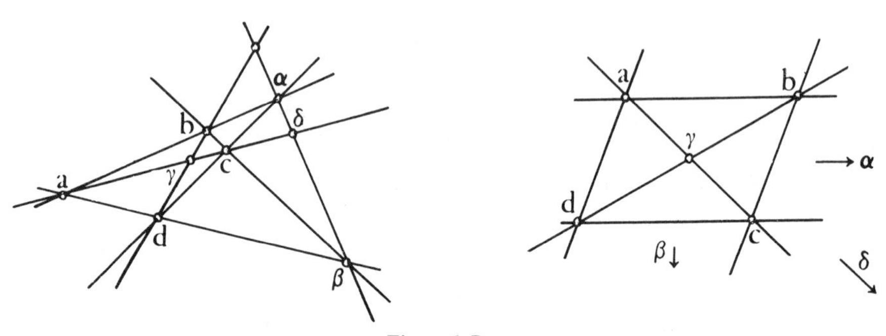

Figure 6.B.

The proof consists in sending points α and δ to infinity (see 5.D). We obtain a parallelogram, whose diagonals are known to cross at their midpoint.

6.C Duality ([B, 6.5])

Using the ideas in 4.B, consider in a projective space $P(E)$ four hyperplanes $H_i (i = 1, 2, 3, 4)$ which belong to the same pencil, and let D be a line in $P(E)$ which does not intersect the codimension-two subspace that defines this pencil. Then the cross-ratio $[H_i]$ of the four points on the projective line in $P(E^*)$ which constitutes this pencil satisfies

$$[H_i] = [H_i \cap D].$$

We deduce the following useful results. Let m, m' be two points on a projective plane, and let m^*, m'^* be the pencils of lines defined by them (cf. 4.B). Let $f: m^* \to m'^*$ be a homography taking $\langle m, m' \rangle$ to $\langle m', m \rangle$. Then, as

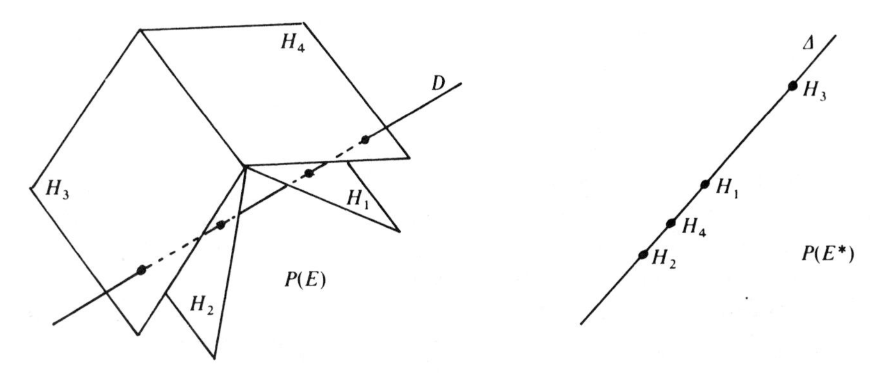

Figure 6.C.1.

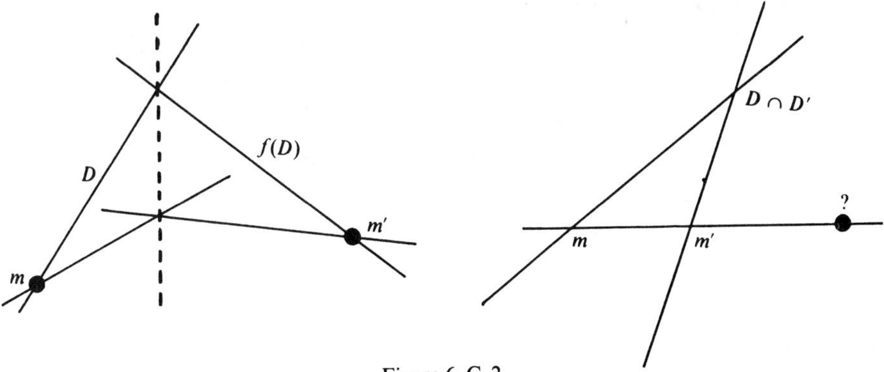

Figure 6.C.2.

the line D runs through m^*, the intersection point $D \cap f(D)$ describes a straight line. In the case when $f(\langle m, m' \rangle) \neq \langle m', m \rangle$ the point $D \cap f(D)$ describes a conic; see 16.C.

The dual statement is: If two points m, m' run through two lines D, D' of a projective plane, in such a way that the mapping $m \mapsto m'$ is a homography preserving the point $D \cap D'$, then the line $\langle m, m' \rangle$ goes through a fixed point.

6.D Homographies of a Projective Line ([B, 6.6, 6.7])

We want to study the elements of the projective group of a projective line, which is essentially $\tilde{K} = K \cup \infty$. In terms of a chart, such a homography f can be written

$$z \mapsto \frac{\alpha z + \beta}{\gamma z + \delta}, \quad \text{where} \quad \alpha, \beta, \gamma, \delta \in K,$$

the convention being that

$$f(\infty)=\frac{\alpha}{\gamma} \quad \text{and} \quad f\left(-\frac{\delta}{\gamma}\right)=\infty$$

(unless $\gamma=0$, in which case $f(\infty)=\infty$).

If the field K is algebraically closed, f is studied by means of its fixed points. If there is only one of these, we put it at infinity; then f is a translation of the affine line obtained by taking infinity away. Otherwise, let a, b be the two distinct fixed points; then f satisfies $[a, b, m, f(m)]=$ constant for all m.

This constant value is λ/μ, where λ and μ are the two eigenvalues of the matrix of f, namely

$$A=\begin{pmatrix}\alpha & \beta\\ \gamma & \delta\end{pmatrix}.$$

An *involution* of a projective line is by definition a homography different from the identity, whose square is the identity. The analytical condition for a homography is

$$\operatorname{trace}(A)=\alpha+\delta=0.$$

An involution is determined by its value on two points. Every homography is the product of at most three involutions. If K is algebraically closed and a, b are the two fixed points of an involution, then $[a, b, m, f(m)]=-1$ for every m (cf. 3.B).

Problems

6.1 RELATION BETWEEN THE CROSS-RATIOS OF FIVE POINTS ([B, 6.8.1]). Let x, y, z, u, v be five points on the same projective line. Show that the following always holds:

$$[x, y, u, v][y, z, u, v][z, x, u, v]=1.$$

6.2 RICATTI DIFFERENTIAL EQUATIONS ([B, 6.8.12]). If $a, b, c:[\alpha, \beta] \to \mathbf{R}$ are continuous functions, we consider the differential equation (called a Ricatti equation): $y'(t)=a(t)y^2+b(t)y+c(t)$; show that if $y_i (i=1,2,3,4)$ are four solutions of this equation, the cross-ratio $[y_i(t)]$ is independent of t.

6.3 DUALITY IN A TETRAHEDRON ([B, 6.8.21]). Let T be a tetrahedron in a three-dimensional projective space and let D be a line. Show that the cross-ratio of the four intersection points of D with the faces of T is equal to the cross-ratio of the four planes passing through D and the vertices of T.

6.4 FIVE-POINT SETS ON A PROJECTIVE PLANE ([B, 6.8.17]). Let $(a_i)_{i=1,\ldots,5}$ be points on a projective plane, so that the first four form a projective base. We *denote* by d_{ij} the line $\langle a_i, a_j\rangle$. Show that the following holds:

$$[d_{12}, d_{13}, d_{14}, d_{15}][d_{23}, d_{21}, d_{24}, d_{25}][d_{31}, d_{32}, d_{34}, d_{35}] = 1.$$

Show that a necessary and sufficient condition for the existence of a homography taking the $(a_i)_{i=1,\ldots,5}$ into new points $(a_i')_{i=1,\ldots,5}$ is that the following two equalities be satisfied:

$$[d_{12}, d_{13}, d_{14}, d_{15}] = [d_{12}', d_{13}', d_{14}', d_{15}']$$

and

$$[d_{23}, d_{21}, d_{24}, d_{25}] = [d_{23}', d_{21}', d_{24}', d_{25}'],$$

where we have put $d_{ij}' = \langle a_i', a_j' \rangle$. Generalize for the case of a projective space of arbitrary dimension.

6.5 EIGENVALUES OF A HOMOGRAPHY ([B, 6.8.7]). Let f be a homography with two distinct fixed points a, b; show that the pair $\{k, 1/k\}$, where $k = [a, b, m, f(m)]$ for every m, depends only on f and not on the choice of the order of a, b. If f has matrix $M = \begin{pmatrix} \alpha & \beta \\ \gamma & \delta \end{pmatrix}$ show that $\{k, 1/k\}$ are the roots of the equation

$$(\alpha\delta - \beta\gamma)k^2 - (\alpha^2 + 2\beta\gamma + \delta^2)k + (\alpha\delta - \beta\gamma) = 0.$$

6.6 CLASSIFICATION OF COMPLEX HOMOGRAPHIES ([B, 6.6.8]). The data are those of 6.5, and moreover $K = \mathbf{C}$. We say that f is *elliptic* if the complex number k (or $1/k$) has absolute value 1, *hyperbolic* if k is positive real, and *loxodromic* otherwise. Show that, normalizing $M(f)$ by $\alpha\delta - \beta\gamma = 1$, we can characterize these three cases by using the trace t of f, given by $t = \alpha + \delta$:

$$f \text{ elliptic} \iff t \text{ is real and } |t| < 2;$$

$$f \text{ hyperbolic} \iff t \text{ is real and } |t| > 2;$$

$$f \text{ loxodromic} \iff t \text{ is not real.}$$

Study, for the three cases considered, the nature of the iterates $f^n (n \in \mathbf{Z})$.

Chapter 7
Complexifications

Here we treat only explicit (i.e. non-intrinsic) complexifications of the different objects in question; intrinsic complexifications can be found in [B, 7]. The objects to be complexified are vector spaces, affine and projective spaces, preserving the operations of vectorialization and projectivization of affine spaces.

It is also fundamental to be able to return from the complex to the real case; this can only be correctly done if the complexifications are endowed with a *conjugation*, as for the usual complex numbers. A subset of the complexification can only be brought down to the real case if it is invariant by conjugation.

We will not treat in this chapter the *morphisms* between the objects considered, but they too can be complexified without problems. *Polynomials* can also be complexified.

7.A Complexification of a Vector Space ([B, 7.1, 7.2, 7.3, 7.4])

If E is a real vector space, a *complexification* of E, denoted by E^C, is a complex vector space containing E and such that $E^C = E \oplus iE$ (where $i = \sqrt{-1}$). Each z in E^C is written as $x + iy$, with $x, y \in E$. For instance, given a basis of E, we can take the embedding $(x_1, \ldots, x_n) \mapsto (x_1, \ldots, x_n)$ of E into $\mathbf{C}^n$, where the x_i on the right-hand side are complex numbers.

The important thing is to introduce in E^C the *conjugation* σ (which is not a complex automorphism, since it is only semilinear), defined by $\sigma(x+iy) = x - iy$. A subspace F of E has a *complexification* F^C in E^C, which is a complex vector subspace. On the other hand, a complex vector subspace Z of E^C is an F^C for some F if and only if $\sigma(Z) = Z$, in which case $Z = F^C$ for $F = E \cap Z$.

Using the preceding remarks, one can easily show that every endomorphism of a finite-dimensional real vector space leaves invariant a line or a plane.

7.B Complexification of a Projective Space ([B, 7.5])

This is now easy: we define a complexification of $P(E)$ as $P(E^C)$, and we *denote* it by $(P(E))^C$. It is a complex projective space containing $P(E)$. The conjugation, still *denoted* by σ, is obtained from the conjugation in E^C by passing to the quotient. A projective subspace $\mathscr{X}$ of $P(E^C)$ is the complexification of a projective subspace of $P(E)$ if and only if $\sigma(\mathscr{X}) = \mathscr{X}$, and then we have $\mathscr{X} = (\mathscr{X} \cap P(E))^C$.

7.C Complexification of an Affine Space ([B, 7.6])

This is a more delicate operation. If X is an affine space, one way to complexify it is to take a vectorialization of X (cf. 2.A) and apply 7.A to it, thus obtaining a complex vector space X^C which gives rise to a complex affine space (see 2.A). Here too we obtain a conjugation σ which gives a characterization for the complex affine subspaces of X^C which are complexifications of affine subspaces of X, just as in 7.A and 7.B. Observe also that $(\vec{X})^C = \overrightarrow{(X^C)}$.

7.D Adding Up ([B, 7.6])

Adding up all of the above, we obtain a way to complexify the operation of projective completion of a real affine space X, which gives $\tilde{X} = X \cup \infty_X = X \cup P(\vec{X})$ (cf. 5.A). We find a space $(\tilde{X})^C = \widetilde{(X^C)}$, which can be written as $\tilde{X}^C = X^C \cup \infty_X^C$, with $\infty_X^C = (P(\vec{X}))^C = P(\overrightarrow{(X)^C})$. The conjugation map σ on $\tilde{X}^C$ is the union of the conjugations on X^C and in ∞_X^C.

Analytically, we start from an n-dimensional real affine space, then take an affine frame in which to express the points $x = (x_1, \ldots, x_n)$ of X, and we embed X into $P^n(\mathbf{C})$, the n-dimensional complex projective space, by means of

$$(x_1, \ldots, x_n) \mapsto p((1, x_1, \ldots, x_n)),$$

where the $n+1$ quantities $1, x_1, \ldots, x_n$ are considered as complex numbers. Conjugation in $P^n(\mathbf{C})$ is given by

$$\sigma(p((z_0, z_1, \ldots, z_n))) = p((\bar{z}_0, \bar{z}_1, \ldots, \bar{z}_n)).$$

The points of ∞_X^C are those of the form $p((0, z_1, \ldots, z_n))$.

An example of return to the reals. Suppose we have obtained, by complexifying a certain situation, four points a, b, c, d in ∞_X^C (in other words, directions of complex lines in X^C), which are permuted by σ (but none of them is left fixed). Then we claim that, among the six lines joining these points pairwise, exactly two come from lines in ∞_X. In fact, we can write our four points as $a, \sigma(a), b, \sigma(b)$, and the two lines will be $\langle a, \sigma(a)\rangle, \langle b, \sigma(b)\rangle$.

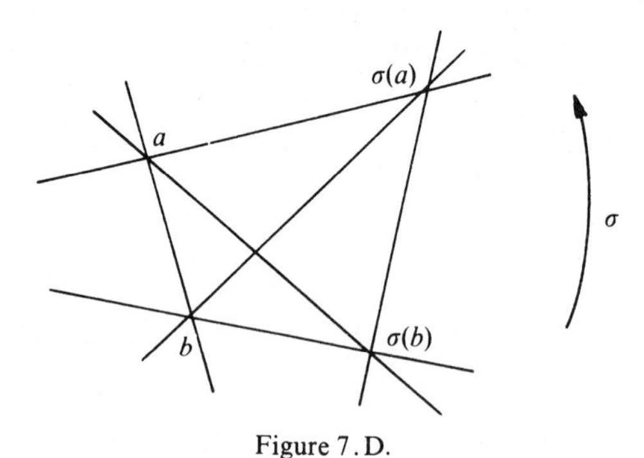

Figure 7.D.

Typical examples of the complexification techniques developed above are the construction of the *cyclical points* of a Euclidean space, and the applications stemming from this construction. See for instance 8.H, 9.D, 17.C, problem 16.3.

Problems

7.1 A NATURAL COMPLEXIFICATION ([B, 7.7.5]). Show that $\mathrm{Hom}_{\mathbf{R}}(\mathbf{C}; E)$, the set of all $\mathbf{R}$-linear maps from $\mathbf{C}$ into the real vector space E, is a natural complexification of E.

Chapter 8

More about Euclidean Vector Spaces

8.A Definitions ([B, 8.1])

A *Euclidean vector space* E is a real, finite-dimensional vector space (whose dimension will always be *denoted* by n), endowed with a positive definite symmetric bilinear form φ (i.e. $\varphi(x, x) > 0$ for all $x \neq 0$). We write $\varphi(x, y)$ as $(x|y)$, and we call this number the *scalar product* (or *inner product*) of x and y. The number

$$\|x\| = \sqrt{\varphi(x, x)}$$

is called the *norm* of x. Two vectors x and y are called *orthogonal* if $(x|y) = 0$.

It is always true that $|(x|y)| \leq \|x\| \, \|y\|$, and $\mathrm{d}(x, y) = \|x - y\|$ makes E into a metric space, which will be studied in Chapter 9. A subset $\{e_i\}_{i=1,\ldots,p}$ is called *orthonormal* if $\|e_i\| = 1$ for every i and $(e_i|e_j) = 0$ for every $i \neq j$.

The linear isomorphisms f of E that preserve the Euclidean structure, i.e. those for which $(f(x)|f(y)) = (x|y)$ for all x, y, are called (vector) *isometries*. They form the *orthogonal group* of E, denoted by $O(E)$. The elements of $O^+(E) = O(E) \cap \mathrm{GL}^+(E) = \{f \in O(E) : \det f > 0\}$ (cf. 2.G) are called *rotations*. For f to belong to $O(E)$ it is necessary and sufficient that its matrix A, relative to an orthonormal basis, satisfy ${}^tAA = I$, where ${}^tA$ is the transpose of A.

The *standard example* of a Euclidean space is $\mathbf{R}^n$, endowed with the scalar product $((x_1, \ldots, x_n)|(y_1, \ldots, y_n)) = \Sigma_i x_i y_i$. Every Euclidean space of dimension n is isomorphic to $\mathbf{R}^n$.

8.B Duality and Orthogonality ([B, 8.1])

We say that two subsets A, B of E are *orthogonal* if $(a|b)=0$ for every a in A and every b in B. To study this relation, it is convenient to introduce the *canonical isomorphism* $x \mapsto \{y \mapsto (x|y)\}$ between E and its dual E^*. For then Euclidean orthogonality is identical with orthogonality between E and E^* under the natural pairing. ("Orthogonal sets" in E and E^*, of course, are so called because of this fact.)

For example, we shall denote by $A^\perp$ the *orthogonal* set to subset A, defined by $A^\perp = \{x \in A : (x|a)=0,\ \forall a \in A\}$. This set is always a vector subspace of E, and for A, B vector subspaces of E, the following relations always hold:

$$(A^\perp)^\perp = A,\ A \oplus A^\perp = E, (A+B)^\perp = A^\perp \cap B^\perp, (A \cap B)^\perp = A^\perp + B^\perp.$$

A direct sum $E = A \oplus B$ is said to be *orthogonal* if $B = A^\perp$.

8.C Reflections ([B, 8.2])

The involutions of $O(E)$ are all of the form $f = \mathrm{Id}_S - \mathrm{Id}_T$, where $E = S \oplus T$ is an orthogonal direct sum (cf. 2.D and 8.B). Since T is uniquely determined by S, we say that f is the (orthogonal) *reflection*, or *symmetry*, through S.

Every $f \in O(E)$ is the product of at most n reflections through hyperplanes, and for $n \geq 3$, every $f \in O^+(E)$ is the product of at most n codimension-2 reflections.

8.D Structure of $O(E)$ for $n=2$ ([B, 8.3])

In this case, every element of $O^-(E) = \{f \in O(E) : \det f < 0\}$ is a reflection through some line. Moreover, $O^+(E)$ is commutative, and is formed by maps f whose matrix (in an orthonormal basis) has the form

$$\begin{pmatrix} \cos\vartheta & -\sin\vartheta \\ \sin\vartheta & \cos\vartheta \end{pmatrix}$$

where ϑ is a real number (the "angle" of f), determined modulo 2π provided we have chosen an orientation for E.

8.E Structure of an Element of $O(E)$ ([B, 8.4])

For every element f in $O(E)$ there is an orthogonal direct sum decomposition of $E = I_+ \oplus I_- \oplus R_1 \oplus \ldots \oplus R_r$ satisfying the following conditions:

(i) f leaves invariant each factor of the decomposition;
(ii) f is the identity on I_+ and minus the identity on I_-;

(iii) each R_i is two-dimensional, and on each R_i the matrix of f has the form

$$\begin{pmatrix} \cos\vartheta_i & -\sin\vartheta_i \\ \sin\vartheta_i & \cos\vartheta_i \end{pmatrix}$$

This result can be proven by recurrence using 8.D and the following fact: every element f of $O(E)$ leaves invariant some line or plane in E (see 7.A or problem 8.2).

EXAMPLE. In dimension 3, every rotation is a *rotation around a* (fixed) *line* in E, i.e. it is given by a matrix of the form

$$\begin{pmatrix} \cos\vartheta & -\sin\vartheta & 0 \\ \sin\vartheta & \cos\vartheta & 0 \\ 0 & 0 & 1 \end{pmatrix}$$

On the other hand, in dimension 4, a matrix like

$$\begin{pmatrix} \cos\vartheta & -\sin\vartheta & 0 & 0 \\ \sin\vartheta & \cos\vartheta & 0 & 0 \\ 0 & 0 & \cos\eta & -\sin\eta \\ 0 & 0 & \sin\eta & \cos\eta \end{pmatrix}$$

defines, for generic values of ϑ and η, a rotation which leaves nothing fixed (other than the origin).

8.F Angles and Oriented Angles ([B, 8.6, 8.7])

If Δ, Δ' are two oriented straight lines (or half-lines), their *angle*, *denoted* by $\overline{\Delta\Delta'}$, is the real number in the interval $[0, \pi]$ defined by

$$\cos(\overline{\Delta\Delta'}) = \frac{(x|x')}{\|x\|\,\|x'\|},$$

for any $x \in \Delta$, $x' \in \Delta'$ different from zero.

If D, D' are two lines, their *angle* $\overline{D, D'}$ is the element of $[0, \pi/2]$ defined by

$$\cos(\overline{DD'}) = \frac{|(x|x')|}{\|x\|\,\|x'\|}$$

for every $x \in D$, $x' \in D'$ different from zero.

On a plane it is possible to define a finer notation than the above, for oriented lines and lines. This is due to the fact that in this case $O^+(E)$ is commutative (cf. 8.D). The *oriented angle* between two oriented lines Δ, Δ', denoted by $\widehat{\Delta, \Delta'}$, is an element of $O^+(E)$ (the one that takes Δ to Δ'); we will use *additive* notation for angles and the group of oriented angles will be *denoted* by $\tilde{\mathfrak{U}}(E)$. Since we also have an orientation for E, we can identify $O^+(E)$, hence $\tilde{\mathfrak{U}}(E)$, with the multiplicative group $\mathbf{U}$ of complex numbers of absolute value 1, which in turn can be identified with $\mathbf{R}/2\pi\mathbf{Z}$ by means of the complex exponential map.

Oriented angles satisfy the following properties:

$$\widehat{f(\Delta)f(\Delta')}=\widehat{\Delta\Delta'} \quad \text{if } f \text{ is a rotation;}$$

$$\widehat{f(\Delta)f(\Delta')}=-\widehat{\Delta\Delta'} \quad \text{if } f \text{ is a reflection;}$$

$$\widehat{\Delta\Delta'}+\widehat{\Delta'\Delta''}=\widehat{\Delta\Delta''} \quad \text{for all } \Delta, \Delta', \Delta'';$$

$$\widehat{\Delta\Delta'}=\widehat{\Delta_1\Delta_1'} \quad \text{is equivalent to } \widehat{\Delta,\Delta_1}=\widehat{\Delta'\Delta_1'}$$

The *bisectors* of Δ, Δ' are the oriented lines Σ such that $\widehat{\Delta\Sigma}=\widehat{\Sigma\Delta}$; there are two of them, opposite to each other.

For non-oriented lines, it is still possible to define an oriented angle $\widehat{DD'}$, which verifies the four properties above. The corresponding group is *denoted* by $\mathfrak{U}(E)$. The lines S such that $\widehat{DS}=\widehat{SD'}$, which are also called *bisectors* of D and D', are again two, but they are orthogonal. This finer notion of angle will play an essential role in chapter 10; see 10.D, problems 10.3 and 10.7.

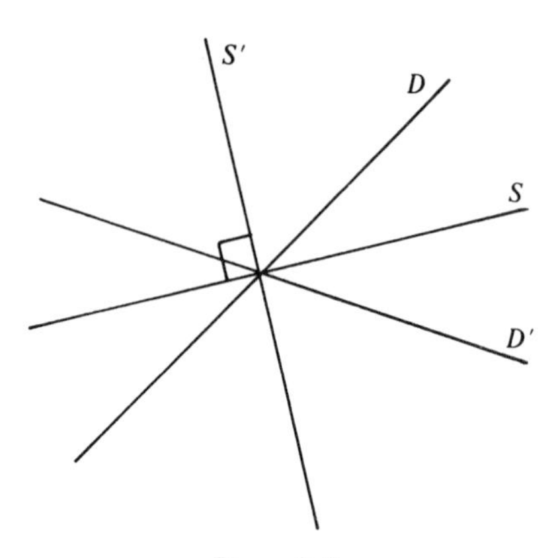

Figure 8.F.

8.G Similarities ([B, 8.8])

Similarities are the elements of $GL(E)$ that are the product of an isometry and a vector homothety; the group of similarities is *denoted* by $GO(E)$. We put $GO^+(E)=GO(E)\cap GL^+(E)$; this is the group of *orientation-preserving* similarities.

There are many equivalent definitions for similarities. Essentially, they are the linear maps which preserve ratios between norms, i.e. the maps f such that there is a real μ with $\|f(x)\|=\mu\|(x)\|$ for every x. Also, f is a similarity if and only if it always takes two orthogonal lines into orthogonal lines. It is certainly true that similarities preserve angles between lines or oriented lines.

8.H Isotropic Cone, Isotropic Lines, Laguerre Formula ([B, 8.8])

With the notion of complexification (Chapter 7) we obtain some nice and useful facts about similarities. Let $N: x \mapsto \|x\|^2 = \varphi(x, x)$ be the quadratic form that defines the Euclidean structure on E. Let E^C and $N^C: E^C \to \mathbf{C}$ be the complexifications of E and N, and introduce the kernel $(N^C)^{-1}(0)$ (observe that it is invariant under the conjugation σ; see Chapter 7). This kernel is a cone, called the *isotropic cone* of E (although it sits in E^C; in E it does not exist!)

If, in particular, E is a plane, the isotropic cone consists of two lines I, J called the *isotropic lines* of E (again, they are actually lines in E^C); they satisfy $\sigma(I) = J$. Observe that I, J form a *pair* of which neither element is distinguished *a priori*. On the other hand, if E is oriented, we can *call* I the line whose slope relative to an orthonormal basis is $i = \sqrt{-1}$.

The similarities of E can be characterized as the linear automorphisms which leave invariant the isotropic cone. For the case of the plane, we have $f \in \mathrm{GO}^+(E)$ if and only if $f^C(I) = I$ (this automatically implies $f^C(J) = J$), and $f \in \mathrm{GO}^-(E)$ if and only if $f(I) = J$.

Given two lines D, D' on a Euclidean plane E, we can consider the cross-ratio of the four lines D^C, D'^C, I, J in E^C (see 6.A). The Laguerre formula relates this cross-ratio to the angle $\overline{DD'}$ in the following way:

$$\overline{DD'} = \tfrac{1}{2}\left|\log([D^C, D'^C, I, J])\right|.$$

For instance, orthogonality of D, D' is equivalent to the harmonic conjugation $[D^C, D'^C, I, J] = -1$; see 17.C.

There is a finer Laguerre formula for $\widehat{DD'}$, the oriented angle between D and D'; it says that $[D^C, D'^C, I, J] = e^{2it}$, where t is any measure of the angle $\widehat{DD'}$. Together with 6.D, this shows that the condition "the angle $\widehat{DD'}$ is constant" can be interpreted as saying that D' can be obtained from D by a fixed homography.

8.I Quaternions and Rotations

Due to lack of space, we can only mention here, for the sake of completeness, that quaternions provide a very nice way to study $O^+(3)$ and $O^+(4)$; see [B, 8.9].

8.J Orientation, Vector Products, Gram Determinants ([B, 8.11])

In this section we consider an *oriented* n-dimensional Euclidean vector space. Such a space possesses a *canonical volume form* λ_E, defined by the conditions that it is an exterior n-form (i.e. an alternating multilinear n-form), and its value $\lambda_E(e_1,\ldots,e_n)$ on any positive orthonormal basis is 1. In dimension 3 this form is often called the *mixed product* and is *denoted* simply by $\lambda_E(x, y, z) = (x, y, z)$.

The *vector product* of $n-1$ vectors $x_1,\ldots,x_{n-1}$ in E is the vector $x_1 \times \cdots \times x_{n-1}$ in E defined by the following duality relation:

$$(x_1 \times \cdots \times x_{n-1}|y) = \lambda_E(x_1,\ldots,x_{n-1}, y) \quad \text{for all } y \in E.$$

The vector product is zero if and only if the x_i are linearly dependent. Otherwise, it is orthogonal to all the x_i, and added to them it forms a positive basis; finally, its norm is given by

$$\|x_1 \times \cdots \times x_{n-1}\|^2 = \operatorname{Gram}(x_1,\ldots,x_{n-1}),$$

where $\operatorname{Gram}(x_1,\ldots,x_p)$ *denotes*, in general, the determinant

$$\operatorname{Gram}(x_1,\ldots,x_p) = \det\big((x_i|x_j)\big) = \begin{vmatrix} \|x_1\|^2 & (x_1|x_2) & \cdots & (x_1|x_p) \\ (x_2|x_1) & \|x_2\|^2 & \cdots & (x_2|x_p) \\ \vdots & \vdots & & \vdots \\ (x_p|x_1) & (x_p|x_2) & \cdots & \|x_p\|^2 \end{vmatrix}.$$

Finally, observe the following relation, very useful in calculating volumes: $\operatorname{Gram}(x_1,\ldots,x_n) = (\lambda_E(x_1,\ldots,x_n))^2$.

Problems

8.1 AN IRREDUCIBLE GROUP CAN LEAVE INVARIANT AT MOST ONE EUCLIDEAN STRUCTURE ([B, 8.12.1]). Let E be a finite-dimensional real vector space, φ and ψ two Euclidean structures on E, and $G \subset \mathrm{GL}(E)$ a subgroup of the linear group of E; we assume G is irreducible (cf. [B, 8.12.2]). Show that if $G \subset O(E, \varphi) \cap O(E, \psi)$ (i.e. if every element of G leaves φ and ψ invariant, cf. 13.E), then φ and ψ are proportional.

8.2 A VECTOR ISOMETRY ALWAYS POSSESSES AN INVARIANT LINE OR PLANE ([B, 8.12.2]). To prove that $f \in O(E)$ always leaves some line or plane invariant, consider some $x \in S(E)$ such that $\|f(x)-x\|$ is minimal, and show that $x, f(x), f^2(x)$ lie in the same plane.

8.3 DIVIDING AN ANGLE BY n ([B, 8.12.7]). Show that, for every $n \in \mathbf{N}^*$ and every $a \in \tilde{\mathfrak{U}}(E)$, the equation $nx = a$ has exactly n solutions in $\tilde{\mathfrak{U}}(E)$. Draw the solutions on a circle for a few values of a, with $n = 2, 3, 4, 5$.

8.4 FIND THREE LINES WHOSE BISECTORS ARE GIVEN ([B, 8.12.20]). We call a *bisector* of two lines A, B in a Euclidean vector space any bisector of A and B in the plane generated by the two lines (cf. 8.F).

Let S, T, U be three lines in a 3-dimensional Euclidean vector space. Find three lines A, B, C such that line S is a bisector of A, B, line T is a bisector of B, C, and line U is a bisector of C, A. Study the possible generalizations of this problem: replacing lines by half-lines, considering more than three lines, or considering higher-dimensional spaces.

8.5 AUTOMORPHISMS OF **H** ([B, 8.12.11]). Show that every automorphism of **H** is of the form $a \mapsto \mathscr{R}(a) + \rho(\mathscr{P}(a))$, where $\rho \in O^+(3)$.

8.6 VECTOR PRODUCTS IN R^3 ([B, 8.12.9]). For every $a, b, c \in \mathbf{R}^3$, prove the following formulas:

$$a \times (b \times c) = (a|c)b - (a|b)c, \tag{1}$$

$$(a \times b, a \times c, b \times c) = (a, b, c)^2, \tag{2}$$

$$(a \times b) \times (a \times c) = (a, b, c)a. \tag{3}$$

Show that $\mathbf{R}^3$, endowed with the operations of addition and vector product, is an anticommutative algebra which, instead of being associative, satisfies the *Jacobi identity*

$$a \times (b \times c) + b \times (c \times a) + c \times (a \times b) = 0.$$

Such an algebra is called a *Lie algebra*.

If p, q, r denote the projections from $\mathbf{R}^3$ onto the three coordinate planes, show that

$$\|a \times b\|^2 = \operatorname{Gram}(p(a), p(b)) + \operatorname{Gram}(q(a), q(b)) + \operatorname{Gram}(r(a), r(b)).$$

Find a geometrical interpretation for this result (see the definition of the Gram determinant in 8.J).

Study the equation $x \times a = b$ (for a and b given); find whether there is a solution and whether it is unique.

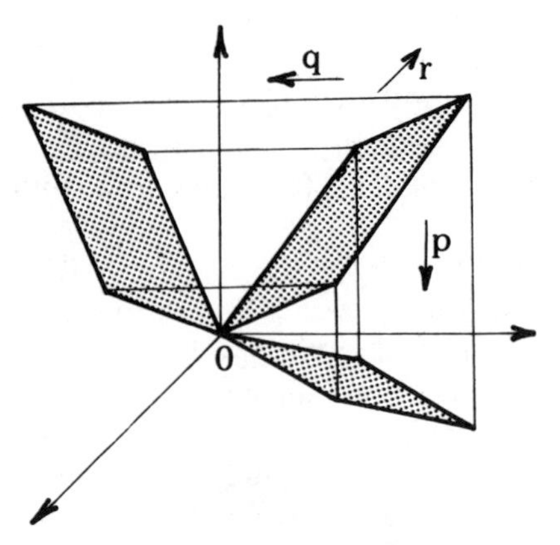

Figure 8.6.

Chapter 9

Euclidean Affine Spaces

9.A Definitions ([B, 9.1])

We consider a real affine space X of finite dimension (which is always *denoted* by n), and whose underlying vector subspace $\vec{X}$ (see 2.A) is endowed with a Euclidean structure; we say that X is a *Euclidean affine space*. The standard example is $\mathbf{R}^n$, considered as an affine space.

We make X into a metric space by taking the *distance function* $d(x, y) = \|\overrightarrow{xy}\|$, which is generally written simply $d(x, y) = xy$. The triangle inequality is *strict*, which means that $xz = xy + yz$ implies that y belongs to the segment $[x, z]$ (cf. 3.C). A Euclidean affine space possesses a canonical topological structure (see 2.G), whose compact sets are the closed sets bounded in the metric d.

The group of *isometries* of X, i.e. the bijections of X such that $f(x)f(y) = xy$ for every x, y in X, is *denoted* by $\mathrm{Is}(X)$. We have the following fundamental fact: an isometry is necessarily an affine map, or, more precisely,

$$\mathrm{Is}(X) = \{ f \in \mathrm{GA}(X) \text{ and } \vec{f} \in O(X) \} \text{ (cf. 2.B).}$$

We can then define (see 2.G) $\mathrm{Is}^{\pm}(X) = \mathrm{Is}(X) \cap \mathrm{GA}^{\pm}(X)$; the elements of $\mathrm{Is}^{+}(X)$ are called (proper) motions, and those of $\mathrm{Is}^{-}(X)$ are sometimes called improper motions.

9.B Subspaces ([B, 9.2])

According to 2.D and 8.C, the involutive isometries of a Euclidean affine space are exactly the (orthogonal) reflections through (affine) subspaces of X. Here, in contrast with the case of vector spaces, a subspace S has no well-defined orthogonal complement; in fact, for any point x in X there is exactly one orthogonal subspace to S passing through x. Call T this subspace. Then the intersection point $y = S \cap T$ is the unique point in S whose distance to x is minimal (among points in S):

$$d(x, y) = \inf\{d(x, z) : z \in S\} = d(x, S).$$

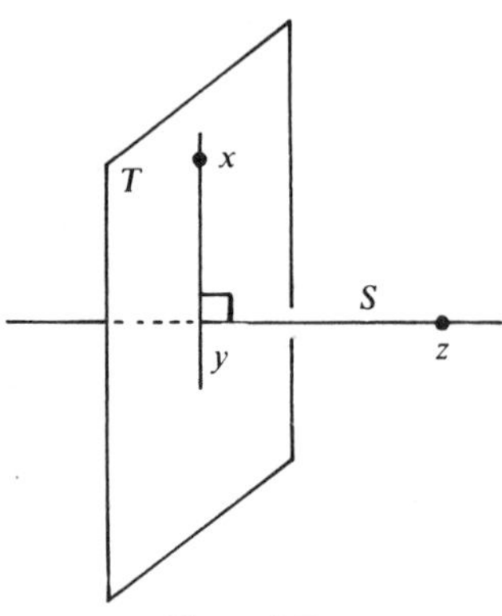

Figure 9.B.

The distance from x to a hyperplane H is easily computed if we know an affine form f such that $H = f^{-1}(0)$ (cf. 2.B); in fact, we have

$$d(x, H) = \frac{|f(x)|}{\|\vec{f}\|}.$$

9.C Structure of an Element of Is(X) ([B, 9.3])

Every element of Is(X) (see 2.A) is the composition of a translation of X and a vector isometry of some vectorialization of X (cf. 8.A). But there is a finer decomposition, as follows: let $f \in \mathrm{Is}(X)$. There exists a *unique pair* consisting of an isometry g of X and a translation t_{ξ} satisfying the following conditions: The set G of points fixed by g is non-empty, and we have $\vec{\xi} \in \vec{G}$ and $f = t_{\xi} \circ g$. Moreover, the following holds:

$$\vec{G} = \mathrm{Ker}(\vec{f} - \mathrm{Id}_{\vec{X}}) \text{ and } t_{\xi} \circ g = g \circ t_{\xi}.$$

Here are some particular cases of this theorem:

If $f \in \mathrm{Is}^+(X)$ and if f is not a translation of the Euclidean affine plane X, then f possesses a fixed point a, so $\vec{f}$ is a rotation of the Euclidean vectorialization X_a; this rotation has a well-defined *angle* in $\tilde{\mathfrak{U}}(\vec{X})$ (cf. 8.F).

Every orientation-reversing isometry of the Euclidean affine plane can be uniquely written as $f = t_{\xi} \circ g$, where $g = \sigma_D$ is the reflection through a line D such that $\vec{\xi} \in \vec{D}$.

Every orientation-preserving isometry of three-dimensional Euclidean affine space, except for translations, is a *screw motion*, i.e. the product (or composition) of a translation t_{ξ} by a vector rotation which fixes all points in a line parallel to $\vec{\xi}$ (see 8.E).

9.D Similarities ([B, 9.5])

The *similarities* of a Euclidean affine space X are the affine maps of X such that $\vec{f} \in \mathrm{GO}(\vec{X})$, the group of (vector) similarities of $\vec{X}$ (cf. 8.G). The group of similarities is *denoted* by $\mathrm{Sim}(X)$. A similarity is *direct* (resp. *inverse*) if $\vec{f} \in \mathrm{GO}^+(X)$ (resp. $\vec{f} \in \mathrm{GO}^-(X)$); the corresponding sets are *denoted* by $\mathrm{Sim}^+(X)$ and $\mathrm{Sim}^-(X)$. Every similarity which is not an isometry possesses a unique fixed point, called its *center*. Similarities have some nice characteristics: first of all, they preserve ratios between distances:

$$\frac{f(x')f(y')}{f(x)f(y)} = \frac{x'y'}{xy}, \quad \text{for any } x, y, x', y'.$$

They are also characterized by taking spheres into spheres, and they preserve angles of lines and oriented lines (see 8.F).

The procedures of complexification (8.H) and completion (7.D) can be applied here as follows: We introduce the complexified projective completion $\tilde{X}^C = X^C \cup \infty_X^C$, where $\infty_X^C = P(\vec{X}^C)$. In $P(\vec{X}^C)$, we use the terminology of 14.A and define the *umbilical* (*locus*) of X as the quadric $\Omega = p((N^C)^{-1}(0))$, which is the image of the isotropic cone of X in the projective space ∞_X^C. (Observe the umbilical of X does not sit in X, but the terminology is convenient.) When $n = 3$, the umbilical is a conic in the projective plane ∞_X^C; when $n = 2$ it is two points $p(I)$, $p(J)$ in the projective line ∞_X^C, the image of the isotropic lines of X. We will still denote these two points by $I = p(I)$ and $J = p(J)$, and call them *cyclical points* of X (for a justification for the name, see 17.C). Choosing either cyclical point is equivalent to choosing an orientation for X (see 8.H).

In order that $f \in \mathrm{GA}(X)$ be a similarity, it is necessary and sufficient that its projective completion $\tilde{f}^C$ satisfy the relation $\tilde{f}^C(\Omega) = \Omega$. When X is a plane, f belongs to $\mathrm{Sim}^+(X)$ if $\tilde{f}^C$ fixes each cyclical point, and to $\mathrm{Sim}^-(X)$ if $\tilde{f}^C$ switches the two.

Orthogonality of two lines D, D' in X corresponds here to the following formula (cf. 8.H):

$$[\infty_{D^C}, \infty_{D'^C}, I, J] = -1,$$

the cross-ratio being taken in the projective line at infinity ∞_X^C. The Laguerre formulas also hold.

9.E Plane Similarities ([B, 9.6])

Plane similarities can be neatly expressed by identifying a Euclidean affine plane X with $\mathbf{C}$ via an orthonormal frame; we associate with $x=(a,b)$ the complex number $a+ib$. Then every direct (resp. inverse) similarity can be written as $z \mapsto \alpha z+\beta$ (resp. $z \mapsto \alpha \bar{z}+\beta$).

A direct plane similarity possesses an *angle* $\alpha \in \tilde{\mathfrak{U}}(\vec{X})$ (cf. 8.H), which is the argument of the complex number α when the similarity is written as above. We have $\overrightarrow{\Delta f(\Delta)}=\alpha$ for any line (oriented or not).

The group $\operatorname{Sim}^{+}(X)$ contains interesting one-real-parameter subgroups which mix homotheties and rotations; expressed in a vectorialization X_a, they are sets of vector similarities of X_a whose matrices (relative to an orthonormal basis) are given by

$$\begin{pmatrix} e^{kt}\cos ht & -e^{kt}\sin ht \\ e^{kt}\sin ht & e^{kt}\cos ht \end{pmatrix},$$

where k, h are fixed real numbers and t is the parameter of the group. The orbits of such a group are called, by definition, *logarithmic spirals*; they have the property that their tangent at an arbitrary point x makes a constant angle with the line $\langle a, x\rangle$.

9.F Metric Properties ([B, 9.7])

The group $\operatorname{Is}(X)$ is as transitive as can be: let k be an arbitrary integer and $\{x_i\}_{i=1,\dots,k}, \{y_i\}_{i=1,\dots,k}$ two k-tuples of points in X such that

$$d(x_i, x_j)=d(y_i, y_j)$$

for all $i, j=1,\dots,k$. Then there exists an isometry f of X taking x_i to y_i for every $i=1,\dots,k$.

But the distances $d_{ij}=d(x_i, x_j)$ are not arbitrary; apart from the triangle inequality and its generalizations, we have the following universal relation, when $k=n+2$:

$$\begin{vmatrix} 0 & 1 & 1 & \cdots & 1 \\ 1 & 0 & d_{1,2}^2 & \cdots & d_{1,n+2}^2 \\ 1 & d_{2,1}^2 & 0 & \cdots & d_{2,n+2}^2 \\ \vdots & \vdots & \vdots & & \vdots \\ 1 & d_{n+2,1}^2 & d_{n+2,2}^2 & \cdots & 0 \end{vmatrix}=0.$$

The *Apollonius formula* uses the barycenter g of the $\{(x_i, y_i)\}$ (see 3.A) and says that for any z we have

$$\sum_i \lambda_i zx_i^2 = \sum_i \lambda_i gx_i^2 + \Big(\sum_i \lambda_i\Big) gz^2.$$

This formula allows the solution of numerous problems involving loci. For example, the set of points z such that $\Sigma_i \lambda_i zx_i^2 = \text{constant}$ is in general a sphere, and so is in particular the locus of points whose ratio of distances to two fixed points is a constant!

9.G Length of Curves ([B, 9.9, 9.10])

Let M be an arbitrary metric space, with distance function d. A *curve* f in M is a map $f:[a,b]\to M$ from an interval $[a,b]$ into M. Its *length*, length (f), is the supremum of the sums

$$\sum d(f(t_i), f(t_{i+1})), \quad a = t_0 < t_1 < \cdots < t_{n-1} < t_n = b.$$

The length can be infinite. A *segment* of M is a curve $[a,b]\to M$ such that $d(f(t), f(t')) = t' - t$ for all $a \le t \le t' \le b$. In particular

$$\text{length}(f) = d(f(a), f(b)).$$

Not all metric spaces have segments. A metric space is called *excellent* if for every pair of points (x, y) in it there is a segment joining them. Euclidean affine spaces are excellent (see also 18.B, 19.A, 19.B).

In a Euclidean affine space X, the distance function $d: X \times X \to \mathbf{R}$ is differentiable at a pair (x, y) such that $x \neq y$. Its derivative at that point is given by

$$\begin{aligned} d'(x,y)(u,v) &= \frac{1}{\|\overrightarrow{xy}\|}(\overrightarrow{xy}|\vec{v} - \vec{u}) \\ &= \|\vec{v}\|\cos\big(\overline{\vec{v},\overrightarrow{xy}}\big) - \|\vec{u}\|\cos\big(\overline{\vec{u},\overrightarrow{xy}}\big). \end{aligned}$$

This is called the *first variation formula*, and the angles are between oriented lines (8.F).

9.H Canonical Measure, Volumes ([B, 9.12])

A finite-dimensional real affine space has a canonical measure only up to a scalar (cf. 2.G); but if it is a Euclidean space, the canonical measure is well-defined. It is obtained by pulling back the Lebesgue measure from $\mathbf{R}^n$ to X by any isometry $X \to \mathbf{R}^n$, and it will be *denoted* by μ.

Then we can talk about the volume $\mathscr{L}(K)$ of a compact K in X, defined as

$$\mathscr{L}(K) = \int_X \chi_K m\mu,$$

where χ_K is the characteristic function of K.

The reader can find in [B, 9.12; 9.13] some important results about volumes: the Stein symmetrization and the isodiametric inequality.

Problems

9.1 SYLVESTER'S THEOREM ([B, 9.14.25]). Prove that if a set of n points $(x_i)_{i=1,\dots,n}$ on an affine plane has the property that every straight line containing two of the x_i also contains a third, then all the points are on the same line.

9.2 BISECTORS ([B, 9.14.3]). Let D, D' be two non-parallel lines in a Euclidean plane X; show that $\{x \in X : d(x, D) = d(x, D')\}$ is formed by the two bisectors of D, D'. What is this set when X is higher-dimensional?

9.3 LIGHT POLYGONS ([B, 9.14.33]). Let C be a plane curve in a Euclidean plane, of class C^1 (cf. 3.6) and strictly convex. Show that for any integer $n \geq 3$ there is at least one n-sided *light polygon* inscribed in C, i.e. a polygon for each vertex of which the exterior bisector of the two sides meeting at this vertex coincides with the tangent to C at this point.

9.4 FINDING THE CENTER OF A SIMILARITY ([B, 9.14.40]). Given four points a, b, a', b' in $\mathbf{R}^2$, construct the centers of the similarities taking a to a' and b to b'.

9.5 INSCRIBING A SQUARE IN A TRIANGLE ([B, 9.14.16]). Given a triangle, inscribe a square inside it: see figure below.

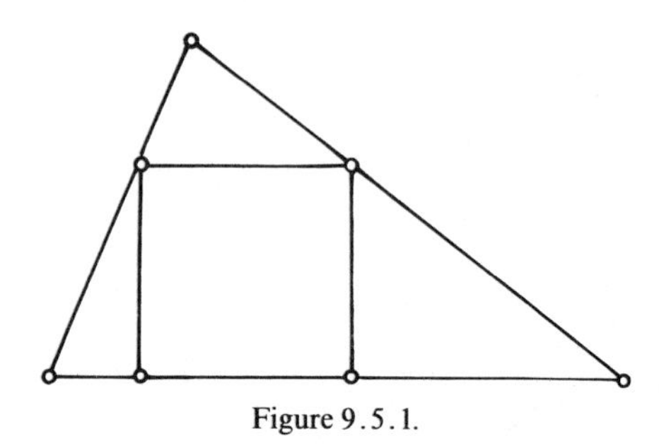

Figure 9.5.1.

9.6 CONVEXITY OF PASCAL LIMAÇONS ([B, 9.14.18]). Given a circle and a point, we call a *Pascal limaçon* the curve obtained by projecting this point to all the tangents to the circle (in general, a curve obtained from another by this procedure is called its "pedal curve"). Study the convexity characteristics of Pascal limaçons as a function of the position of the point relative to the circle.

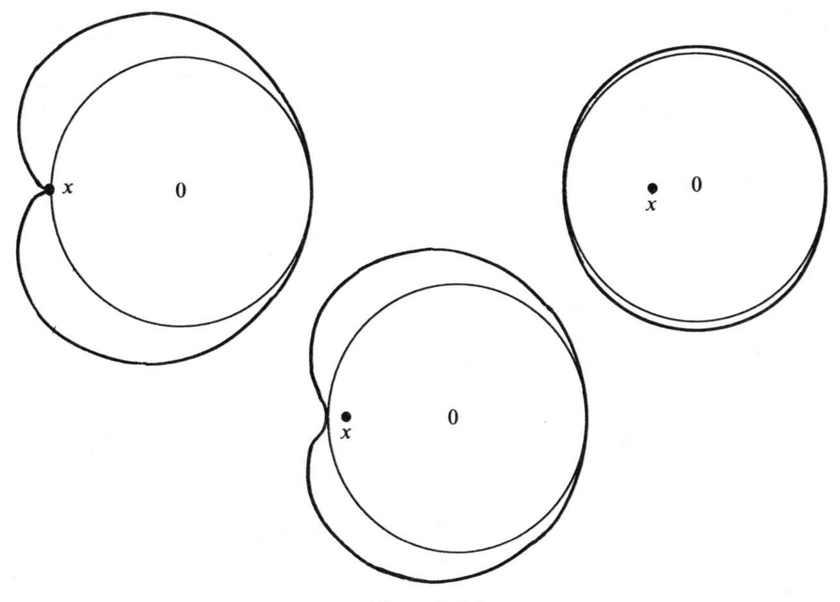

Figure 9.6.1.

Chapter 10

Triangles, Spheres, and Circles

10.A Triangles ([B, 10.1, 10.2, 10.3])

A *triangle* is a simplex (see 2.E) in a Euclidean affine plane, or, in other words, three non-collinear points x, y, z; we can write $\mathscr{T}=\{x, y, z\}$. These three points are the *vertices* of $\mathscr{T}$; we call *sides* the segments $[y, z],[z, x],[x, y]$, and also their lengths, which are conventionally written $a=yz$, $b=zx$, $c=xy$. The *angles* of $\mathscr{T}$ are elements of $]0, \pi[$, measured between oriented lines (cf. 8.F), and written $A=\overrightarrow{xy},\overrightarrow{xz}$, $B=\overrightarrow{yz},\overrightarrow{yx}$, $C=\overrightarrow{zx},\overrightarrow{zy}$. We also have the *semi-perimeter* $p=(a+b+c)/2$.

Triangles satisfy a long list of formulas ([B, 10.3, 10.13.2]), but they are all based on a few essential ones:

$$A+B+C=\pi;$$

$$a^2=b^2+c^2-2bc\cos A;$$

$$\frac{\sin A}{a}=\frac{\sin B}{b}=\frac{\sin C}{C}=\frac{1}{2R};$$

$$S=\sqrt{p(p-a)(p-b)(p-c)}\,;$$

where R denotes the radius of the circle circumscribed around $\mathscr{T}$, and S is the area of $\mathscr{T}$ (cf. 9.H). Observe that $\sin A$, as opposed to $\cos A$, is not enough to determine A uniquely.

Two triangles $\mathscr{T}, \mathscr{T}'$ are called *similar* if there is a similarity (cf. 9.D) taking the vertices of $\mathscr{T}$ to those of $\mathscr{T}'$. Either of the two conditions below characterizes two similar triangles (primes denote values related to $\mathscr{T}'$):

$$\frac{a'}{a} = \frac{b'}{b} = \frac{c'}{c}; \tag{i}$$

$$A = A', \quad B = B', \quad C = C'. \tag{ii}$$

10.B Spheres ([B, 10.7])

The *sphere of radius a and center r* (in an n-dimensional Euclidean affine space) is the set $S(a, r) = \{x \in X : ax = r\}$; when $n = 2$ spheres are generally called *circles*. If Y is a subspace of X, the intersection $Y \cap S(a, r)$ falls into three cases: it is empty if $d(a, Y) > r$, it consists of one point if $d(a, Y) = r$ (in which case we say that Y is *tangent* to $S(a, r)$ at this point), and when $d(a, Y) < r$ it is the sphere in Y of radius $\sqrt{r^2 - ax^2}$ and center x, where x is the point of Y such that $d(a, Y) = ax$ (cf. 9.B).

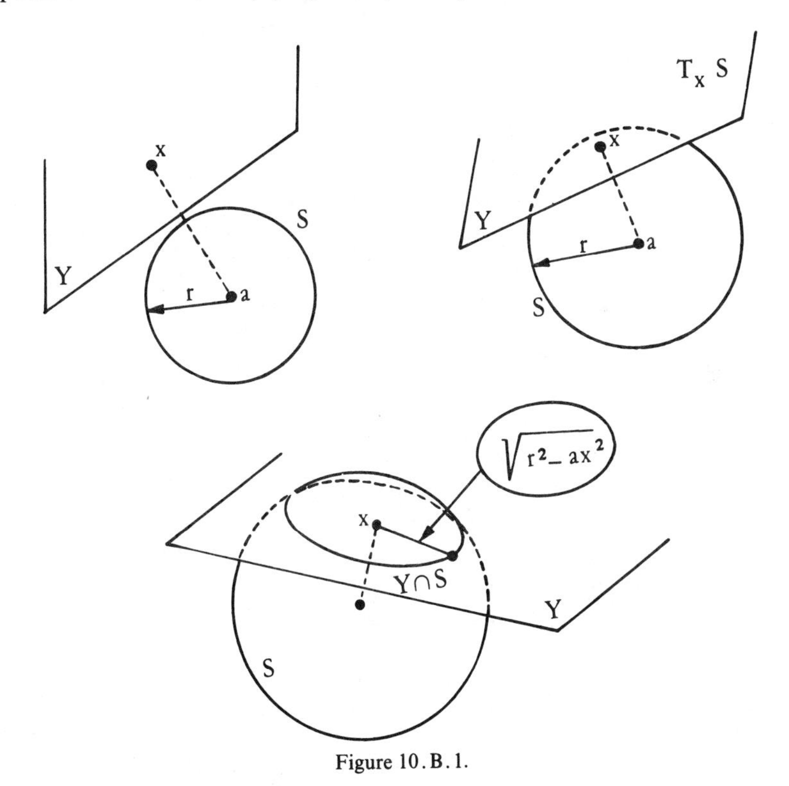

Figure 10.B.1.

The intersection of two spheres $S(a, r)$ and $S(a', r')$ can also be the empty set, a single point or a sphere in a hyperplane. The *angle* of two such spheres is defined as the number φ such that

$$\cos\varphi = \frac{r^2 + r'^2 - aa'^2}{2rr'}.$$

When $\varphi = 0$ or π, the spheres are said to be *tangent*, and when $\varphi = \pi/2$, they are *orthogonal.* It is also easy to define the angle between a sphere and a hyperplane, and between two hyperplanes.

The *power of the point x relative to* $S(a,r)$ is the real number $xa^2 - r^2$; its importance stems from the fact that $xa^2 - r^2 = \pm\, xt.xt'$ whenever a straight line passing through x intersects the sphere at t and t'. (The sign is positive if x is outside and negative if x is inside the sphere.) The locus of points that have same power relative to two non-concentric spheres is a hyperplane, called *radical.*

Polarity with respect to $S(a,r)$ is the geometric transformation that associates with every point $x \neq a$ in X the hyperplane $H(x)$ that intersects $\langle a, x\rangle$ orthogonally and such that $D(a, H(x)) = r^2/ax$.

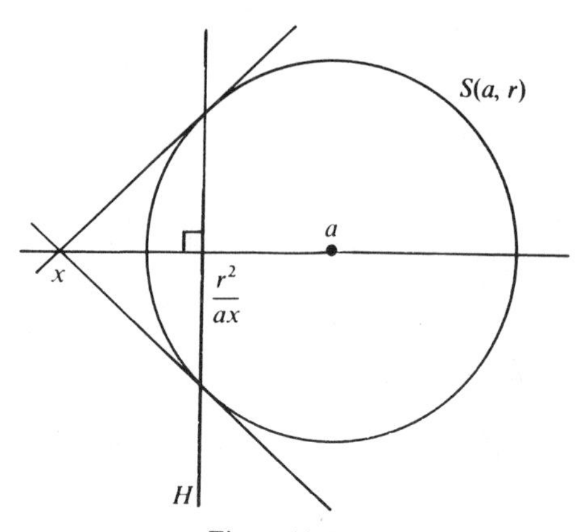

Figure 10.B.2.

This is a very useful duality transformation, which we will encounter again in 11.B and problem 12.4. The image of a subset of X is the *envelope* of the hyperplanes associated with its points by polarity. Note that if $x \in S(a,r)$ then $H(x)$ is the hyperplane tangent to $S(a,r)$ at x. Observe finally that this is a particular case of polarity relative to a quadric (14.E).

10.C Inversion ([B, 10.8])

Let c be a point in X, and $\alpha \in \mathbf{R}^*$; the *inversion with pole c and power* α is the map

$$i = i_{c,\alpha}\colon X\backslash c \to X\backslash c$$

defined by $i(x) = (\alpha/\|x\|^2)\cdot x$ in the vectorialization X_c. This map is involutive.

Inversions transform any sphere or hyperplane into a sphere or hyperplane. They also preserve the angle between two spheres (or a sphere and a hyperplane, or two hyperplanes).

Inversions can be very useful in simplifying certain proofs of properties which only involve angles; taking as pole a point common to several spheres, each sphere is transformed into a hyperplane. See for example 10.D and problem 18.7.

10.D Circles on the Plane and Oriented Angles between Lines ([B, 10.9])

The most important property of circles on the plane is that "the central angle is twice the internal angle"; in other words, if C is a circle of radius ω, and a, b, x are two points on C, we have

$$\widehat{\overrightarrow{\omega a}, \overrightarrow{\omega b}} = 2\widehat{\overrightarrow{xa}, \overrightarrow{xb}}.$$

Here the angles are oriented, and the lines are oriented on the left-hand side and non-oriented on the right-hand side. The factor 2 makes the equality work, since the two angles belong to different groups.

The essential corollary of this result is the following condition for four points to be on the same line or circle: *four distinct points a, b, c, d are collinear or cocyclic if and only if*

$$\widehat{\overrightarrow{ca}, \overrightarrow{cb}} = \widehat{\overrightarrow{da}, \overrightarrow{db}}.$$

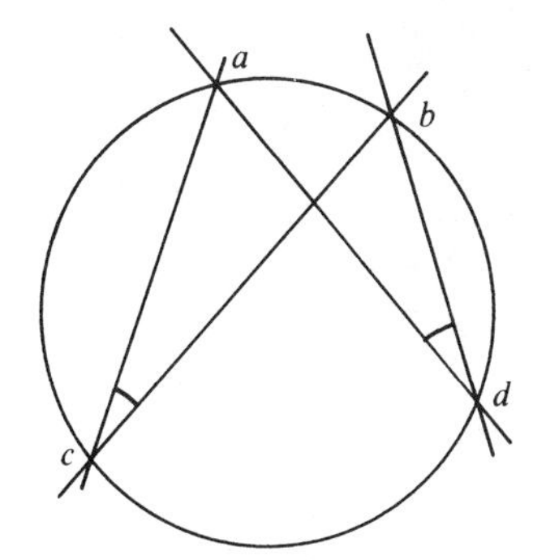

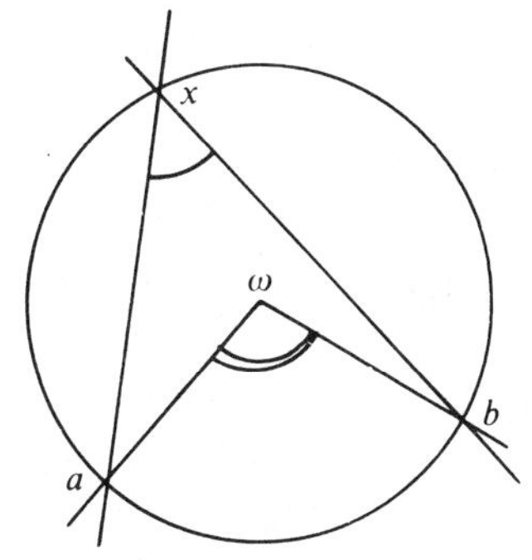

Figure 10.D.1.

Two classical corollaries of this are the following:

The Simson line. Given a triangle $\{a, b, c\}$, a point x belongs to the circle circumscribed around this triangle if and only if the three projections of x on the sides of the triangle are collinear.

The six circles of Miguel. Let C_i be four circles, and let their intersections be $C_1 \cap C_2 = \{a, a'\}$, $C_2 \cap C_3 = \{b, b'\}$, $C_3 \cap C_4 = \{c, c'\}$, $C_4 \cap C_1 = \{d, d'\}$; then a, b, c, d are cocyclic if and only if a', b', c', d' are.

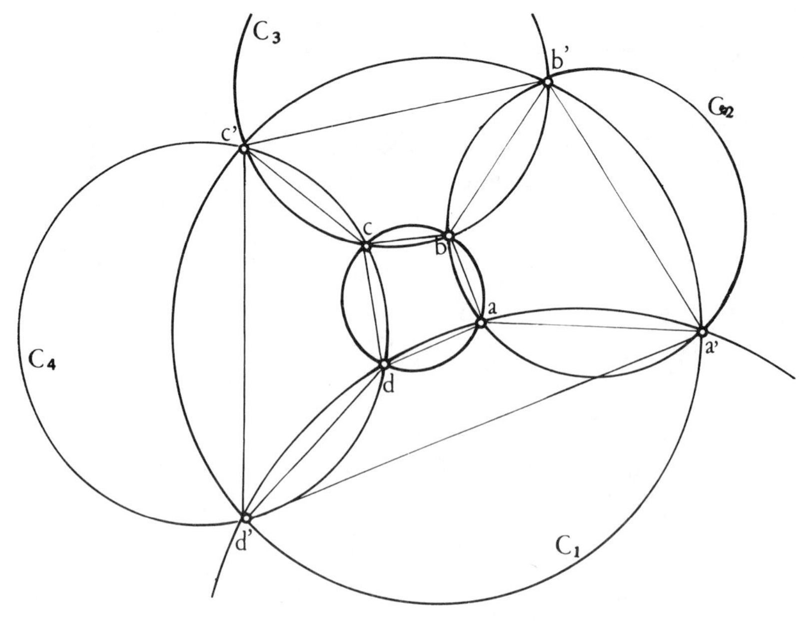

Figure 10.D.2.

Finally, we observe that a pair C, C' of circles in the plane can always be transformed by inversion into a very simple figure: if they intersect, the figure is two lines (cf. 10.C), and if not, it is two concentric circles. In the latter case there are two possible poles for the inversion, and they are called the *limit points* of the pair C, C'. They can be obtained by considering *pencils* of circles ([B, 10.10]); they are the two points common to all circles which are orthogonal to both C and C'.

Problems

10.1 THE HEIGHTS OF A TRIANGLE ARE CONCURRENT ([B, 10.13.1]).

(i) Let T be a triangle; form a triangle T' by taking a parallel line to each side of T passing through the opposite vertex. Deduce from this construction that all three heights of a triangle intersect in a point.

(ii) Deduce the same fact from the following property (to be shown below): for every four points $\{a, b, c, d\}$ in a Euclidean plane, we have

$$(\overrightarrow{ab}|\overrightarrow{cd}) + (\overrightarrow{ac}|\overrightarrow{db}) + (\overrightarrow{ad}|\overrightarrow{bc}) = 0.$$

(iii) Deduce it now from the Theorem of Ceva (problem 2.1).

(iv) Deduce it by considering cyclical points (9.D) and using the Desargues theorem on pencils of conics (14.D).

10.2 TRIANGLES INSCRIBED IN A CIRCLE AND CIRCUMSCRIBED AROUND ANOTHER ([B, 10.13.3]). Given a triangle $\mathscr{T}$, show that the radius R of the circumscribed circle C, the radius r of the inscribed circle Γ and the distance d between the two centers satisfy the relation $R^2 - 2Rr = d^2$. Show that if, conversely, two circles C, Γ of radii R, r and whose centers are d units apart satisfy the condition $R^2 - 2Rr = d^2$, then for any $x \in C$ there is a triangle $\mathscr{T}$ inscribed in C and circumscribed around Γ, and having x as a vertex. Deduce from this fact the theorem of Poncelet for the case of triangles (16.H).

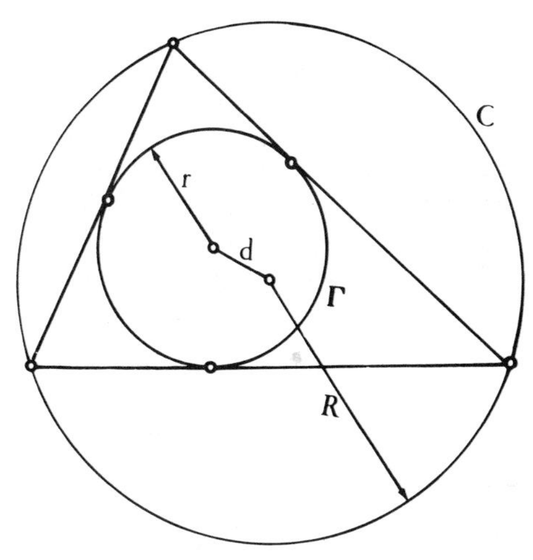

Figure 10.2.

10.3 LINE OF THE IMAGES ([B, 10.13.16]). Let $\mathscr{T}$ be a triangle and D a line. Show that the reflections of D through each side have a point in common if and only if D goes through the intersection of the heights of $\mathscr{T}$. What happens when D rotates around this intersection point?

10.4 MINIMIZING THE SUM OF THE DISTANCES TO FOUR FIXED POINTS ([B, 10.13.8]). Given a convex quadrilateral, find the minimum of the function "sum of the distances from a variable point to the vertices of the quadrilateral". Do the same for a convex hexagon circumscribed around an ellipse.

10.5 CONSTRUCTING A SQUARE WHOSE SIDES PASS THROUGH FOUR POINTS ([B, 10.13.28]). Given four points a, b, c, d in a Euclidean plane, construct a square $ABCD$ such that $a \in AB$, $b \in BC$, $c \in CD$, $d \in DA$.

10.6 PROBLEM OF NAPOLEON-MASCHERONI ([B, 10.11.2]). Find the center of a given circle using just the compass (no ruler!). More generally, show the following theorem of Mohr-Mascheroni: "Every construction that can be done with ruler and compass is also possible with the compass alone."

10.7 A CHAIN OF THEOREMS ([B, 10.13.19]). In this problem, the objects are supposed to be "in general position". Let $(D_i)_{i=1,2,3,4}$ be four lines in a Euclidean plane, and let C_i be the circle circumscribed around the triangle formed by the three lines D_j $(j \neq i)$; show that the four circles C_i have a point in common. (See another explanation in [B, 17.4.3.5].) Draw an illustrative picture.

Let $(D_i)_{i=1,\dots,5}$ be five lines in a Euclidean plane, and let p_i be the point associated in the preceding theorem with the four lines D_j $(j \neq i)$; show that the five points p_i belong to the same circle. Draw a picture.

State and prove a chain of theorems whose two first elements are the results above.

10.8 FORD CIRCLES ([B, 10.13.29]). Consider three pairwise tangent circles, and also tangent to a line D, in the situation shown in figure 10.8.1. If r, s are the radii of γ, δ, and ρ is the radius of the small circle, find ρ, $\overline{ax}$, $\overline{bx}$ as functions of $\overline{ab}$, r, s, where a, b, x are the tangency points of γ, δ and the small circle with D.

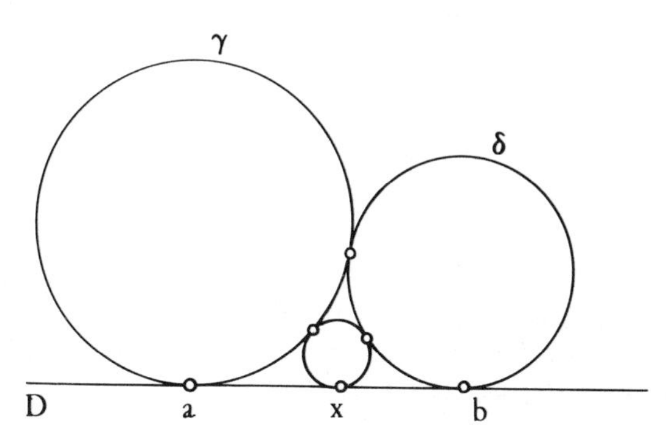

Figure 10.8.1.

Starting now from two circles whose equations, in an orthonormal frame, are $x^2 + y^2 - y = 0$ and $x^2 + y^2 - 2x - y + 1 = 0$, use the induction to construct circles as shown in figure 10.8.2. Show that the tangency points of these circles with the x-axis always have rational abscissas. Are all rational numbers in $[0, 1]$ obtained in this way?

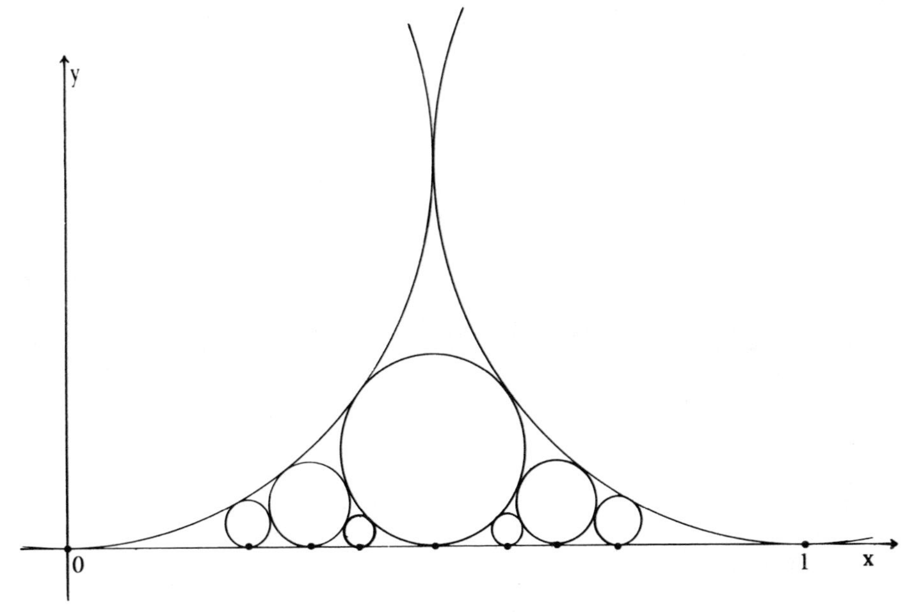

Figure 10.8.2.

Chapter 11

Convex Sets

We'll be working in a real affine space X of real dimension d.

11.A Definition; First Properties ([B, 11.1, 11.2])

A subset S of X is called *convex* if the segment $[x, y]$ (cf. 3.C) is contained in S for every $x, y \in S$. A weaker condition is that defining star-shaped sets: S is *star-shaped with center* $a \in X$ if $[a, x] \subset S$ for every $x \in S$. In $\mathbf{R}$ convex sets are the same as intervals.

An arbitrary intersection of convex sets is convex; in particular every subset A of X gives rise to a smallest convex set containing it, which we'll *denote* by $\mathscr{E}(A)$ and call *convex hull* of A. The convex hull $\mathscr{E}(A)$ of A is characterized as the set of barycenters of families $\{(x_i, \lambda_i)\}$, where each x_i is in A and each $\lambda_i \geq 0$. A theorem of Caratheodory asserts that this holds even if we take only families of at most $d+1$ elements.

If S is convex, so are its closure $\bar{S}$ and its interior $\mathring{S}$. The *dimension* of a convex set is that of the smallest affine subspace $\langle S \rangle$ containing it; saying that $\dim S = d$ is the same as saying that S has non-empty interior.

11.B The Hahn-Banach Theorem. Supporting Hyperplanes ([B, 11.4, 11.5])

The Hahn-Banach theorem states the following: Let A be a non-empty open convex set in X, and let L be an affine subspace of X such that $A \cap L \neq \phi$. Then there is a hyperplane of X which contains L and does not meet A.

We immediately conclude that through each point in the boundary of a closed convex set S there passes at least one *supporting hyperplane* of S, i.e. a hyperplane H intersecting the boundary of S and such that S is entirely contained in one of the closed subspaces defined by H (cf. 2.G).

Next, we can study the following polarity operation for the convex sets (compare with 10.B and 14.E): We assume X is a Euclidean vector space with origin O. We associate to every subset A of X its *polar reciprocal* A^*, defined by

$$A^* = \{\, y \in X : (x|y) \leq 1 \quad \text{for every } x \in A\,\}.$$

The correspondence $A \mapsto A^*$ is a good duality whenever A is a compact convex set containing O in its interior; we have $(A^*)^* = A$ and the support hyperplanes of A are the polar hyperplanes of the points of A^*, relative to the unit sphere $S(0,1)$.

Finally, the Hahn-Banach theorem implies Helly's theorem: For X a (d-dimensional) affine space, let $\mathcal{F}$ be a family of convex compact subsets of X such that the intersection of any $d+1$ elements of $\mathcal{F}$ is non-empty; then the intersection of all the elements of $\mathcal{F}$ is non-empty as well.

11.C Boundary Points of a Convex Set ([B, 11.6])

They can be classified in many ways; one of the most useful is the following: A point x in the boundary of a convex set A is called *extremal* if whenever $x = (y+z)/2$ with $y, z \in A$ we have $y = z$. The theorem of Krein and Milman says that a convex compact set is the convex hull of its extremal points.

Problems

11.1 PARTITION OF THE PLANE INTO CONVEX SETS ([B, 11.9.22]). Find all partitions of the plane into two convex sets.

11.2 EXTREMAL POINTS IN TWO DIMENSIONS ([B, 11.9.8]). Prove that the extremal points of a convex set in the plane form a closed set.

11.3 HILBERT GEOMETRY ([B, 11.9.8]). Let A be a convex compact set of X whose interior is non-empty. Given two distinct points x, y of $\mathring{A}$, put $d(x,y) = |\log[x,y,u,v]|$, where u, v are the two points where the line $\langle x, y\rangle$ meets the boundary of A, and $[.,.,.,.]$ denotes the cross-ratio (cf. 6.A or [B, 6]). Show that $d: \mathring{A} \times \mathring{A} \to \mathbf{R}$, defined by the equation above and $d(x,x) = 0$, $\forall x \in \mathring{A}$, is a metric. Show that this metric is excellent (9.G). Study the relation between the strict triangle inequality (9.A) and the nature of the boundary points of A.

11.4 THE LUCAS THEOREM ([B, 11.9.21]). Let P be a polynomial with complex coefficients, and let P' be its derivative. Show that in the affine space **C**, all the roots of P' belong to the convex hull of the roots of P. When P has degree three and distinct roots a, b, c, show that there is an ellipse inscribed in the triangle $\{a, b, c\}$ and whose foci are the roots u, v of P'.

11.5 STAR-SHAPED SETS ([B, 11.9.20]). Given a subset A of an affine space X, consider the set $N(A)$ of points a such that A is star-shaped with center a. Show that $N(A)$ is convex. Find $N(A)$ for a number of shapes of A.

Chapter 12

Polytopes; Compact Convex Sets

12.A Polytopes ([B, 12.1, 12.2, 12.3])

We'll be working in a d-dimensional real affine space X, for d finite. A *polytope* is a convex compact set with non-empty interior, which can be realized as the intersection of a *finite* number of closed half-spaces of X (cf. 2.G). We shall assume there are no *superfluous* half-spaces in the intersection. For $d = 2$ we use the word *polygon*.

The *faces* of a polytope P are the intersections of its boundary with the hyperplanes that define P. A face is itself a polytope (of dimension $d-1$) inside the hyperplane which contains it. We define by induction the k-faces of P as the faces of all the $(k+1)$-faces of P; 1-faces are called *edges* and 0-faces are called *vertices*.

From now on X is Euclidean affine.

Give a polytope P, we can define its *volume* $\mathscr{L}(P)$ and its *area* $\mathfrak{A}(P)$; the area is the sum of the volumes of all the faces (considered as $(d-1)$-polytopes).

12.B Convex Compact Sets ([B, 12.9, 12.10, 12.11])

A convex compact set whose interior is non-empty can be approximated, in the best sense of the word, by polytopes. This allows us to define, by passing to the limit, the *volume* $\mathscr{L}(C)$ and the *area* $\mathfrak{A}(C)$ of a convex compact set C. These two notions possess very nice properties. For example, for every real positive λ we put

$$B(C,\lambda)=\{x\in X: d(x,C)\leq\lambda\}.$$

Then

$$\mathfrak{A}(C)=\lim_{\lambda\to 0}\frac{\mathscr{L}(B(C,\lambda))-\mathscr{L}(C)}{\lambda}$$

Next, the isoperimetric inequality:

$$\frac{\mathfrak{A}(C)}{\alpha(d)}\geq\left(\frac{\mathscr{L}(C)}{\beta(d)}\right)^{(d-1)/d},$$

which holds for every convex compact set with non-empty interior; here $\alpha(d)$ (resp. $\beta(d)$) denote the area (resp. the volume) of the unit sphere in $\mathbf{R}^d$. Moreover, equality only holds if C is a sphere.

Notice that the two above results can be generalized for subsets of X which are "nice" enough though not convex: for example differentiable manifolds (cf. problem 12.3).

12.C Regular Polytopes ([B, 12.4, 12.5, 12.6])

A polygon is called *regular* if all its sides have same length and all its angles are equal. For $n\geq 3$ there are always n-sided regular polygons, and all n-sided regular polygons are similar. A regular polygon can always be inscribed in a circle. Its symmetry group possesses $2n$ elements, and is called the *dihedral group* of order $2n$; it acts simply transitively (cf. 1.D) on the pairs formed by a vertex and a side that ends at this edge. See problem 12.1.

The easiest way to generalize this notion for $d\geq 3$ is the following: consider the d-tuples $(F_0, F_1,\ldots, F_{d-1})$ such that the F_i are i-faces and $F_{i-1}\subset F_i$, $1\leq i\leq d-1$. Then the polytope P is *regular* if its isometry group $G(P)$ acts transitively on the set of such d-tuples, in which case the action is also *simply* transitive (cf. 1.D). A regular polytope can always be inscribed in a sphere.

The following are examples of regular polytopes: the *cocube* Coc_d whose vertices are $\pm e_i$ $(i=1,\ldots,d)$ in an orthonormal basis of X; it has $2d$ vertices and 2^d faces. The *cube* Cub_d has as vertices the 2^d points whose coordinates are given by $(\pm 1,\ldots,\pm 1)$ (in an orthonormal basis), and it has $2d$ faces. And the *regular simplex* Sim_d which, considered in the hyperplane $\Sigma_{i=1}^{d+1}x_i=1$, has as vertices the $d+1$ points e_i of an orthonormal basis; it also has $d+1$ faces.

It can be shown that for $d\geq 5$ these are the only regular polytopes (up to a similarity, of course). On the other hand, for $d=3$ there are two exceptional regular polytopes, the dodecahedron and the icosahedron (Figure 1.F), and for $d=4$ there are three exceptions. The existence of these exceptional polytopes is not obvious; the problem can be reduced to the case $d=3$. Problem 19.1 gives a construction of the regular icosahedron; another one consists in cleverly placing regular pentagons on a cube ([B, 12.5.5]), and finally one can leave aside all subtlety and just give the coordinates for the vertices: $(0, \pm\tau, \pm 1), (\pm 1, 0, \pm\tau), (\pm\tau, \pm 1, 0)$, where $\tau = (\sqrt{5}+1)/2$.

Problems

12.1 REGULAR PENTAGON ([B, 12.12.4]). Justify the following two constructions for the regular pentagon:

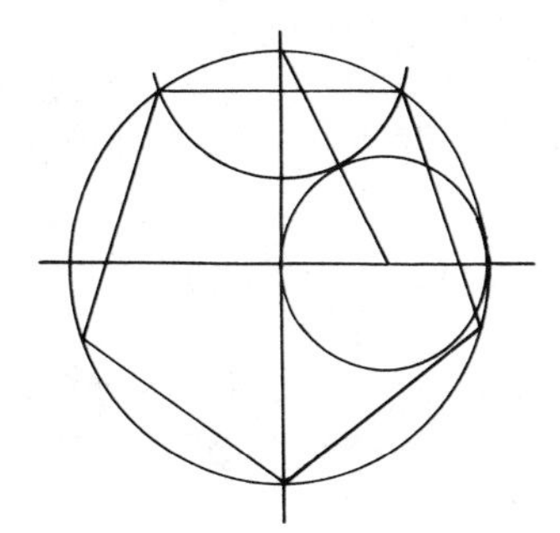

Figure 12.1.1.

Figure 12.1.2.

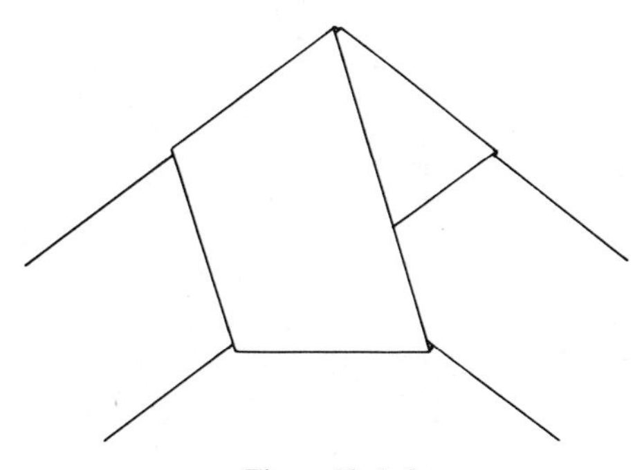

Figure 12.1.3.

12.2 ORDER OF THE GROUP OF A REGULAR POLYHEDRON ([B, 12.12.10]). Show that, for a regular polyhedron (i.e. a 3-dimensional polytope), the order of its group of isometries is equal to four times the number of its edges.

12.3 THEOREMS OF GULDIN ([B, 12.12.10.9]). Consider a compact set K of a plane P in the 3-dimensional Euclidean space E. Show that the volume of the compact set of C of E, generated by rotating K around a line D of P which does not intersect K, is given by the formula

$$\mathscr{L}_E(C)=2\pi\cdot d(g,D)\cdot\mathscr{L}_P(K),$$

where $g=\operatorname{cent}(K)$ denotes the centroid of K (see 2.G).

If the boundary of K is considered as a homogeneous wire and h is the center of mass of this wire (in the usual sense), show that the area of C is given by the formula

$$\mathfrak{A}_E(C)=2\pi\cdot d(h,D)\cdot\mathfrak{A}_P(K).$$

(Both areas are understood in the sense of differentiable manifolds.)

Find applications of this formula, as well as special cases of volumes or areas already known.

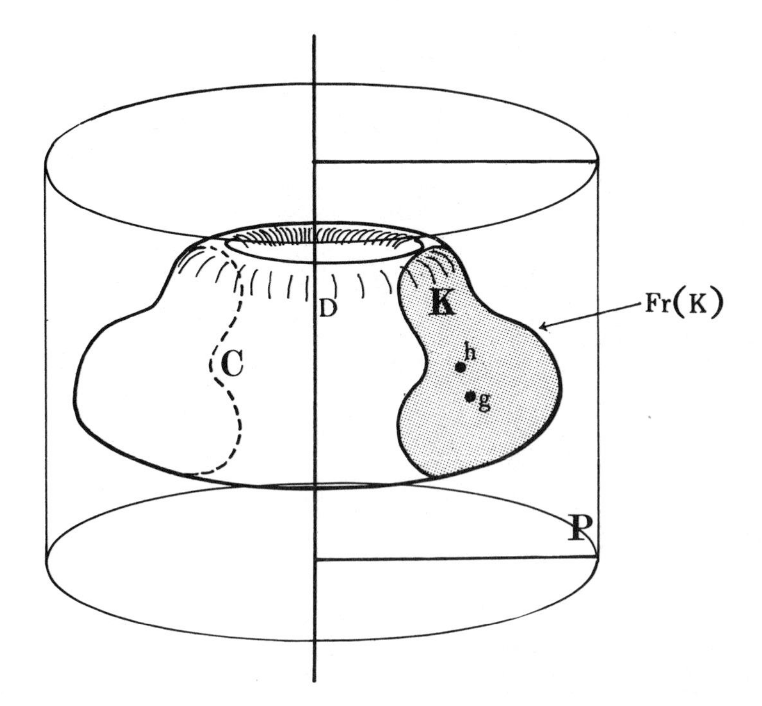

Figure 12.3.1.

12.4 VOLUME OF POLAR RECIPROCALS OF ELLIPSOIDS ([B, 12.12.2]). Show that if E is an ellipsoid in the Euclidean vector space X, containing O in its interior, then its polar reciprocal E^* (see 11.B) is an ellipsoid with the same property. Their volumes satisfy $\mathscr{L}(E)\mathscr{L}(E^*) \geq (\beta(d))^2$, and equality takes place if and only if O is the center of E (for the definition of $\beta(d)$, see 12.B or [B, 9.12.4]).

12.5 THE BLASCHKE ROLLING THEOREM ([B, 12.12.4]). Let C be a compact convex set in the plane whose boundary is a biregular curve (cf. M. Berger and B. Gostiaux, *Géométrie Différentielle*, Armand Colin, 1972, p. 309) of class C^2. Let A (resp. a) be a point on the boundary of C where the curvature is maximal (resp. minimal). Show that the osculating circle γ at a can roll all around the boundary, always staying inside C, and the boundary can roll all around the osculating circle Γ at A. Is this still true if we replace γ by the largest circle contained in C or Γ by the smallest circle containing C?

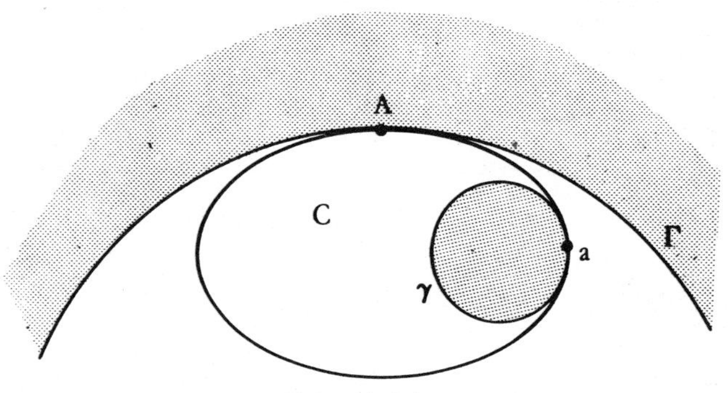

Figure 12.5.1.

Chapter 13
Quadratic Forms

Here E is a vector space of finite dimension n over a commutative field of characteristic different from 2.

13.A Definitions ([B, 13.1])

A *quadratic form* q on E is a map of the form

$$x \mapsto q(x) = P(x, x)$$

where P is a symmetric bilinear form over E; such a P is well determined by q by the formula

$$P(x, y) = \tfrac{1}{2}(q(x+y) - q(x) - q(y)),$$

and is called the *polar form* of q. Over a subspace F of E, the restriction of q is still a quadratic form, *denoted* by $q|_F$.

In a basis $\{e_i\}$ of E, the form q is determined by its matrix

$$A = (a_{ij}) = \big(P(e_i, e_j)\big).$$

If f is an automorphism of E, and q has matrix A in the basis $\{e_i\}$, the matrix of q in the new basis $\{f(e_i)\}$ will be ${}^tSAS$ (where ${}^tS$ denotes the transpose of S).

Let E, E' be two vector spaces (over the same field), and $f: E \to E'$ a linear map. If q' is a quadratic form over E', the *pull-back* of q' by f is the quadratic form q, written f^*q', defined by

$$q(x) = (f^*q')(x) = q'(f(x))$$

for every $x \in E$.

13.B Equivalence, Classification ([B, 13.1, 13.4, 13.5])

Two quadratic forms q and q', over E and E', respectively, are called *equivalent* if there is an isomorphism $f: E \to E'$ such that $q = f^*q'$. *Classifying* quadratic forms (over a field K in dimension n) means finding the equivalence classes of quadratic forms over K^n. This problem has not been solved, except for special fields K.

In dimension 1, the problem is easy because the formula above, $A \mapsto {}^tSAS$, shows that the determinant $\det A$ has an image in the quotient $K/(K^*)^2$ (called the *discriminant*) which is invariant under equivalence; this is true in general, but in dimension 1 the converse is also true, so the equivalence classes are in one-to-one correspondence with the elements of $K/(K^*)^2$. See problems 13.2 and 13.3.

Over $\mathbf{C}^n$ (or, more generally, any vector space over an algebraically closed field), there are exactly $n+1$ classes of quadratic forms, each represented by one of the forms $\sum_{i=1}^k z_i^2$, for $k = 0, 1, \ldots, n$ (the classifying invariant k is called the *rank* of the form).

Over $\mathbf{R}^n$, there are $[(n+1)(n+2)]/2$ classes of quadratic forms, represented by the following forms:

$$\sum_{i=1}^{p} x_i^2 - \sum_{i=p+1}^{n} x_i^2.$$

This result is called "Sylvester's law of inertia".

A form is called *neutral* if it is equivalent to the form $q = 2\sum_{i=1}^p x_i y_i$ over K^{2p}; a space endowed with a neutral form is called an *Artin space* and is *denoted* by Art_{2p}. Examples are $\sum_{i=1}^{2p} z_i^2$ over $\mathbf{C}^{2p}$, and $\sum_{i=1}^p x_i^2 - \sum_{i=p+1}^{2p} x_i^2$ over $\mathbf{R}^{2p}$.

We mention that every quadratic form, for an arbitrary field, can always be *diagonalized*, i.e. we can find a basis in which its matrix A is diagonal: in this basis q is written as $\sum_i a_i x_i^2$, for some $a_i \in K$. If E is moreover a *Euclidean* space, we can choose this basis to be orthonormal.

13.C Rank, Degeneracy, Isotropy ([B, 13.2])

Given a quadratic form q, we *denote* by φ the linear map from E into its dual E^* defined by $\varphi(y) = \{P(x, y)\}$, where P is the polar form. The *rank* of q is defined as the rank of φ; the *radical* $\mathrm{rad}(q)$ is the kernel of φ, i.e. the subspace of E containing the elements y such that $P(x, y) = 0$ for every $x \in E$. Thus we have $\operatorname{rank}(q) + \dim(\operatorname{rad}(q)) = \dim E$. We say that q is *non-degenerate* if its rank is equal to n, and *degenerate* otherwise.

An important difference between the general and the Euclidean case is that here the restriction of a non-degenerate form q to a subspace can be degenerate. A subspace F of E is called *singular* (resp. *non-singular*) if $q|_F$ is degenerate (resp. non-degenerate); it is called a *null subspace* if $q|_F = 0$, i.e. if q is identically zero over F.

The *isotropic cone* of q is the subset $q^{-1}(0)$ of E; its elements are called *isotropic* vectors. An *anisotropic* form is one for which the isotropic cone consists of the zero vector only.

From now on q will be assumed non-degenerate.

13.D Orthogonality ([B, 13.3])

Let E be a vector space endowed with a quadratic form, and F a subspace; the *orthogonal complement* of F, *denoted* by $F^\perp$, is the subspace of E defined by $F^\perp = \{P(x, y) = 0 \text{ for all } x \in F\}$. In general, unlike the Euclidean case, the sum $F + F^\perp$ is not always direct; but the following properties hold for all subspaces F, F':

$$(F^\perp)^\perp = F; \quad \dim F + \dim F^\perp = \dim E;$$

$$\operatorname{rad} F = F^\perp \cap F; \quad F \text{ is a null subspace} \Leftrightarrow F \subset F^\perp,$$

$$(F \cap F')^\perp = F^\perp + F'^\perp, \quad (F + F')^\perp = F^\perp \cap F'^\perp.$$

$$E = F \oplus F^\perp \text{ (direct sum)} \Leftrightarrow F \text{ is non-singular}$$

$$\Leftrightarrow F^\perp \text{ is non-singular} \Leftrightarrow F \cap F^\perp = 0.$$

13.E The Group of a Quadratic Form ([B, 13.6, 13.7])

The *orthogonal group* of (E, q), *written* $O(E)$ or $O(q)$, is defined as

$$O(E) = \{f \in \operatorname{GL}(E) : f^*q = q\}.$$

For a basis in which the matrix of q is A, the matrices S of the elements f of $O(E)$ are those fulfilling the condition ${}^tSAS = A$. In particular one always has $\det f = \pm 1$ and as usual we put

$$O^\pm(E) = \{f \in O(E) : \det f = \pm 1\}.$$

The *involutions* of $O(E)$ are exactly the maps of the form $f = \operatorname{Id}_S - \operatorname{Id}_T$, where

$E = S \oplus T$ is an *orthogonal* direct sum; according to 13.D they correspond exactly to non-singular subspaces S. Involutions are also called *symmetries* or *reflections* (through S).

As in the Euclidean case, we can show that every element of $O(E)$ is the product of at most n reflections through hyperplanes, but the proof in this case is much more difficult (Cartan-Dieudonné); see problem 13.5. One essential result is Witt's theorem: let F, F' be two subspaces of E and f be a linear map from F into F' such that $q|_F = f^*(q|_{F'})$. Then f can be extended to an element of $O(E)$.

A useful tool for the results above is the following: let F be a subspace of E, and suppose its radical $\mathrm{rad}(F)$ has dimension s. Let G be such that $G + \mathrm{rad}(F) = F$, and let $\{x_i\}_{i=1,\ldots,i}$ be a basis for $\mathrm{rad}(F)$. Then there are s planes P_i in E such that each P_i contains x_i and is an Artin space under $q|_{P_i}$ (see 13.B), the P_i and G are pairwise orthogonal, and the orthogonal direct sum $\bar{F} = G \oplus P_1 \oplus \cdots \oplus P_s$ is non-singular (we say that $\bar{F}$ is a *non-singular completion* of F). This lemma shows immediately that the null subspaces of E have dimension at most $n/2$; moreover, if E has a null subspace of dimension $s = n/2$, then E is necessarily an Artin space Art_{2s}. See also problem 13.1.

The Witt theorem shows in addition that the maximal null subspaces of a pair (E, q) are all conjugate under elements of $O(E)$, and in particular have the same dimension.

13.F The Two-dimensional Case ([B, 13.8])

In dimension 2, the subgroup $O^+(E)$ is always commutative, and $O^-(E)$ is formed by reflections through lines; see problem 13.4.

In E there are either two distinct isotropic lines or no isotropic lines; in the first case we necessarily have $E = \mathrm{Art}_2$, and we can study $O(E)$ in the same way we did for the Euclidean case (cf. [B, 13.8]).

Problems

13.1 ANISOTROPIC FORMS ([B, 13.9.4]). Reduce the classification of quadratic forms to that of anisotropic forms.

13.2 FORMS IN DIMENSION 1 OVER A FINITE FIELD ([B, 13.9.10]). Show that if $n = 1$ and K is finite, there are exactly three classes of quadratic forms.

13.3 FORMS IN DIMENSION 1 OVER THE RATIONALS ([B, 13.9.9]). Show that if $K = \mathbf{Q}$ and $n = \dim E = 1$ there are an infinite number of non-isometric forms (E, q).

13.4 AN EXCEPTIONAL PLANE ([B, 13.9.16]). Show that $O(E)$ is never commutative unless $E = \mathrm{Art}_2$ over the field with three elements.

13.5 EXCEPTIONAL ISOMORPHISMS ([B, 13.9.15]). Show there are vector spaces E possessing quadratic forms q such that (E, q) admits isomorphisms f (i.e. $f \in O(q)$) satisfying the following condition: $f(x) - x$ is non-zero and isotropic for any non-isotropic vector x.

Chapter 14
Projective Quadrics

In this chapter E denotes a vector space of dimension $n+1$ over a commutative field K of characteristic different from 2; the associated projective space is *denoted* by $P(E)$, and $p: E\backslash 0 \to P(E)$ is the canonical projection. We *denote* by $Q(E)$ the vector space of quadratic forms over E, and by $\mathrm{PQ}(E)$ the associated projective space $P(Q(E))$; the polar form of $q \in Q(E)$ is *denoted* by P. We will always have $n \geq 1$.

14.A Definitions ([B, 14.1])

A (*projective*) *quadric* in $P(E)$ is a non-zero element α of $\mathrm{PQ}(E)$, i.e. a quadratic form q, over E, considered up to a non-zero scalar. Such a form q representing the class α is called an *equation* of α.

The *image* of α, *denoted* by $\mathrm{im}(\alpha)$, is defined as $p(q^{-1}(0)\backslash 0)$; it is the image in $P(E)$ of a cone of E, and it may be empty. The *rank* of α is the rank of one of its equations; the quadric is called *degenerate* if q is degenerate, and *proper* otherwise. When $n=2$ we use the term *conic* instead of quadric.

14.B Notation, Examples ([B, 14.1])

If α is a quadric in $P(E)$ and $S = P(F)$ is a (projective) subspace of $P(E)$, we define the *intersection* of α and S as the quadric having $q|_F$ as an equation (where q is an equation of α); *denoting* this intersection by $\alpha \cap S$ we have, naturally enough, $\mathrm{im}(\alpha \cap S) = \mathrm{im}(\alpha) \cap S$.

Case $n=1$: If α has rank 1 (degenerate case), im(α) consists always of one point, whereas when α is proper its image can either be empty or consist of two distinct points. By the paragraph above, this means that a projective line of $P(E)$ intersects the image of a quadric in 0, 1 or 2 points (never three or more, unless the line is entirely contained in im(α), which means the plane defining this line is a null subspace; cf. 13.C). When the intersection has one point or is the whole line, the line is said to be *tangent* to α.

If α is a quadric and f is a homography from $P(E)$ into $P(E')$, we can define the image quadric $f(\alpha)$ in $P(E')$. In particular the projective group operates on PQ(E); *classifying* the quadrics of $P(E)$ means finding the orbits of this action.

One example of a quadric is the umbilical locus of a Euclidean space (9.D).

Taking homogeneous coordinates for $P(E)$ (see 4.C), the image of a quadric α such that the matrix of q is $A=(a_{ij})$ will be

$$\sum_{i,j} a_{ij}x_i x_j = 0;$$

for example, a conic has as equation

$$ax^2+a'y^2+a''z^2+2byz+2b'zx+2b''xy=0.$$

Observe that a proper quadric is a "nice" submanifold of $P(E)$, for $K=\mathbf{R}$ or $\mathbf{C}$: it is of class C^∞, and even real or complex analytic. At each point m of its image it has a tangent (projective) hyperplane, which is the projection in $P(E)$ of the orthogonal hyperplane to a vector x such that $p(x)=m$ (see 13.D and 14.E for the equation of this hyperplane).

14.C Classification ([B, 14.1, 14.3, 14.4])

From 13.B we see that all proper quadrics of $P^n(\mathbf{C})$ are isomorphic, so we can talk about *the* complex quadric in dimension n, the one defined by $\Sigma_i z_i^2=0$. Note that the image im(α) determines α for $K=\mathbf{C}$, whether α is proper or degenerate.

For $P^n(\mathbf{R})$, there are $[(n+1)/2]+1$ types of proper quadrics (where $[s]$ denotes the largest integer less than or equal to s). The following have a non-empty image: two points for $n=1$, the "ellipse" for $n=2$, the "ellipsoid" and the "hyperboloid" for $n=3$. The neutral case (for $n=4$) is interesting (cf. 13.B): The image im(α) contains two distinct families of lines; see [B, 14.4] for details. Examples are $z_1^2+z_2^2+z_3^2+z_4^2$ over $\mathbf{C}$, and $x^2+y^2-z^2-t^2$ over $\mathbf{R}$.

The image of a degenerate conic is in general composed of two distinct lines. When $n=3$ the image is in general a cone whose directrix is a conic.

14.D Pencils of Quadrics ([B, 14.2])

We now study the projective space of quadrics, $PQ(E)$, for its own sake. To begin with, its dimension is $[n(n+3)]/2$. Given $m \in P(E)$, the subset $H(m) = \{\alpha \in PQ(E) : m \in \operatorname{im}(\alpha)\}$ is a hyperplane; this implies (cf. 4.B) that given $[n(n+3)]/2$ points of $P(E)$ there is always at least one quadric going through them. For conics in a plane, this number is 5.

Following the nomenclature of 4.B, we shall call a line in $PQ(E)$ a *pencil of quadrics*; a pencil is determined by two distinct quadrics or, for example, by $(n^2+3n-2)/2$ points $P(E)$ (four for conics in a plane). A pencil of conics contains in general $n+1$ degenerate quadrics, occurring at the values of (λ, λ') determined by the equation $\det(\lambda A + \lambda' A') = 0$, where A, A' are the matrices of the equations of two quadrics defining the pencil.

The theorem of Desargues says that, given a fixed line D, the quadrics of a pencil intersect it in pairs of points $\{m, f(m)\}$ which correspond under an *involution* of D (cf. 6.D).

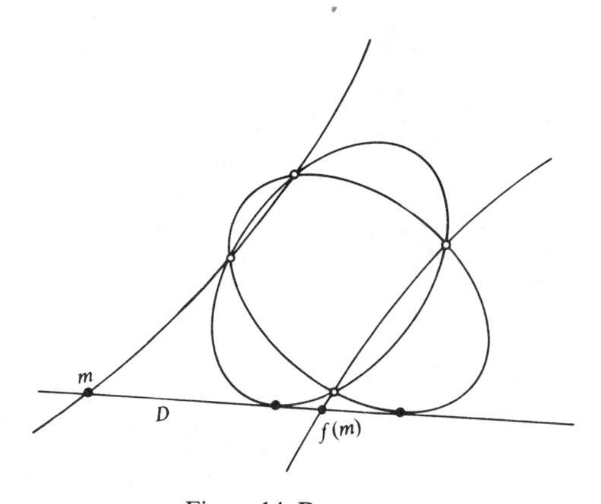

Figure 14.D.

From now on α is a proper quadric.

14.E Polarity ([B, 14.5])

Polarity relative to a proper quadric α with equation q is the projectivization of orthogonality relative to q in E (cf. 13.D). We know that q, via its polar form P, determines a linear isomorphism between E and its dual E^*, *denoted* by φ (cf. 13.C). Since $P(E^*)$ is identified with $\mathscr{H}(E)$, the set of (projective) hyperplanes of E (see 4.A), the quadric α determines a bijection between the points m of $P(E)$ and the hyperplanes of $P(E)$; we will denote by $m^\perp$ the hyperplane associated to m in this way, and we will call it the *polar* hyperplane of m. The point m such that $m^\perp = H$ is called the *pole* of the hyperplane H.

If $\sum_{i,j} a_{ij} x_i x_j = 0$ is an equation for α, and if $m = (\xi_i)$, the polar hyperplane $m^\perp$ has as equation $\sum_{i,j} a_{ij} \xi_i x_j = 0$; a practical rule is to replace x_i^2 by $\xi_i x_i$ and $x_i x_j$ by $\frac{1}{2}(\xi_i x_j + \xi_j x_i)$ (watch out for the factor $\frac{1}{2}$). In this way we can for example find the equation of the tangent hyperplane at an arbitrary point of the image, which in this case coincides with the polar hyperplane.

More generally, two points $m = p(x)$ and $n = p(y)$ of $P(E)$ will be called *conjugate relative* to α if $P(x, y) = 0$ (i.e. x and y are orthogonal relative to q). Taking the points conjugate to all points of a subspace S of $P(E)$, we obtain a subspace $S^\perp$, called *conjugate* to S. The relations in 13.D have various geometric consequences, like the following: If m, n are two points of a projective plane, the pole of the line $\langle m, n\rangle$ is the intersection $m^\perp \cap n^\perp$ of the polar lines of m and n. Or again: the tangents to a quadric drawn from a given point m are obtained by intersecting $\mathrm{im}(\alpha)$ and the polar hyperplane $m^\perp$ of m.

When $\mathrm{im}(\alpha)$ is non-empty, we have the following geometrical construction for conjugate points (and hence, in general, for the polar hyperplane of a point): If the line $\langle m, n\rangle$ intersects $\mathrm{im}(\alpha)$ in a, b, then m, n are conjugate relative to α if and only if they are harmonic conjugates relative to a, b:

$$[m, n, a, b] = -1 \text{ (cf. 6.B)}.$$

The property of the complete quadrilateral (cf. 6.B) gives the rest of the construction:

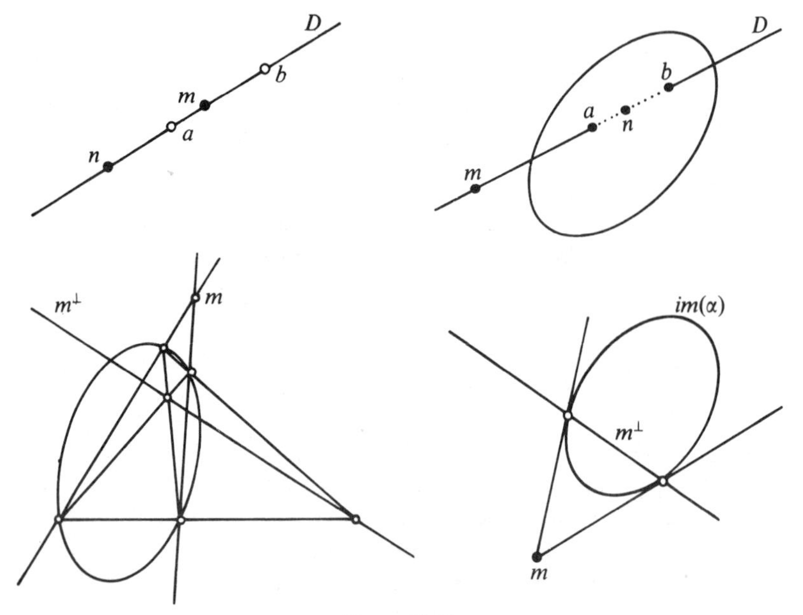

Figure 14.E.

A *simplex* $\{m_i\}$ of $P(E)$ is composed of $n+1$ linearly independent points (cf. 4.B); it is called *self-polar* relative to α if for every i the polar hyperplane of m_i is precisely the hyperplane spanned by the $m_j (j \neq i)$. Observe that an associated base will diagonalize q (cf. 13.B).

Let α, α' be two proper quadrics, with equations q, q', respectively, and let φ, φ' be the associated isomorphisms. Then the condition trace $(\varphi^{-1}\varphi')=0$ can be interpreted as follows: the trace is zero if $\mathrm{im}(\alpha')$ contains a self-polar simplex with respect to α. Conversely, if K is algebraically closed (for instance), we can construct an infinite number of simplices inscribed in α' and self-polar relative to α, and even choose the vertex m_1 arbitrarily on $\mathrm{im}(\alpha')$. We say that α' is *harmonically circumscribed* around α; this condition on α', expressed in $\mathrm{PQ}(E)$, is linear.

14.F Duality; Envelope Equation ([B, 14.6])

The set of hyperplanes $m^{\perp}$, when m varies in $\mathrm{im}(\alpha)$, is a subset of $\mathscr{H}(E) = P(E^*)$. What is this set? It is a proper quadric, in essence α considered as the envelope of its tangent hyperplanes. It shall be *denoted* by α^*. Writing an equation of α in a basis of E as the matrix A, we get A^{-1} as the matrix of an equation of α^* in the dual basis; we call this the *envelope equation*. See for example problem 16.6.

In general, we shall call any quadric of $P(E^*)$ a *tangential quadric*; a *tangential pencil* of quadrics will be a line in $\mathrm{PQ}(E^*)$. One way to obtain tangential quadrics is to take two proper quadrics α, β and consider the set of polar hyperplanes of points of $\mathrm{im}(\beta)$ with respect to α; we obtain a tangential quadric *denoted* by β_α^*, called the *polar* quadric of β relative to α. In the case when α is the unit sphere of a Euclidean vector space, we fall back into the situation of 10.B, 11.B. Taking bases as above, we get that a matrix for β_α^* is $AB^{-1}A$, where A (resp. B) is a matrix for α (resp. β).

14.G The Group of a Quadric ([B, 14.7])

If q is an equation for α, the *group* of α is the image in $P(E)$ of the group $O(q)$; it is *denoted* by $\mathrm{PO}(\alpha)$. The elements of $\mathrm{PO}(\alpha)$, of course, leave $\mathrm{im}(\alpha)$ invariant. The theorem of Cartan-Dieudonné (cf. 10.E) shows that $\mathrm{PO}(\alpha)$ is generated by the homographies f of $P(E)$ defined in the following way: For m in the complement of $\mathrm{im}(\alpha)$, associate with $t \in P(E)$ the point $f(t)$ that lies in the line $\langle m, t\rangle$ and such that $[m, \langle m, t\rangle \cap m^{\perp}, t, f(t)] = -1$. The group $\mathrm{PO}(\alpha)$ in the neutral case ($n=4$) studied in 14.C acts transitively on the lines of the two families that form the image of the quadric.

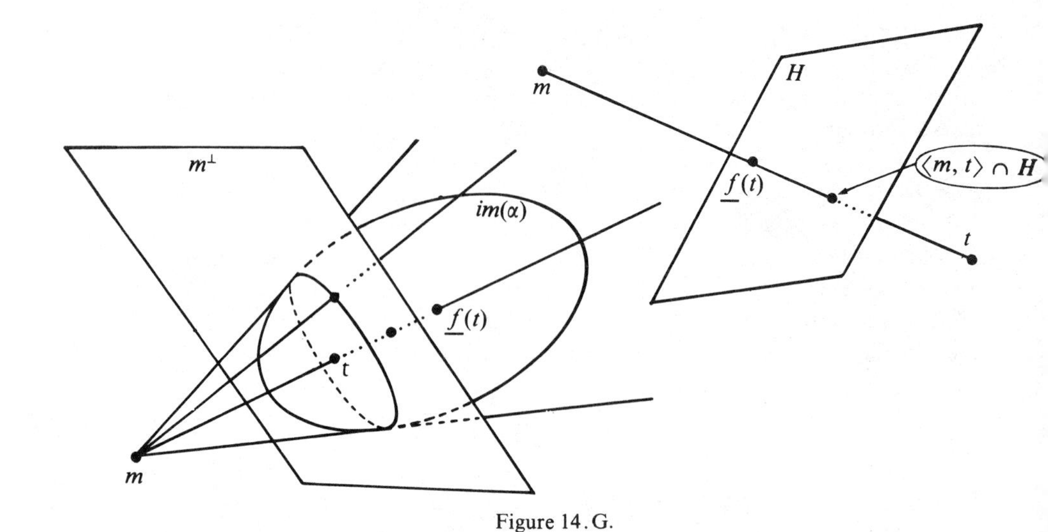

Figure 14.G.

Problems

14.1 THE COMPLEX QUADRIC AND THE GRASSMANN MANIFOLD ([B, 14.8.4]. Let $C(n)$ be the complex quadric in dimension n (cf. 14.C), i.e. the unique non-degenerate quadric of the complex projective space $P^n(\mathbf{C})$, given in homogeneous coordinates by the equation $\Sigma_{k=0}^{n} z_k^2 = 0$.

Show that $C(n)$ is homeomorphic to the Grassmann manifold of oriented lines of $P^n(\mathbf{R})$, which is the same as the set of oriented two-dimensional vector subspaces of $\mathbf{R}^{n+1}$.

14.2 SIX POINTS ON THE SAME CONIC ([B, 14.8.11]). Let α be a proper conic, $\{a, b, c\}$ and $\{a', b', c'\}$ two self-polar triangles with respect to α; show that the six points a, b, c, a', b', c' belong to one single conic.

14.3 HARMONICALLY INSCRIBED QUADRIC ([B, 14.8.10]). We say that the proper quadric α' is *harmonically inscribed* in α if trace $(\varphi'^{-1}\varphi) = 0$. Interpret this condition geometrically.

Deduce that two triangles which are self-polar with respect to the same conic are circumscribed around one single conic. Prove also that if there is a triangle inscribed in a conic C and circumscribed around a conic γ, every point of C from which it is possible to take tangents to Γ is the vertex of a triangle inscribed in C and circumscribed around Γ.

Show finally that the circle circumscribed around a triangle circumscribed around a parabola passes through its focus.

Chapter 15

Affine Quadrics

In all of this chapter X will be an affine space of finite dimension $n \geq 1$ over a commutative field of characteristic $\neq 2$. We shall use (cf. chapter 5) the projective completion $\tilde{X} = X \cup \infty_X$ of X, where the hyperplane at infinity is $\infty_X = P(\vec{X})$.

15.A Definitions ([B, 15.1])

An *affine quadratic form* over X is a polynomial over X whose degree is less than or equal to 2 (cf. 3.E); we *denote* by $Q(X)$ the vector space of such polynomials. The *symbol* $\vec{q}$ of $q \in Q(X)$ is a polynomial of degree 2 over $\vec{X}$. In every vectorialization of X, we can write $q = q_2 + q_1 + q_0$, where $q_0 \in K$, q_1 is a linear form, and $q_2 = \vec{q}$.

An (affine) *quadric* in X is an element α of the projective space $\mathrm{QA}(X) = P(Q(X))$ such that, for $\alpha = p(q)$, we have $\vec{q} \neq 0$. If $\alpha = p(q)$, we say that q is an *equation* of α; if $n = 2$ we use the term *conic* instead of *quadric*. The *image* of α is $\mathrm{im}(\alpha) = q^{-1}(0)$.

By passing from X to $\tilde{X}$, we see that there is a bijection $\tilde{}$ between the affine quadrics of X and the projective quadrics $\beta = \tilde{\alpha}$ of $\tilde{X}$ such that $\mathrm{im}(\beta)$ does not contain ∞_X. Under this correspondence it is true that $\mathrm{im}(\alpha) = \mathrm{im}(\tilde{\alpha}) \cap X$ and $\tilde{\alpha} \cap \infty_X = \vec{\alpha}$ if $\vec{\alpha} = p(\vec{q})$ for an equation q of α.

We say that α is *proper* if $\tilde{\alpha}$ is; the *rank* of α is that of $\tilde{\alpha}$ and the *index* of $\vec{\alpha}$ is the rank of α.

Expressed in an affine frame, an equation q of α will be:

$$\sum_{i,j} a_{ij} x_i x_j + 2 \sum_i b_i x_i + c, \quad \text{where} \quad \vec{q} = \sum_{i,j} a_{ij} x_i x_j. \tag{1}$$

The equation of $\tilde{\alpha}$ is obtained by introducing the $(n+1)$-th variable t to make (1) homogeneous:

$$\sum_{i,j} a_{ij}x_i x_j + 2\sum_i b_i x_i t + ct^2.$$

We associate with $\tilde{\alpha}$ the matrices

$$\vec{A} = (a_{ij}) \quad \text{and} \quad \hat{A} = \begin{pmatrix} & & & b_1 \\ & \vec{A} & & \vdots \\ & & & b_n \\ b_1 & \cdots & b_n & c \end{pmatrix}.$$

15.B Reduction of Affine Quadratic Forms ([B, 15.2, 15.3])

Every quadratic form is of one of the following types, after an appropriate choice of basis:

$$q = \sum_{i=1}^{r} a_i x_i^2 \quad (\text{all the } a_i \neq 0); \qquad \text{type I}$$

$$q = \sum_{i=1}^{r} a_i x_i^2 + 1 \quad (\text{all the } a_i \neq 0); \qquad \text{type II}$$

$$q = \sum_{i=1}^{r} a_i x_i^2 + 2x_n \quad (\text{all the } a_i \neq 0); \qquad \text{type III}$$

This fact permits the classification of affine quadratic forms when $K = \mathbf{C}$ or $\mathbf{R}$. In particular, if $K = \mathbf{R}$ and $n = 2$, we obtain three types of proper conics whose image is non-empty, called *ellipse*, *hyperbola* and *parabola*. When $K = \mathbf{R}$ and $n = 3$, we obtain five types of proper quadrics with a non-empty image, called *ellipsoid*, *one-sheeted hyperboloid*, *two-sheeted hyperboloid*, *elliptic paraboloid* and *hyperbolic paraboloid*.

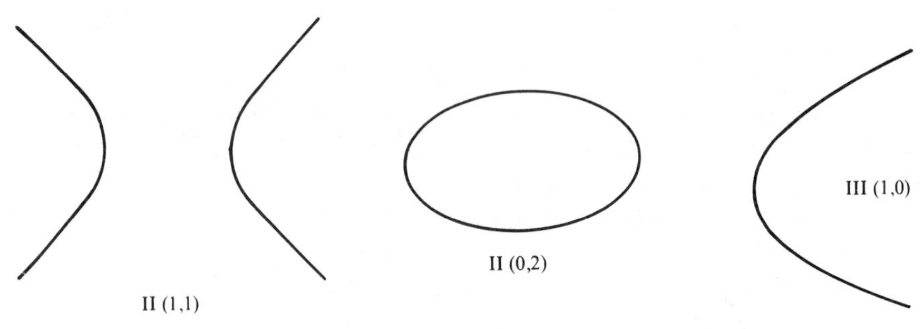

Figure 15.B.1.

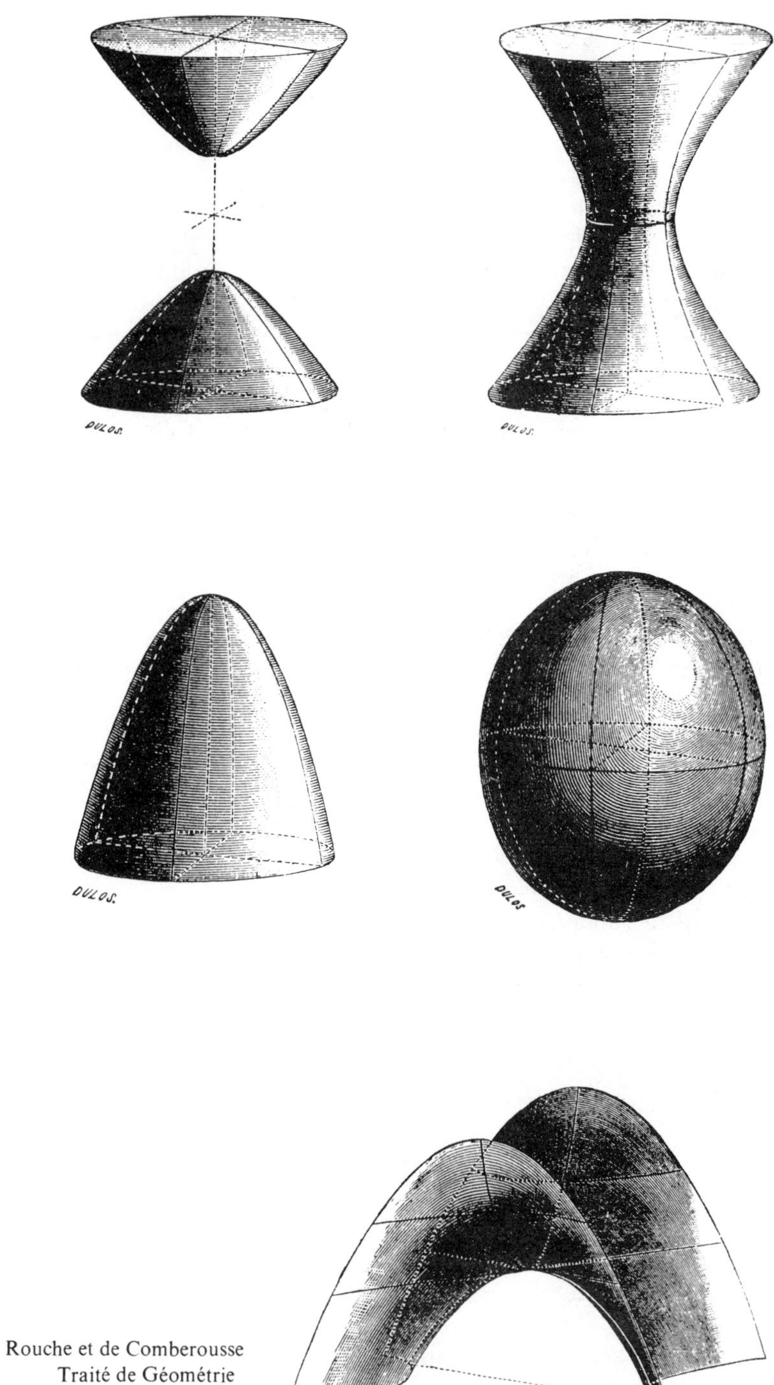

Rouche et de Comberousse
Traité de Géométrie
Gauthier-Villars Paris

Figure 15.B.2.

15.C Polarity ([B, 15.5])

Polarity relative to a *proper* affine quadric α is by definition the same as polarity relative to $\tilde{\alpha}$; it is a relation in X, or if necessary in $X \cup \infty_X$ (cf. 14.E and chapter 5).

It is easy to see that the following three conditions are equivalent: the quadric α is of type II, the hyperplane ∞_X is not tangent to X, and the pole $C = \infty_X^\perp$ of the hyperplane at infinity is not at infinity. In this situation we say that α is a *central quadric*, for in effect $c = \infty_X^\perp$ is a center of symmetry for $\operatorname{im}(\alpha)$.

If $(\xi_1, \ldots, \xi_n)$ is a point in X, its polar hyperplane relative to the quadric of equation (1) (15.A) has the following equation (cf. 14.E):

$$\sum_{i,j} a_{ij}\xi_i x_j + \sum_i b_i(\xi_i + x_i) + c = 0.$$

The equations of the center $(x_1, \ldots, x_n)$ are then

$$\sum_j a_{ij}x_j + b_i = 0 \quad (i = 1, \ldots, n).$$

An interesting case of polarity is when we take a point $a \in \infty_X$; if α is a central quadric, the polar hyperplane $a^\perp$ of a is an affine hyperplane passing through the center c, and the affine reflection through the hyperplane $a^\perp$ and parallel to the direction a (cf. 2.D) leaves invariant the image of α. We say that $a^\perp$ is a *diametral* hyperplane for α; for $n = 2$ we call it a *diameter*. If we take n points $a_1, \ldots, a_n$ in ∞_X such that the simplex $\{c, a_1, \ldots, a_n\}$ is self-polar relative to α (cf. 14.E) (which here means that the points a_i are pairwise conjugate with respect to α), then the lines going through c and whose directions are the a_i are said to form a set of *conjugate diameters* of α.

We extend the notion of diameter to quadrics of type III (typified by the parabola) in the following way: instead of passing through the center, the diameters are all parallel and their direction is that of the point at infinity of α.

15.D Euclidean Affine Quadrics ([B, 15.6])

Because of 13.B, every proper affine quadric in a Euclidean affine space will have, in some appropriate orthonormal frame, one of the following equations:

$$\sum_{i=1}^{r} a_i x_i^2 - \sum_{i=r+1}^{n} a_i x_i^2 + 1 \quad \text{or} \quad \sum_{i=1}^{r} a_i x_i^2 - \sum_{i=r+1}^{n-1} a_i x_i^2 + 2x_n,$$

where all the a_i are strictly positive. The quadric is called an *ellipsoid* if $\operatorname{im}(\alpha) = \mathscr{E}$ can be written as

$$\mathscr{E} = \mathscr{E}(q) = \{x \in X : q(x) = 1\},$$

where q is a positive definite quadratic form over the vectorialization X_a, and a is the center of $\mathscr{E}$.

The *theorem of Appolonius* says the following: let $\mathscr{E}$ be an ellipsoid with center a, and let $\{m_i\}_{i=1,\ldots,n}$ be a set of points of $\mathscr{E}$ such that the directions am_i $(i=1,\ldots,n)$ are conjugate (which means that the lines $\langle a, m_i\rangle$ form a set of conjugate diameters of $\mathscr{E}$, in the sense of 15.C). Then, for each k, the sum $\sum_{i_1<\cdots<i_k} \mathrm{Gram}(m_{i_1},\ldots,m_{i_k})$ of the Gram determinants of all the k-element subsets of the m_i is a constant, depending only on q and not on the choice of the m_i. The cases $k=1$ and $k=n$ are particularly interesting; see problems 15.3 and 15.4, and 17.D.2.

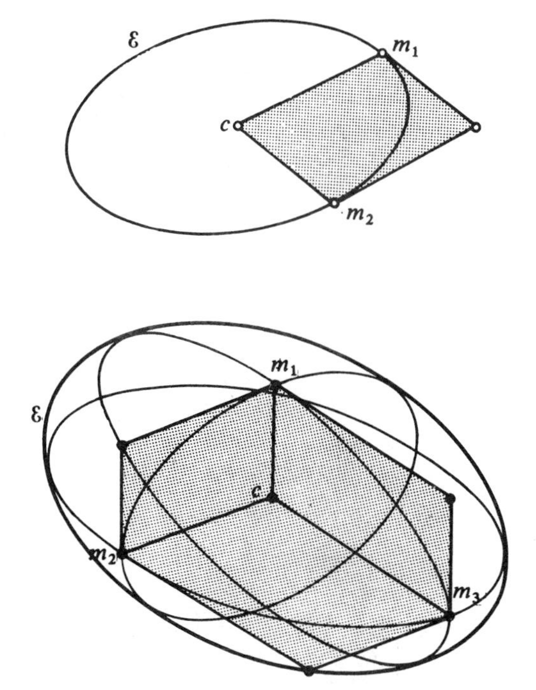

Figure 15.D.

Problems

15.1 ARCHIMEDES' METHOD FOR THE AREA OF THE PARABOLA ([B, 15.7.6]). Let $C=\mathrm{im}(\alpha)$ be the non-empty image of a proper plane conic, let m be a point on the plane, and $\langle m, a\rangle$ and $\langle m, b\rangle$ two distinct tangents to C at a and b, passing through m. Show that the line $D=\langle m,(a+b)/2\rangle$ is a diameter of α, and that the tangents to α at the points of $C \cap D$ are parallel to $\langle a, b \rangle$. When α is a parabola, show that D always intersects C, and the intersection is the midpoint of m and $(a+b)/2$. Deduce a geometric construction for a sequence of points on an arc of parabola, given two points and the tangents at these points.

Observing that the area of the triangle $\{m, a', b'\}$ in figure 15.1 is $1/4$ of the area of $\{m, a, b\}$, deduce that the shaded area is $2/3$ of the area of $\{m, a, b\}$ (this limiting process is due to Archimedes).

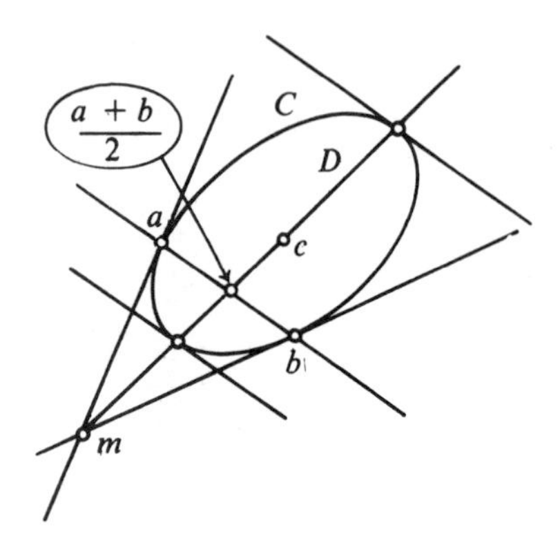

Figure 15.1.1.

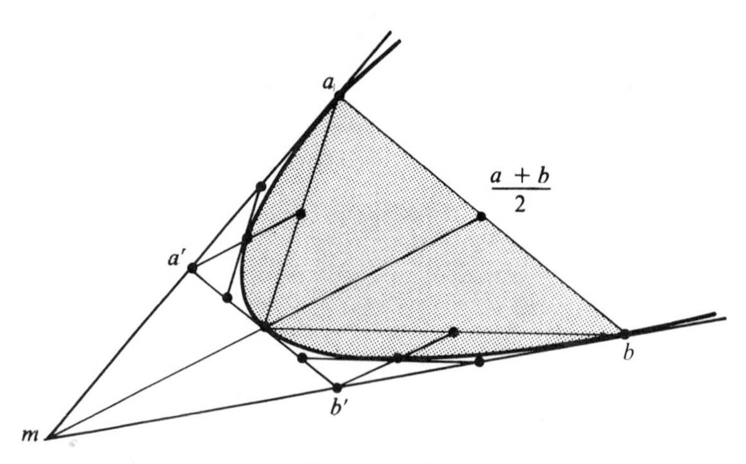

Figure 15.1.2.

15.2 ELLIPSES AND PARALLELOGRAMS Given a parallelogram, show that there are ellipses inscribed in them so that the tangency points are at the middle of each side. With the notation of figure 15.2.1, show that such an ellipse always satisfies $\overrightarrow{c\alpha} = \sqrt{2}\,\overrightarrow{c\beta}$.

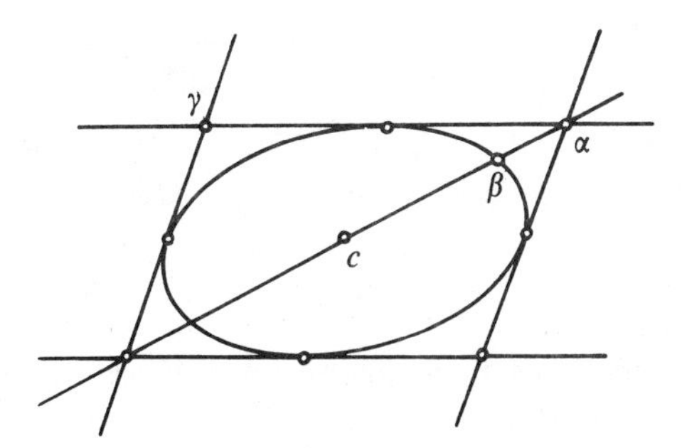

Figure 15.2.1.

15.3 METRIC RELATIONS IN ELLIPSOIDS: I ([B, 15.7.9]). Let Q be the image of an ellipsoid in a three-dimensional Euclidean affine space, and let x be a fixed point in the interior of Q (cf. 15.D). Take three orthogonal lines D, E, F through x, and let the intersection points of Q with D be a, b, with E be c, d and with F be e, f. Show that the sum

$$\frac{1}{\overline{xa}\cdot\overline{xb}}+\frac{1}{\overline{xc}\cdot\overline{xd}}+\frac{1}{\overline{xe}\cdot\overline{xf}}$$

is constant. Give examples and generalize.

Now consider three lines D, E, F through x whose directions are pairwise conjugate relative to Q, and let the intersection points of Q with D be a, b, with E be c, d and with F be e, f. Show that the sum

$$\overline{xa}\cdot\overline{xb}+\overline{xc}\cdot\overline{xd}+\overline{xe}\cdot\overline{xf}$$

is constant.

15.4 METRIC RELATIONS IN ELLIPSOIDS: II ([B, 15.7.20]). Let $\mathscr{E}$ be an ellipsoid with center O in an n-dimensional Euclidean affine space (cf. 15.D). We consider the sets $\{a_i\}_{i=1,\ldots,n}$ of points of $\mathscr{E}$ such that the vectors $\overrightarrow{Oa_i}$ are orthogonal. Show that

$$\sum_{i=1}^{n}\frac{1}{(Oa_i)^2}$$

is a constant.

Use this fact to find the envelope of the hyperplanes containing the a_i.

Using polarity with respect to a sphere centered at O (cf. 10.B), show that the preceding result implies that the locus of the points which are the intersection of n orthogonal hyperplanes tangent to an ellipsoid is a sphere (called the *orthoptic sphere* of that ellipsoid).

15.5 NORMALS TO A QUADRIC FROM A GIVEN POINT ([B, 15.7.15]). Let Q be a quadric in a 3-dimensional Euclidean affine space, and let m be a point. Show that the number of *normals* to Q that pass through m is "in general" equal to six. Show that the feet of all normals to Q from m are contained in a second-degree cone with vertex m and containing the center of Q and the parallels to the axes of Q which go through m.

15.6 HOMOFOCAL QUADRICS ([B, 15.7.17]). We consider in $\mathbf{R}^3$ the family of quadrics $Q(\lambda)$ whose equations are

$$\frac{x^2}{a^2+\lambda}+\frac{y^2}{b^2+\lambda}+\frac{z^2}{c^2+\lambda}-1=0,$$

with $a>b>c$. Find how many quadrics $Q(\lambda)$ pass through a fixed point (x_0, y_0, z_0); show that if three quadrics $Q(\lambda)$ pass through a point, their tangent planes at that point are orthogonal (cf. 17.B, 17.D).

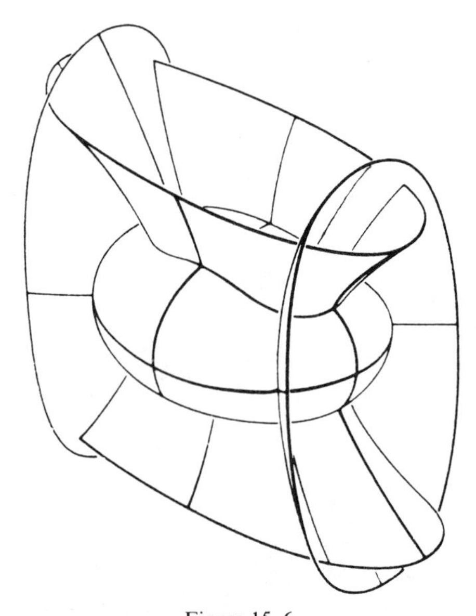

Figure 15.6.

Chapter 16

Projective Conics

In all of this chapter, $P = P(E)$ is a projective plane over a commutative field K of characteristic $\neq 2$; we *put* $P^* = P(E^*)$. We will often identify a point $m \in P$ with its homogeneous coordinates (x, y, z).

We will generally fix a conic $\alpha \in \mathrm{PQ}(E)$ and its image $C = \mathrm{im}(\alpha)$ (in most cases, α will be proper and have non-empty image), as well as one equation q for α. For a a point of C, the tangent to C at a will sometimes be *denoted* by $\langle a, a \rangle$.

16.A Notation ([B, 16.1])

The general equation of a conic will be written

$$q = ax^2 + a'y^2 + a''z^2 + 2byz + 2b'zx + 2b''xy.$$

Depending on whether the triangle $p(1,0,0)$, $q(0,1,0)$, $r(0,0,1)$ is inscribed in C, self-polar relative to C (cf. 14.E) or "bitangent to C", we have three simplified equations for the conic, shown in the figure below:

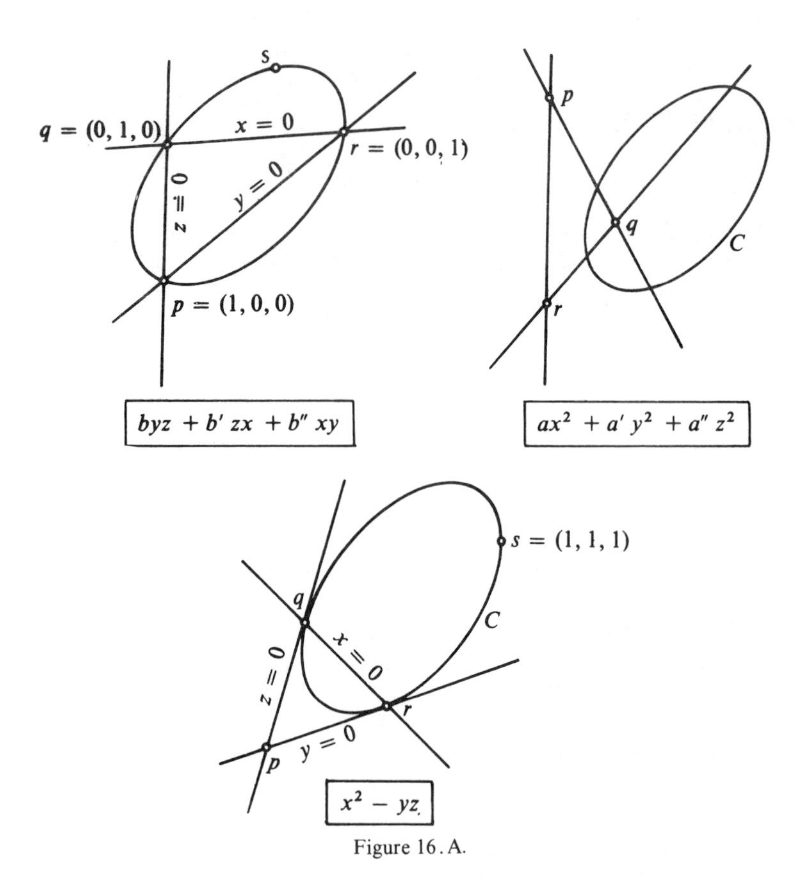

Figure 16.A.

Recall (cf. 14.D) that five points determine in general exactly one conic. Recall also that the tangent to α at point (x_0, y_0, z_0), or more generally the polar line of (x_0, y_0, z_0) (cf. 14.E), has the following equation:

$$ax_0x + a'y_0y + a''z_0z + b(y_0z + z_0y) + b'(z_0x + x_0z) + b''(x_0y + y_0x) = 0.$$

Finally, recall that a degenerate conic is formed by either two distinct lines, or a "double" line ("to the square").

16.B Good Parametrizations ([B, 16.2])

The idea is to use the figure below to parameterize a proper conic C. We obtain a bijection between the elements of m^* (where $m \in C$; see 4.B) and the points of C, by associating with a line $D \in m^*$ the point in $D \cap C$ other than m. Next we identify m^* with $\tilde{K} = K \cup \infty$ (cf. 5.A) to get a bijection $\tilde{K} \to C$; such a bijection is called a *good parametrization* (of a proper conic). The essential result here is that two good parametrizations differ only by a homography of $\tilde{K}$.

An equivalent definition for good parametrizations of C is that they are maps of the second degree with image C, i.e. they have the form

$$(\lambda, \mu) \mapsto (u\lambda^2 + u'\lambda\mu + u''\mu^2, v\lambda^2 + v'\lambda\mu + v''\mu^2, w\lambda^2 + w'\lambda\mu + w''\mu^2).$$

We can simplify this formula by putting $\lambda/\mu = t$, as long as we include the value $t = \infty = \infty_K$. For instance, a good parametrization of $x^2 - yz = 0$ is $(\lambda\mu, \lambda^2, \mu^2)$, or, alternatively, $(t, t^2, 1)$.

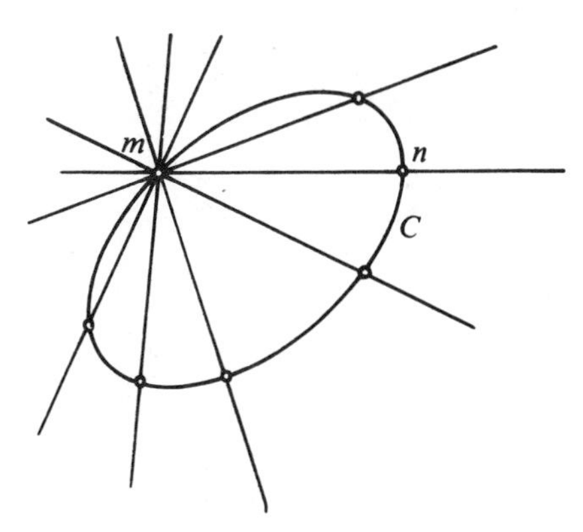

Figure 16.B.

16.C Cross-ratios ([B, 16.2])

Since the notion of the cross-ratio of four distinct points makes sense in $\tilde{K}$ (cf. 6.A), and it is invariant under homographies, good parametrizations allow us to transfer this notion over to C, and to define the *cross-ratio* of four points $m_i (i = 1, 2, 3, 4)$, which will still be *denoted* by

$$[m_i] = [m_1, m_2, m_3, m_4]$$

(or, if necessary, $[m_i]_C$). This number is equal to the cross-ratio of the four lines $\langle m, m_i \rangle$, and is consequently independent of m.

From this property a number of results follow: the theorem of Pascal (see also 16.2 and [B, 16.2.11]) and, by duality, the theorem of *Brianchon*: If a hexagon is circumscribed around a conic, its three diagonals are concurrent. These two theorems have many degenerate cases; see the figure below. Finally, we see that if m and m' are two points in P, and $f: m^* \to m'^*$ is a homography such that $f(\langle m, m' \rangle) \neq \langle m', m \rangle$, then the point $D \cap f(D)$, for D ranging over m^*, describes a conic passing through m and m' (cf. 6.C).

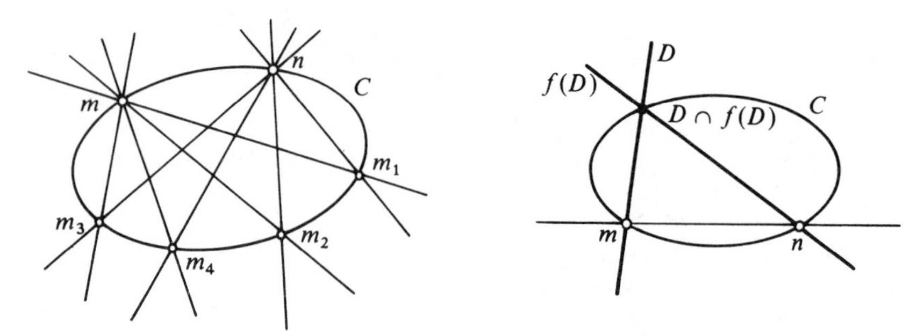

Figure 16.C.1.

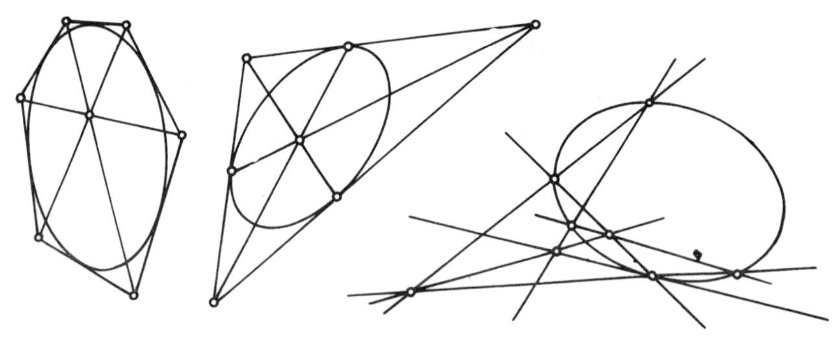

Figure 16.C.2.

16.D Homographies of a Conic ([B, 16.3])

Using good parametrizations, we can also transfer *homographies* from $\tilde{K}$ to C; they will be the bijections $C \to C$ which preserve cross-ratios. They form a group *denoted* by $\mathrm{GP}(C)$. We show (using 6.D and 14.G) that $\mathrm{GP}(C)$ coincides with the restriction to $C = \mathrm{im}(\alpha)$ of the projective group $\mathrm{PO}(\alpha)$.

The essential result for homographies $f \in \mathrm{GP}(C)$ is that, as the points m, n run through C, the points $\langle m, f(n)\rangle \cap \langle n, f(m)\rangle$ stay on a line that depends only on f, called its *homography axis*. The case when f is an involution is a very nice one: the line $\langle m, f(m)\rangle$ passes through a fixed point, called the *Frégier point* of f which is the pole of the axis of f (see the construction of the polar hyperplane in 14.E).

See problem 16.5 to find out what $\langle m, f(m)\rangle$ does when f is not an involution.

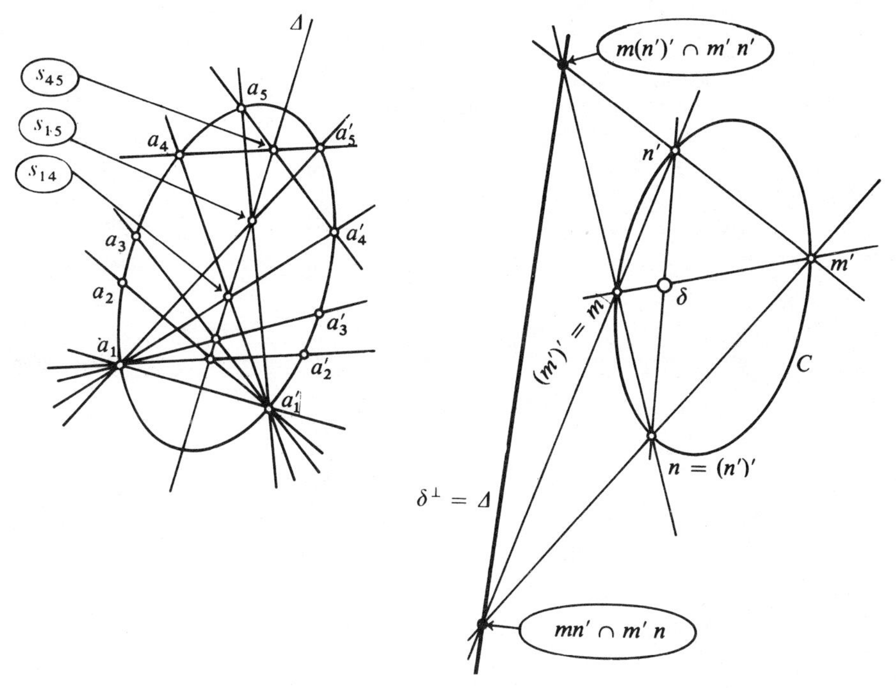

Figure 16.D.

16.E Intersection of Two Conics; Theorem of Bezout ([B, 16.4])

Let α be a proper conic with non-empty image $\mathrm{im}(\alpha) = C$, and α' a second conic, with image C' and not necessarily proper. Using for instance good parametrizations, we can define the *order* $\omega(m)$ of a point $m \in C \cap C'$, which is an integer taking the values 1, 2, 3 or 4. If φ is an equation for the tangent to C at m, we have $\omega(m) \geq 2$ if C' is *tangent to C at m*, i.e. if an equation q' of α' can be written as $q' = kq + \varphi\psi$, where $k \in K^*$ and ψ is a line. We have $\omega(m) \geq 3$ if C' *osculates* C at m, i.e. $q' = kq + \varphi\psi$, where ψ is a line through m; and finally $\omega(m) = 4$ if $q' = kq + \varphi^2$, in which case we say that C' *superosculates* C at m.

The theorem of Bezout says that if K is algebraically closed and if $\{(m_i, \omega_i)\}$ is the set of points of $C \cap C'$ counted according to their orders, we have

$$\sum_i \omega_i = 4.$$

So there are only five cases: (1,1,1,1), (2,1,1), (2,2), (3,1), (4), and they are numbered I, II, III, IV and V, accordingly. In case III we say that C and C' are *bitangent* (see also problem 16.5).

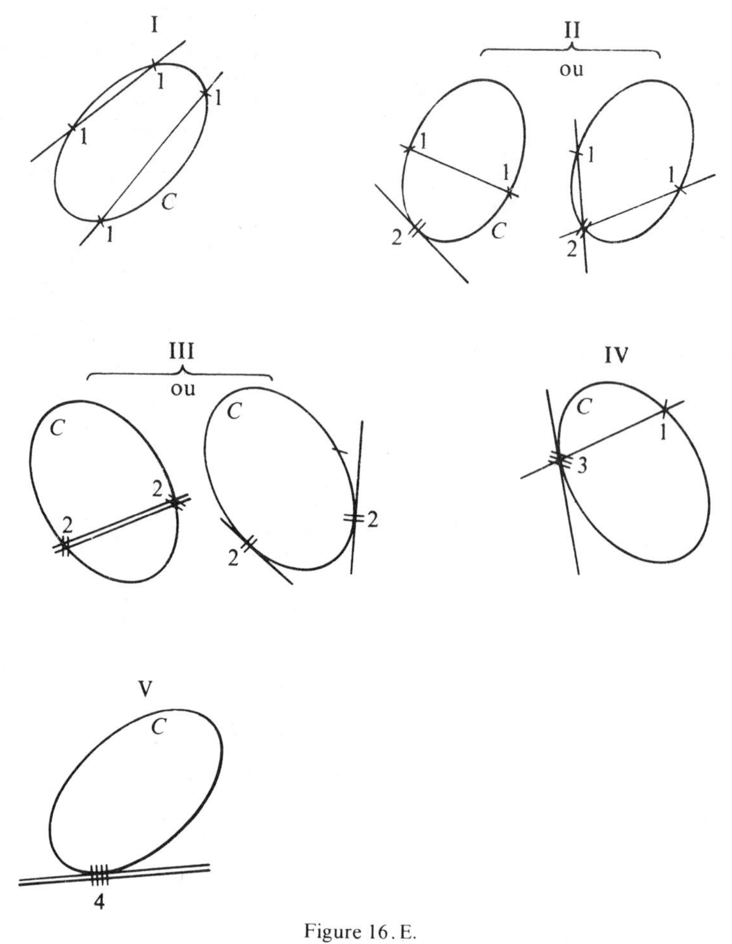

Figure 16.E.

16.F Pencils of Conics ([B, 16.5])

Recall (cf. 14.D) that a pencil of conics is a line in PQ(E), or, analytically, the set of conics with equations $\lambda q + \lambda' q'$, where q, q' are two equations of conics (and we assume that at least one of them is proper), and (λ, λ') runs through $\tilde{K}$. In the algebraically closed case, a pencil of conics is the set of conics that intersect a given proper conic C in a fixed set $\{(m_i, \omega_i)\}$ of points of C satisfying $\Sigma_i \omega_i = 4$. So there are five types, I, II, III, IV and V, of pencils (see drawings in [B, 16.5]).

The degenerate conics in these pencils are: three pairs of lines in case I, two pairs of lines in case II, one pair of lines and a double line in case III, one pair of lines in case IV, and a double line in case V; see figure 16.E.

16.G Tangential Conics

Recall 14.F. We will have here five types I*, II*, III*, IV* and V* of tangential pencils of conics. Geometrically, III = III* and V = V* (bitangent and superosculating conics. For case I*, see problem 16.6.) Recall that the matrix that gives the envelope equation is the inverse of that for the punctual equation.

16.H The Great Poncelet Theorem ([B, 16.6])

This is a delicate theorem, and its proof is involved. It applies in the algebraically closed case, and it says that if C and Γ are two conics such that there is an n-sided polygon, all of whose vertices are on C and all of whose edges are tangent to Γ, then there are infinitely many such polygons, and one of the vertices can be arbitrarily chosen on C. See simple particular cases in problems 10.2, 14.3 and 16.5.

16.I Affine Conics ([B, 16.7])

The equation of an affine conic is

$$ax^2 + 2bxy + cy^2 + 2dx + 2ey + f = 0.$$

The conic is proper if

$$\begin{vmatrix} a & b & d \\ b & c & e \\ d & e & f \end{vmatrix} \neq 0.$$

Its points at infinity are the lines of slope ϑ, where ϑ satisfies the equation

$$a + 2b\vartheta + c\vartheta^2 = 0$$

Problems

16.1 TRIANGLE CIRCUMSCRIBED AROUND A CONIC ([B, 16.8.2]). Show that if a, b, c is a triangle circumscribed around C, and α, β, γ are the tangency points, then the segments $a\alpha, b\beta, c\gamma$ are concurrent. Use only analytic geometry in your proof.

16.2 THEOREM OF PASCAL ([B, 16.8.5]). Let C be any conic and a, b, c, d, e, f any six points on C. Then the points $\langle a, b\rangle \cap \langle d, e\rangle$, $\langle b, c\rangle \cap \langle e, f\rangle$, $\langle c, d\rangle \cap \langle f, a\rangle$ are collinear. Prove this result using calculus, by taking a projective base formed by four of the six points considered.

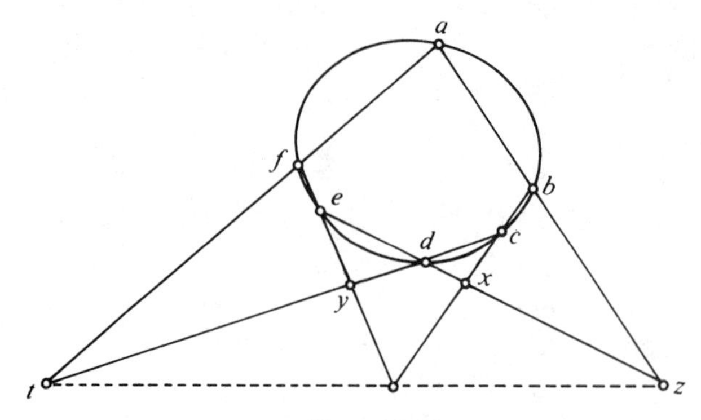

Figure 16.2.1.

16.3 CROSS-RATIOS FOR A CONIC ([B, 16.8.6]). Let C be the non-empty image of a proper conic, and let p, q, r be such that C is tangent to pq at q and to pr at r. Show that, for any $m, n \in C$, the following holds (cf. 16.C):

$$[q, r, m, n]_C^2 = [pq, pr, pm, pn].$$

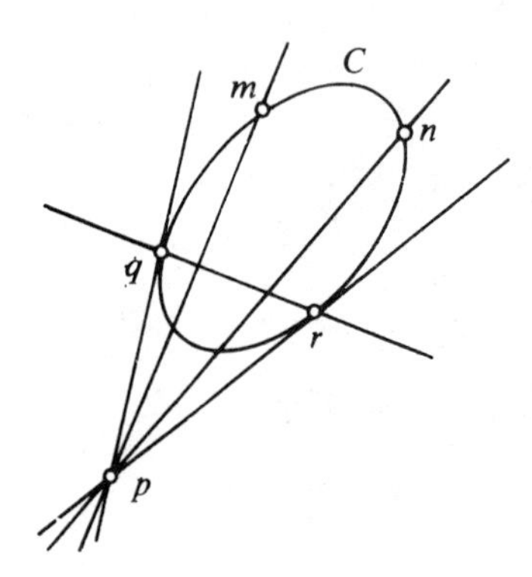

Figure 16.3.1.

16.4 COMMUTING INVOLUTIONS ([B, 16.8.7]). Show that two involutions of a proper conic whose image is non-empty commute if and only if their Frégier points are conjugate (cf. 16.D).

16.5 THE GREAT PONCELET THEOREM FOR BITANGENT CONICS ([B, 16.8.8]). We shall call two conics C, C' *bitangent* if there is a line D, not tangent to C, and a scalar k such that $q' = q + kd^2$, where q, q', d are equations of C, C', D, respectively (cf. 16.E). Notice that if the field K is closed, C and C' are indeed tangent at two distinct points.

Suppose the field K is $\mathbf{R}$ or $\mathbf{C}$ (if not, we must use an algebraic closure of K). Let f be a non-involutive homography (different from the identity) of a proper conic C. Show that the set of lines $\langle m, f(m) \rangle$, for m ranging through C, is the set of tangents to a proper conic which is bitangent to C in the sense above. Prove a converse statement. For two such bitangent conics, prove the great Poncelet theorem (16.H).

16.6 TANGENTIAL PENCILS OF CONICS ([B, 16.8.10]). The figure below shows several conics belonging to the same pencil. Prove rigorously that there are regions of the plane which do not intersect any of the conics of the pencil.

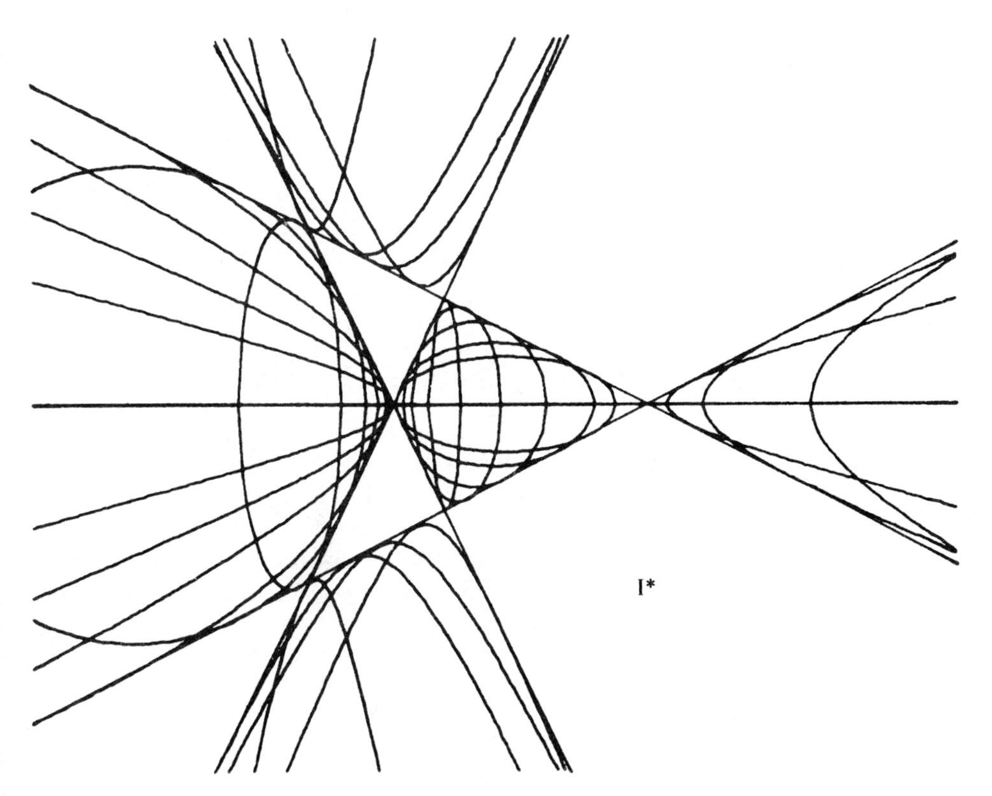

Figure 16.6.1.

16.7 INTERSECTION OF TWO CONICS OVER A FINITE FIELD ([B, 16.8.16]). Study the intersection of $xz - y^2 = 0$ and $xy - z^2 = 0$ over the field K with three elements.

Chapter 17

Euclidean Conics

In this chapter X denotes a Euclidean affine plane, $\tilde{X}$ its projective completion, $\tilde{X}^C$ the complexification of $\tilde{X}$ (see 9.D) and $\{I, J\}$ the cyclical points of X. In general, C will be the non-empty image of a proper conic α in X.

17.A Recapitulation and Notation ([B, 17.1])

From 15.D we know there are three possible cases of Euclidean affine conics: the ellipse, the hyperbola and the parabola. Only the parabola does not have a center; it has a single axis of (orthogonal) symmetry and a vertex. The ellipse and the hyperbola have a center and two axes of orthogonal symmetry; the ellipse has four vertices (with the exception of the circle), and the hyperbola has two. The equations, in an appropriate orthonormal basis, are

$$\frac{x^2}{a^2}+\frac{y^2}{b^2}-1=0 \text{ (ellipse)}, \quad \frac{x^2}{a^2}-\frac{y^2}{b^2}-1=0 \text{ (hyperbola)},$$

$$y^2-2px=0 \text{ (parabola)}.$$

When $a=b$, the ellipse is a circle. A hyperbola for which $a=b$ is called *equilateral*; its points at infinity are orthogonal directions.

The ellipse is obtained from the circle by the affine transformation $(x, y) \mapsto (x,(b/a)x)$; this fact is helpful in the solution of many problems, see for instance 17.2. This affine transformation for the ellipse gives the parametric representation $t \mapsto (a\cos t, b\sin t)$, which we transform into a good parametrization (cf. 16.B) by taking $\vartheta = \tan(t/2)$.

17.B Foci and Directrices ([B, 17.2])

There is a correspondence between the Euclidean conics (apart from circles) and the sets of the plane which can be defined as

$$\{m: fm = e \cdot d(m, D)\},$$

where f is a point and D is a line which does not contain f. We say that f is a focus of the conic C, and D is the associate directrix (which is in fact the polar line of f). The constant e is called the *excentricity* of C, and it has the value 1 for parabolas, greater than 1 for hyperbolas and less than 1 for ellipses. The excentricity depends only on the ratio a/b in the equations of 17.A, or again on the ratio between the eigenvalues of the matrix $\vec{A}$ of 15.A.

For a parabola, the pair (f, D) is unique; for the ellipse and the hyperbola, there are two pairs (f, D) and (f', D'). In this case, when m describes C the sum $mf + mf'$ (resp. $|mf - mf'|$) is constant for an ellipse (resp. a hyperbola) C. The converse is also true. The points f, f' are called the *foci* of the conic.

The first variation formula (cf. 9.G) shows that the oriented lines $\overrightarrow{mf}, \overrightarrow{mf'}$ have as external (resp. internal) bisector the tangent to the ellipse (resp. hyperbola) at m. The "second little theorem of Poncelet" says that for every point m of the plane through which there pass two tangents to the ellipse or

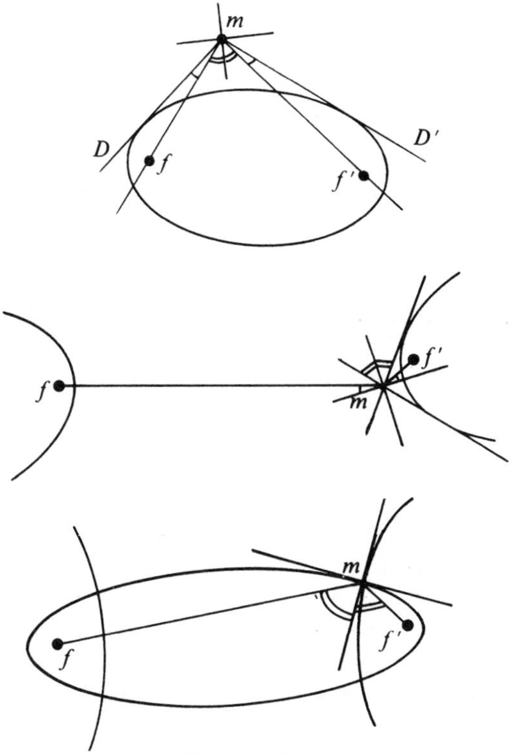

Figure 17.B.

hyperbola C, the two tangents and the lines $\langle m, f\rangle, \langle m, f'\rangle$ have the same bisectors.

17.C Using the Cyclical Points ([B, 17.4, 17.5])

We *denote* by $\bar{\alpha}$ the projective conic in $\tilde{X}^C$ obtained by complexifying the projective completion $\tilde{\alpha}$ of the conic α (cf. 15.A). On the line at infinity ∞_X^C, the position of the points at infinity of $\bar{\alpha}$ can take interesting special values relative to the cyclical points $\{I, J\}$ (cf. 9.D):

— α is an equilateral hyperbola if and only if its points at infinity are conjugate relative to $\{I, J\}$;

— α is a circle if and only if its points at infinity are $\{I, J\}$ themselves.

We can see immediately that the Laguerre formula (8.H) and the fact that the cross-ratio in 16.C is constant imply the condition we found in 10.D for four points to be on the same circle. See also problem 16.3.

Combining the above with the theorem of Desargues for pencils of conics (cf. 14.D), taking D to be the line at infinity of X, one obtains a number of results concerning pencils of Euclidean conics. For instance:

— the directions of the (symmetry) axes of a conic C are the points ∞_X^C which are harmonic conjugates relative to $\{I, J\}$ and also relative to the points at infinity of C;

— a pencil of conics contains a single equilateral hyperbola, unless it contains only such curves;

— a pencil of conics contains a circle if and only if the directions of the axes are fixed; in particular, the common chords have the same inclination relative to the axes, and the axes of the two parabolas in the pencil are orthogonal;

— the centroid of four cocyclical points on a parabola is located on its axis.

The foci of C are the points through which the tangents to C contain the cyclical points: cf. [B, 17.4.3].

17.D Notes

1. The set of plane conics (ellipses and hyperbolas) which share two common foci f, f' satisfies a number of properties. It forms a tangential pencil (cf. 14.F). For details, see [B, 17.6.3], and also problem 15.6.

2. The theorem of Appolonius (cf. 15.D), applied to the special case of conics, says that the area and the sum of the squares of the sides of a parallelogram, two of whose sides are conjugate half-diameters, are both constant quantities for a given ellipse. See [B, 17.9.22] for the geometric construction of an ellipse of which two conjugate diameters are given.

Problems

17.1 CHORDS OF A CONIC ([B, 17.9.20]). Given a fixed point on a conic, consider all the chords whose angle, seen from that point, is a constant. Find the envelope of these chords. Analyze the special case of the right angle.

17.2 COCYCLIC POINTS AND NORMALS TO AN ELLIPSE ([B, 17.7.3, 17.9.15, 17.9.10]). Consider the ellipse parametrized by $(a\cos t, b\sin t)$, where the parameter t is defined modulo 2π, and put $\vartheta = \tan t/2$.

(i) Show that the four points corresponding to values of the parameter $(t_i)_{i=1,2,3,4}$ are on the same circle if and only if

$$t_1 + t_2 + t_3 + t_4 \equiv 0 \,(\mathrm{mod}\, 2\pi).$$

(ii) Consider four points $(m_i)_{i=1,2,3,4}$ on an ellipse, and take the points n_i where the osculating circles at m_i intersect the ellipse (we choose $n_i \neq m_i$ except in the superosculating case). Show that if the m_i are cocyclic, then so are the n_i.

(iii) Show that the normals to the ellipse through the points parametrized by $(t_i)_{i=1,2,3,4}$ are concurrent if and only if the corresponding ϑ_i satisfy the following two conditions:

$$\vartheta_i\vartheta_2 + \vartheta_1\vartheta_3 + \vartheta_1\vartheta_4 + \vartheta_2\vartheta_3 + \vartheta_2\vartheta_4 + \vartheta_3\vartheta_4 = 0,$$

$$\vartheta_1\vartheta_2\vartheta_3\vartheta_4 = -1.$$

In this case we have $t_1 + t_2 + t_3 + t_4 \equiv \pi(\mathrm{mod}\, 2\pi)$.

(iv) If four points in an ellipse have concurrent normals, then the circle passing through three of them also passes through the point diametrically opposite to the fourth (theorem of Joachimstal).

17.3 TANGENT CIRCLES TO TWO CONJUGATE DIAMETERS ([B, 17.9.21]). Show that the tangent circles to two variable conjugate diameters of an ellipse, and whose center is on the ellipse, have constant radius.

17.4 TANGENT ELLIPSES TO A CIRCLE ([B, 17.9.23]). Given a circle C in X and two points a, b on C, consider the ellipses E which are tangent to C, pass through a and b, and whose center is the midpoint of ab. Show that all such ellipses have the same excentricity.

17.5 NORMALS FROM A POINT TO A PARABOLA ([B, 17.9.18.2]). Show that the normals at three points m, m', m'' of a parabola P are concurrent if and only if the barycenter $(m + m' + m'')/3$ belongs to the axis of P. Also if and only if the circle passing through m, m', m'' contains the vertex of P.

Chapter 18
The Sphere for Its Own Sake

In all of this chapter $S = S^d$ denotes the unit sphere in Euclidean space $\mathbf{R}^{d+1}$.

18.A Preliminaries ([B, 18.1, 18.2, 18.3])

The sphere is a compact connected topological space; for $d \geq 2$, it is also *simply connected*, which means that every closed curve in S can be deformed into a point (such is not the case for either S^1 or $P^n(\mathbf{R})$).

A *great circle* of S is the intersection of S with a two-dimensional vector space of $\mathbf{R}^{d+1}$.

There are many possible representations for S, of which we will mention a few. The first is the *stereographic projection* of S^d minus the north pole $(0, 0, \ldots, 0, 1)$ onto the hyperplane $\mathbf{R}^d \subset \mathbf{R}^{d+1}$. This projection is bijective and *conformal*, which means it preserves angles. It is given by the figure below:

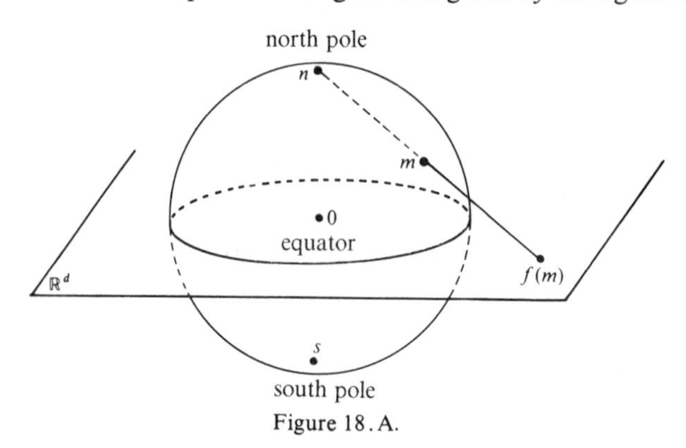

Figure 18.A.

The second representation parametrizes S^2 by its *longitude* and *latitude*, which are the angles φ and ϑ such that

$$x = \cos\varphi\cos\vartheta, \quad y = \sin\varphi\cos\vartheta, \quad z = \sin\vartheta.$$

The sphere possesses a *canonical measure*; its volume for this measure is 2π for S^1, 4π for S^2, $2\pi^2$ for S^3 and so on (see [B, 9.12.4.8] for a general formula). The area of a spherical triangle with angles α, β, γ is $\alpha+\beta+\gamma-\pi$ (see 18.C).

18.B Intrinsic Metric in S ([B, 18.4, 18.5])

The metric $d(x, y) = \|x-y\|$ induced on S^d by the metric of $\mathbf{R}^{d+1}$ is not good, since in it the distance from x to y cannot be realized as the length of a curve in S with endpoints x, y (see 9.G). On the other hand, if we define $\overline{xy}$, for $x, y \in S$, using the scalar product:

$$\cos(\overline{xy}) = (x|y),$$

we see that $\overline{\cdot\,\cdot}$ is a metric on S, called the *intrinsic metric* of S. It has all the desirable properties, and in particular it is excellent (cf. 9.G): the segment joining x to y is the shortest arc of the great circle of S containing x and y (cf. 18.A). This segment is unique, with the only exception of the case $y=-x$, when x and y are called *antipodal points* and any great half-circle joining x and y is a segment.

The strict triangle inequality also holds, i.e. $\overline{xz} = \overline{xy} + \overline{yz}$ if and only if y belongs to a segment whose endpoints are x, z. See also problem 18.3.

The group of *isometries* of S (for the intrinsic metric) coincides with the restriction to S of the orthogonal group $O(d+1) = O(\mathbf{R}^{d+1})$; see 8.A.

18.C Spherical Triangles ([B, 18.6])

A *spherical triangle* in S^2 is a triple $\langle x, y, z\rangle$ consisting of three linearly independent vectors in $\mathbf{R}^3$ and contained in S^2. The *angle* at x, generally *denoted* by α, is the angle (with values in $]0, \pi[$, see 8.F) formed by the vectors at x tangent to arcs of great circle joining x to y and x to z; it is given by

$$\cos\alpha = \left(\frac{y-(x|y)x}{\|y-(x|y)x\|} \middle| \frac{z-(x|z)x}{\|z-(x|z)x\|} \right).$$

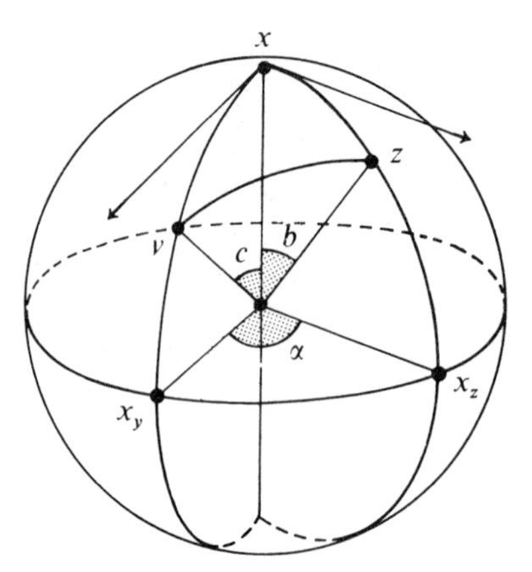

Figure 18.C.1.

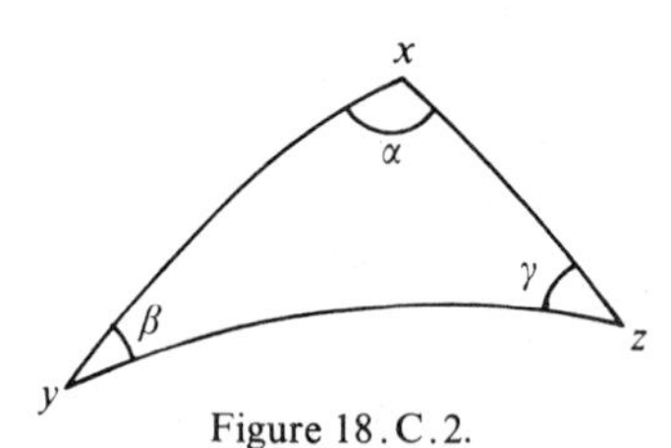

Figure 18.C.2.

We also *denote* by β (resp γ) the angle at y (resp. z) of the spherical triangle $\langle x, y, z\rangle$. The *sides* of $\langle x, y, z\rangle$ refer both to the segments joining x to y, y to z and z to x, and to the distances between each pair of points. The standard *notation* is

$$a=\overline{yz}, \quad b=\overline{zx}, \quad c=\overline{xy}.$$

The *fundamental formula of spherical trigonometry* says that

$$\cos a=\cos b\cos c+\sin b\sin c\cos\alpha. \tag{1}$$

From this we can deduce the strict triangle inequality, and also the formulas

$$\frac{\sin\alpha}{\sin a}=\frac{\sin\beta}{\sin b}=\frac{\sin\gamma}{\sin c}, \tag{2}$$

$$\cos\alpha=-\cos\beta\cos\gamma+\sin\beta\sin\gamma\cos a. \tag{3}$$

Notice the fact that the sines, contrary to the cosines, do not uniquely determine an angle between 0 and π; cf. 10.A. Notice also that the area of a spherical triangle is $\alpha+\beta+\gamma-\pi$ (*Girard's formula*).

In order that two spherical triangles, whose sides and angles are $\{a, b, c, \alpha, \beta, \gamma\}$ and $\{a', b', c', \alpha', \beta', \gamma'\}$, be equivalent under an isometry of S^2, it is sufficient that certain three of the elements be equal for both triangles, for example:

$$a = a', \quad b = b', \quad c = c'; \quad \text{or}$$

$$\alpha = \alpha', \quad \beta = \beta', \quad \gamma = \gamma'; \quad \text{or}$$

$$b = b', \quad c = c', \quad \alpha = \alpha', \quad \text{etc.}$$

Compare with the case of similarity of triangles in $\mathbf{R}^2$ (10.A).

18.D Clifford Parallelism ([B, 10.12, 18.8, 18.9])

This is a phenomenon verified in the sphere S^3. We say that two great circles C, C' of S^3 are *parallel* if, in the intrinsic metric of S^3, the distance from a point $m \in C$ to C', i.e. $D(m, C') = \inf\{\overline{mm'} : m' \in C'\}$, is independent of m. It is easy to see that such parallels exist, for example the orbits of the one-parameter rotation groups acting on $\mathbf{R}^4$ by the matrices of 8.E (for $\vartheta = \eta$ variable).

The essential result is that, given a great circle C of S^3 and a point m' (non-orthogonal to C in $\mathbf{R}^4$), there are exactly two parallels C', C'' to C passing through m', and the circles C', C'' make an angle equal to 2α at m', where $\alpha = d(m', C)$. In particular, this implies that the subset

$$\Delta(\alpha, C) = \{m' \in S^3 : d(m', C) = \alpha\},$$

for $\alpha \in]0, \pi/2[$, which is a topological torus, contains two one-parameter families of great circles parallel to C; the parallels of one family cut those of the other family at a constant angle 2α. Two parallels of the same family do not intersect; they are linked, and parallel to one another.

Applying stereographic projection (18.A) to transfer this picture to $\mathbf{R}^3$, we obtain the following results: a torus of revolution in $\mathbf{R}^3$ possesses four families of circles, namely meridians, longitudes, and two other families (called the *circles of Villarceau*) which arise from Clifford parallels of S^3. Through every point of the torus there passes one representative of each family, and they form constant angles $(\pi/2, \alpha, \pi/2 - \alpha)$. Two circles of Villarceau either intersect or are linked.

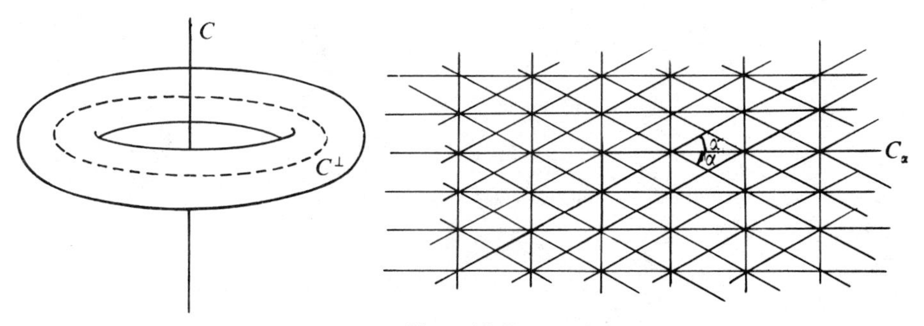

Figure 18.D.

We can obtain the circles of Villarceau in an elementary fashion by cutting the torus by oblique bitangent planes. We can also utilize the theory of cyclids: see problems 18.7 and 20.C, as well as [B, 20.7.2].

18.E The Möbius Group ([B, 18.10])

The spheres S^d is a homogeneous space under a group (cf. 1.B) strictly bigger than $O(d+1)$ (cf. 18.B), which is a Lie group of dimension $[d(d+1)]/2$. This larger group, called *conformal group* or *Möbius group* of S^d and *denoted* by $Möb(d)$, has dimension $[(d+1)(d+2)]/2$. It can be obtained in several equivalent ways.

First, it is the group of conformal transformations of S^d. It is also the group of transformations of S^d generated by $O(d+1)$ and the pull-backs under stereographic projection of the vector homotheties of $\mathbf{R}^d$. Also the group of transformations of S^d which transform every subsphere into a subsphere. Also the group formed by the restrictions to S^d of all inversions (cf. 10.C) and reflections through hyperplanes in $\mathbf{R}^{d+1}$ which leave S^d invariant. Finally, it can be identified with the projective group of the quadric α with equation $-\sum_{i=1}^{d+1} x_i^2 + x_{d+2}^2$ (cf. 14.G) in $P(\mathbf{R}^{d+2})$; the idea is to homogenize the affine equation $\sum_{i=1}^{d+1} x_i^2 = 1$ of S^d (see also chapter 15).

Problems

18.1 THE SPHEROMETER ([B, 18.11.1]). Let A denote the three vertices of an equilateral triangle of side a, B a point of the perpendicular to the triangle passing through its center, and e the distance from B to the plane defined by the triangle. Show that the radius of the sphere passing through B and the three points A has the value $R=(a^2+3e^2)/6e$.

One can build a device to measure the radius of a spherical surface by using the formula above. In the case of the device shown in figure 18.1.2, explain the function of the lever system at the top of the spherometer.

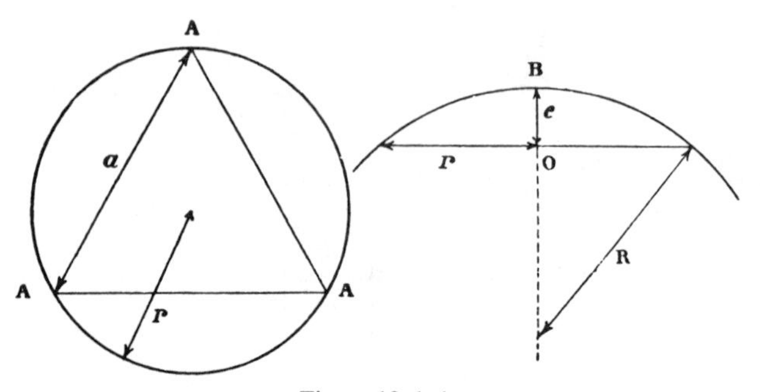

Figure 18.1.1.

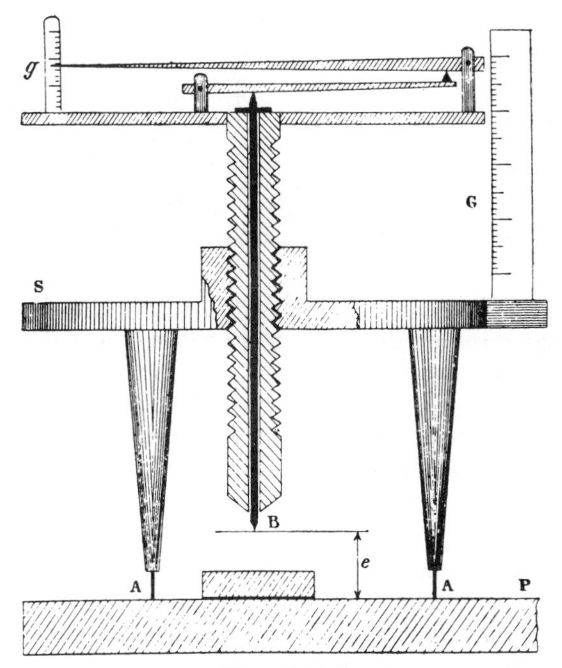

Figure 18.1.2.

H. Bouasse, Appareils de mesure, Delagrave, 1917.

18.2 LOXODROMES ([B, 18.11.3]). Recall that a loxodrome, or rhumb line, of the terrestrial sphere is a curve that makes a constant angle with the meridian at each point. (In the projection of Mercator, loxodromes become straight lines; they represent the trajectories of a ship whose helm is kept fixed. See [B, 18.1.8.2].) Show that, using stereographic projection centered at the north pole (18.A), loxodromes become logarithmic spirals (cf. 9.E). See also problem 6.6.

18.3 THE STRICT TRIANGLE INEQUALITY HOLDS FOR THE SPHERE ([B, 18.11.13]). Prove, using Gram determinants (cf. 8.J), that the distances a, b, c between any three points x, y, z in the sphere verify the inequalities

$$|b-c| \leq a \leq b+c,$$

and equality can only take place if the three points are in the same plane (i.e. they are aligned on a great circle, cf. 18.A).

18.4 UNIVERSAL RELATION BETWEEN DISTANCES OF POINTS IN S^d ([B, 18.11.4]). Show that if $(x_i)_{i=1,\ldots,d+2}$ are $d+2$ points in S^d, their distances $\overline{x_i x_j}$ always satisfy the relation

$$\det\left(\cos\left(\overline{x_i x_j}\right)\right) = 0.$$

18.5 PLANE TRIGONOMETRY AS THE LIMIT OF SPHERICAL TRIGONOMETRY ([B, 18.11.9]). Generalize formulas (1), (2) and (3) of 18.C for the intrinsic metric of a sphere of radius R. Then find out what the formulas become when R approaches infinity.

18.6 HOOKE JOINTS, HOMOKINETIC JOINTS ([B, 18.11.16]). Consider a Hooke joint (figure 18.6.2) whose axes make an angle ϑ. The ratio between the instant angular velocities of the two shafts is a function of the angle between the plane of either fork and the plane containing the axes of the shaft; find the worst possible value for this ratio. To do this, take two great circles C, D in S^2 making an angle ϑ, and two moving points $m(t), n(t)$ on the circles, so that $\overline{m(t)\ n(t)} = \pi/2$. Find the value of the worst possible ratio when $\vartheta = \pi/3$, $\pi/4$, $\pi/6$.

Show that if two shafts A, A' are joined by Hooke joints to a third shaft B whose forks are in the same plane, in such a way that A, B, A' are in the same plane and the angles of B with A and A' are the same (*homokinetic joint*), then A and A' always have the same angular velocity.

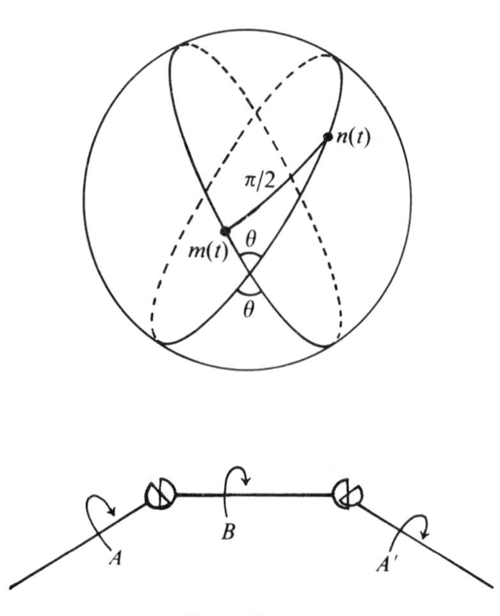

Figure 18.6.1.

18.7 DUPIN CYCLIDS ([B, 18.11.19]). Let $\Sigma, \Sigma', \Sigma''$ be three spheres in $\mathbf{R}^3$; show that, for certain configurations of these spheres, the set of spheres tangent to $\Sigma, \Sigma', \Sigma''$ has for envelope the surface obtained from a torus of revolution by inversion relative to an appropriate point. Deduce several properties of such surfaces, which we will encounter again in 20.C and which are called *Dupin cyclids*.

Chapter 19

Elliptic and Hyperbolic Geometry

19.A Elliptic Geometry ([B, 19.1])

Consider a Euclidean vector space E of dimension $d+1$, and the associated projective space $P = P(E)$ with canonical projection $p: E\backslash 0 \to P$. Recall that P is the set of (vector) lines of E, hence the quotient of the unit sphere $S(E) = \{x \in E: \|x\| = 1\}$ by the equivalence relation $x \sim -x$ (antipodal points).

The *elliptic space* associated with E is P, endowed with the metric $d(D, D') = \overline{DD'}$, where $\overline{DD'}$ denotes an angle between non-oriented lines, ranging from 0 to $\pi/2$ (cf. 8.F). In other words, if $m, n \in P$ and $m = p(x), n = p(y)$, we *define* $\overline{mn}$ by $\cos(\overline{mn}) = |(x|y)|$. As a function of the distance $\overline{xy}$ on S (cf. 18.B), we have

$$\overline{mn} = \inf\left\{\overline{xy}, \overline{x(-y)}\right\}.$$

The distance $\pi/2$ corresponds to orthogonal lines.

The metric thus obtained is *excellent*, and its *segments* (cf. 9.G) are the projections of arcs of great circle of S whose length is less than or equal to $\pi/2$; two points are connected by a unique segment when their distance is less than $\pi/2$, and by two otherwise. The group of isometries of P is the orthogonal projective group $\mathrm{PO}(G)$ (cf. 14.G).

There is a fundamental difference between elliptic geometry on the one hand and Euclidean, spheric (18.C) or hyperbolic (19.B) geometry on the other as regards triangles: in elliptic geometry, two triangles can have the same sides and not be identifiable under an isometry of P. The fundamental reason is that P is not simply connected (see 18.A). There are two types of triangles; type I is formed by those which are projections of triangles of S whose sides are all

$\leq \pi/2$. Type II contains the projections of sets of three segments of S of length $\leq \pi/2$ connecting a point to its antipodal point. For example, there are two triangles of P (up to an isometry) such that the three sides measure $\pi/3$; one is the projection of the spherical triangle of sides $a = b = c = \pi/3$ (and consequently $\alpha = \beta = \gamma$ with $\cos\alpha = 1/3$), and the other has as vertices three *coplanar* concurrent lines forming an angle of $\pi/3$ between each two of them! See problem 19.1.

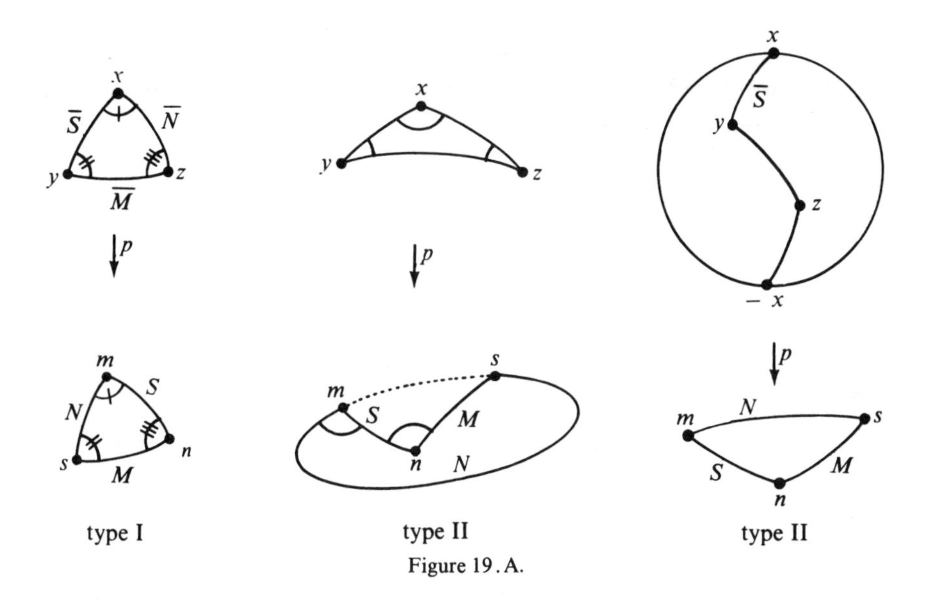

Figure 19.A.

19.B The Hyperbolic Space ([B, 19.2, 19.3])

The *projective model* or *Klein model* for the *hyperbolic space* of dimension n is a metric space whose underlying space is

$$\mathscr{B} = \{ z \in \mathbf{R}^n : \|z\| < 1 \}.$$

The *distance* $d(z, z')$ in $\mathscr{B}$ is defined in the following way: consider first the points $(z, t) \in \mathbf{R}^n \times \mathbf{R} = \mathbf{R}^{n+1}$, and endow $\mathbf{R}^{n+1}$ with the quadratic form q defined by

$$q(z, t) = -\|z\|^2 + t^2,$$

whose polar form is $P((z,t),(z',t')) = -(z|z') + tt'$. Then put

$$\operatorname{ch}(d(z, z')) = \frac{P(\xi, \xi')}{\sqrt{q(\xi)q(\xi')}},$$

where $\xi = (z, 1)$ and $\xi' = (z', 1)$, and ch is the hyperbolic cosine.

If u, v are the intersection points of the line $\langle z, z' \rangle$ with the unit sphere S^{n-1} (the boundary of $\mathscr{B}$), then $d(z, z') = \frac{1}{2}|\log([z, z', u, v])|$ (compare with problem 11.3), where the cross-ratio is taken in the line $\langle z, z' \rangle$; see 6.A.

This metric is excellent (cf. 9.G): there is a unique segment joining z and z', and it is the same as the Euclidean segment $[z, z']$ (but with the hyperbolic metric, of course). The *group of isometries of* $\mathscr{B}$ *is denoted by* $G(n)$; it is the same as the projective group $\mathrm{PO}(\alpha)$ of the projective quadric α in $P^n(\mathbf{R}) = P(\mathbf{R}^{n+1})$ defined by the quadratic form q (cf. 14.G). It is generated by the transformations induced in $\mathscr{B}$ by hyperplane reflections associated with q (see 13.E and 14.G); the latter are the involutive isometries whose fixed points are the points in $H \cap \mathscr{B}$, where H is any hyperplane in $\mathbf{R}^n$ which intersects $\mathscr{B}$. If z, z' are distinct points of $\mathscr{B}$, the set $\{z'' \in \mathscr{B} : d(z'', z) = d(z'', z')\}$ always has the form $H \cap \mathscr{B}$; it is the hyperplane *equidistant* from z and z'. We deduce (cf. 9.F) that if $(z_i)_{i=1,\ldots,k}, (z_i')_{i=1,\ldots,k}$ are two k-tuples of points in $\mathscr{B}$ such that

$$d(z_i, z_j) = d(z_i', z_j')$$

for all i, j, then there exists an isometry $f \in G(n)$ such that $f(z_i) = z_i'$ for all i. In particular, two triangles are "the same" if and only if their sides have same length (by contrast with 19.A; see 18.C).

19.C Angles and Trigonometry ([B, 19.2, 19.3])

It is harder to define angles in $\mathscr{B}$. (The ordinary Euclidean angles in $\mathscr{B} \subset \mathbf{R}^n$ don't work for our hyperbolic metric, since they are not invariant by $G(n)$!) Let $z \in \mathscr{B}$, and consider two vectors $u, v \in \mathbf{R}^n$ which are "tangent to $\mathscr{B}$ at z". We want to work in $\mathbf{R}^{n+1}$, so we take $\xi = (z, 1)$ and replace u (resp. v) by vectors $\bar{u}$ (resp. $\bar{v}$) in $\mathbf{R}^{n+1}$ which are q-orthogonal to ξ (compare with 18.C), and which are linear combinations of ξ and $(u, 0)$ (resp. $(v, 0)$). We get

$$\bar{u} = (u, 0) + \frac{(u|z)}{1 - \|z\|^2}\xi, \quad \bar{v} = (v, 0) + \frac{(v|z)}{1 - \|z\|^2}\xi,$$

and the *angle* formed by u, v at z is defined as $\alpha \in [0, \pi]$ such that

$$\cos\alpha = -\frac{P(\bar{u}, \bar{v})}{\sqrt{q(\bar{u})q(\bar{v})}}.$$

Given a triangle $\mathscr{T} = \{x, y, z\}$ in $\mathscr{B}$, we can now talk about its *sides* $a = d(y, z)$, $b = d(z, x)$, $c = d(x, y)$, and its *angles* α, β, γ at points x, y, z respectively, where α is the angle at x between the vectors $u = \overrightarrow{xy}$, $v = \overrightarrow{xz}$

according to the definition above. We have the following formulas:

$$\operatorname{ch} a = \operatorname{ch} b \operatorname{ch} c - \operatorname{sh} b \operatorname{sh} c \cos \alpha; \tag{1}$$

$$\frac{\sin \alpha}{\operatorname{sh} a} = \frac{\sin \beta}{\operatorname{sh} b} = \frac{\sin \gamma}{\operatorname{sh} c}; \tag{2}$$

$$\cos \alpha = \sin \beta \sin \gamma \operatorname{ch} a - \cos \beta \cos \gamma. \tag{3}$$

The hyperbolic space $\mathscr{B}$ possesses a canonical measure; for that measure the area of a triangle $\mathscr{T}$ of angles α, β, γ is equal to $\pi - \alpha - \beta - \gamma$ (compare with 18.C and 10.A).

19.D The Conformal Models C and H ([B, 19.6, 19.7])

In the projective model, angles had a complicated definition; we would like them in some sense to agree with Euclidean angles. The way to achieve this is the following: define a bijection $\Xi: \mathscr{B} \to \mathscr{B}$ as the composition $\Xi = f \circ g$ of $g: z \mapsto (z\sqrt{1 - \|z\|^2})$ from $\mathscr{B}$ into $\mathbf{R}^{n+1}$ and the stereographic projection f from the north pole of $S^n \subset \mathbf{R}^{n+1}$ onto $\mathbf{R}^n$ (cf. 18.A). Then use Ξ to pull back the metric and obtain a new hyperbolic metric in the ball $\mathscr{B}$; with this new metric the ball will be called the *conformal* or *Poincaré* model, and will be denoted by $\mathscr{C}$. In $\mathscr{C}$ the angles are the same as Euclidean angles; on the other hand, the segments are not Euclidean line segments anymore—they are arcs of circles which intersect the boundary S^{n-1} of $\mathscr{C}$ at right angles.

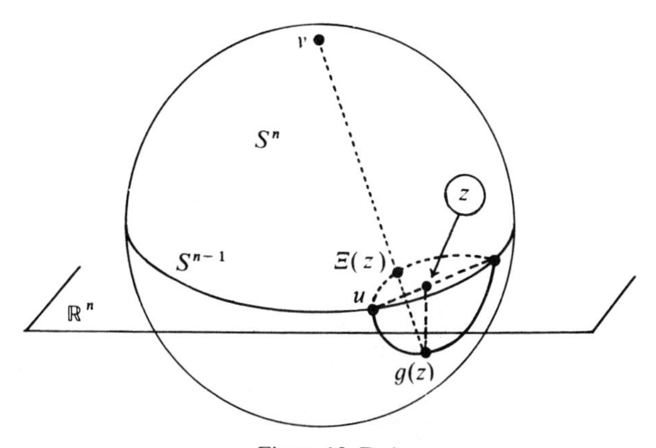

Figure 19.D.1.

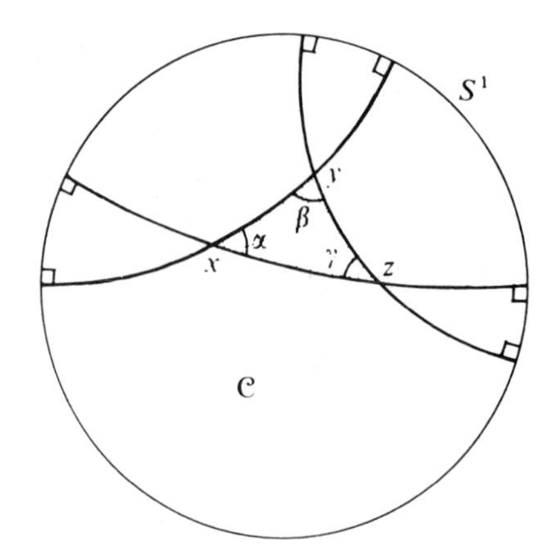

Figure 19.D.2.

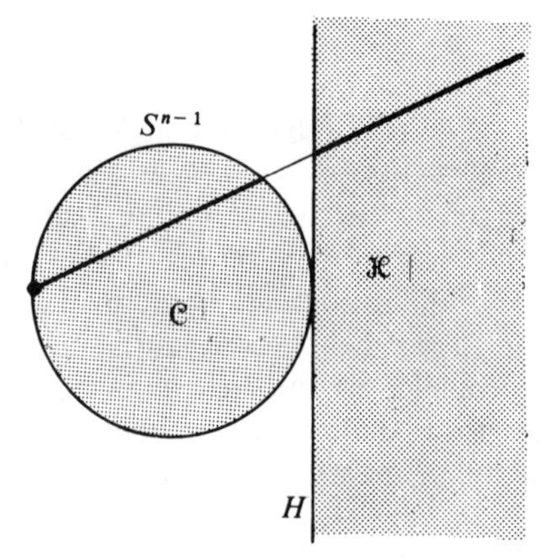

Figure 19.D.3.

Another model $\mathscr{H}$, called the *upper half-space model*, is obtained from $\mathscr{C}$ by applying inversion relative to a point on the boundary of $\mathscr{H}$; it is still conformal, and the segments are still circles.

Problems

19.1 ELLIPTICAL EQUILATERAL SETS ([B, 19.8.24]). An *equilateral set* of a metric space is any set $\{m_i\}_{i=1,\dots,n}$ such that all the distances $d(m_i, m_j)(i < j)$ are equal. Show that the elliptic plane P contains equilateral three-point sets with side lengths ranging from 0 to $\pi/2$; classify them under the action of $\mathrm{Is}(P)$. Show that P contains equilateral four-point sets with side length

$\cos^{-1}(1/\sqrt{3})$ or $\cos^{-1}(1/\sqrt{5})$; study their behavior under the action of Is(P). Show that, up to isometries, P contains exactly one five-point and one six-point equilateral set, and their sides have length $\cos^{-1}(1/\sqrt{5})$.

19.2 HYPERBOLIC QUADRILATERALS WITH THREE RIGHT ANGLES ([B, 19.8.7]). Find the fourth angle of a quadrilateral in the hyperbolic plane such that three of its angles are right and the lengths of the sides which join two right angles are a and b.

19.3 UNIVERSAL DISTANCE RELATION IN HYPERBOLIC SPACES ([B, 19.8.16]). Show that in n-dimensional hyperbolic space every set of $n+2$ points $z_i (i=1,\ldots,n+2)$ obeys the relation

$$\det\left(\mathrm{ch}\left[d(z_i, z_j)\right]\right) = 0.$$

19.4 REGULAR HYPERBOLIC POLYGONS ([B, 19.8.20]). Here n is an integer ≥ 3. We want to study n-sided polygons in the hyperbolic plane, all of whose sides are equal and all of whose angles have the value $2\pi/n$. Are there such polygons for any n? Are they unique up to isometries?

Chapter 20
The Space of Spheres

20.A The Space of Spheres ([B, 20.1])

We fix an n-dimensional Euclidean space E, and *denote* by $Q(E)$ the space of affine quadratic forms q over E; $\vec{q}$, an element of $Q(\vec{E})$, will be the *symbol* of q. A sphere is given by a form q, written as

$$q = k\|\cdot\|^2 + (\alpha|\cdot) + h, \text{ where } \alpha \in \vec{E} \text{ and } k, h \in \mathbf{R};$$

it is an actual sphere if $k \neq 0$ and $\|\alpha\|^2 > 4kh$; if $k \neq 0$ and $\|\alpha\|^2 = 4hk$ the image is a single point (sphere of zero radius), and if $\|\alpha\|^2 < 4hk$ the image is empty, and we say that q represents a sphere "of imaginary radius". For $k = 0$, $h \neq 0$, we obtain a hyperplane (if $\alpha \neq 0$); the case $k = 0$ and $\alpha = 0$, $h \neq 0$ represents *the* point at infinity of E.

Let us now introduce the vector subspace of $Q(E)$ formed by the forms q such that $\vec{q} = k\|\cdot\|^2$; we *denote* it by $\check{S}(E)$, and projectivize it to obtain $S(E) = P(\check{S}(E))$, the canonical projection being $p: \check{S}(E)\backslash 0 \to S(E)$. In this way we can identify the actual spheres of E with a subset Σ of $S(E)$, the affine hyperplanes with a subset Θ, and the spheres of zero radius with E itself. The complement of $\Sigma \cup \Theta \cup E$ in $S(E)$ consists of the spheres of imaginary radius, plus one point which corresponds to the equation $q = 1$; we *denote* this point by ∞ and call it *the point at infinity* of E. Unlike the construction in chapter 5, which was adequate for linear geometry and in which there was a whole hyperplane of points at infinity, our construction here has only one point at infinity, the same in all directions. This new completion of E will be written $\hat{E} = E \cup \infty$.

20.B The Canonical Quadratic Form ([B, 10.2])

What makes the above construction useful is the introduction of a canonical non-degenerate quadratic form in $\check{S}(E)$, from which follows the existence of a proper quadric in $S(E) = P(\check{S}(E))$; see 14.A). This quadratic form, *denoted* by ρ, represents the square of the radius:

$$\rho\big(k\|\cdot\|^2 + (\alpha|\cdot) + h\big) = \frac{\|\alpha\|^2 - 4hk}{4}.$$

Its polar form R is given by

$$R\big(k\|\cdot\|^2 + (\alpha|\cdot) + h,\, k'\|\cdot\|^2 + (\alpha'|\cdot) + h'\big) = \tfrac{1}{4}(\alpha|\alpha') - 2(k'h + kh').$$

The form ρ has a projective quadric, also *denoted* by ρ, associated to it; its image is exactly $\hat{E}$. The associated projective group $\mathrm{PO}(\rho)$ is in fact the Möbius group $\mathrm{Möb}(n)$ (cf. 18.E) if we choose an appropriate identification of $\hat{E}$ and S^n.

It is natural to consider polarity with respect to ρ (cf. 14.E). The tangent hyperplane to $\hat{E}$ at ∞ is exactly Θ (with infinity adjoined). The points of a sphere $s \in S(E)$ are represented by the intersection points of the polar hyperplane of s with $\hat{E}$ (see figure). The center of s is the intersection point (other than ∞) of the line $\langle s, \infty\rangle$ with $\hat{E}$. Two spheres (including hyperplanes) are orthogonal (cf. 10.B) exactly when they represent two conjugate points with respect to ρ. More generally, the *angle between two spheres* (including hyperplanes) is the real number $[s, s']$ in the interval $[0, \pi/2]$ defined by the following equation, where q (resp. q') is an equation for s (resp. s'):

$$\cos([s, s']) = \frac{|R(q, q')|}{\sqrt{\rho(q)\rho(q')}}.$$

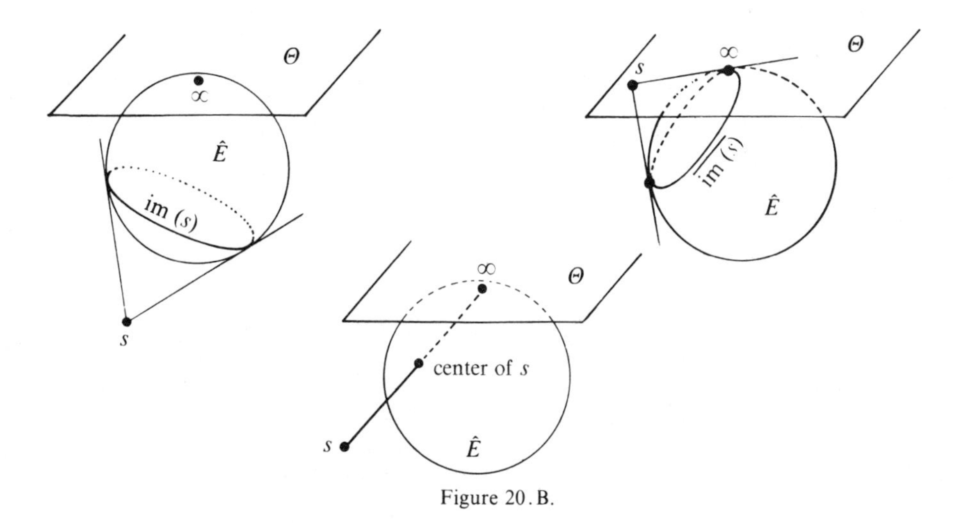

Figure 20.B.

Observe that true spheres are the points of $S(E)$ which lie "outside" $\hat{E}$. Those inside represent imaginary spheres (which, incidentally, are models for hyperbolic geometry; see 19.B).

Pencils of spheres are the (projective) lines of $S(E)$. They differ depending on whether the line does or does not intersect $\hat{E}$; see [B, 20.5.6].

The above construction allows us to extend inversion to the completion $\hat{E}$; the pole c is taken to ∞ and ∞ is taken to c. Inversions are in fact those elements of $\mathrm{PO}(\rho)$ which are reflections through hyperplanes; this means they generate the conformal group $\mathrm{Möb}(n)$ of S^n.

20.C Polyspheric Coordinates ([B, 20.7])

Since E is embedded in $S(E)$, we can represent the points of E using homogeneous coordinates in the projective space $S(E)$ (so there are $n+2$ coordinates for an n-dimensional space E); the chosen basis will preferably be one in which ρ is diagonal. Such coordinates are called *polyspheric*. *Cyclids* are the hypersurfaces of E which are quadrics when expressed in polyspheric coordinates; they are algebraic hypersurfaces of the fourth degree. The cyclids of Dupin are those which have two equal coefficients; the torus of revolution is a special case of a cyclid of Dupin. See problems 18.7 and 20.2. Certain cyclids contain six families of circles: [B, 20.8.7].

Problems

20.1 TANGENT HYPERPLANE TO $\mathrm{im}(s)$ IN THE SPACE OF SPHERES ([B, 20.8.1]). Construct geometrically the tangent hyperplane to $\mathrm{im}(s)$ in $S(E)$.

20.2 THE TORUS IS A CYCLID OF DUPIN ([B, 20.8.5]). Show that the torus is a cyclid of Dupin (cf. 20.C).

20.3 THE THEOREM OF DARBOUX ([B, 20.8.7]). If three points of a line describe three spheres whose centers are collinear, then every point of that line also describes such a sphere, or possibly a plane for one exceptional point. Find a relation between four points of the line and the centers of the four spheres they describe.

Figure 20.3.1.

Gabriel Koenigs, Leçons de cinématique, A. Hermann, 1897.

Suggestions and Hints

Chapter 1

1.1 Use reflection through the midpoints of the sides.

1.3 The five groups of motions can be obtained as extensions of their subgroups of translations, which are free abelian, by adding a rotation. Thus we have the semi-direct product of a free abelian group and a cyclical group; this gives a presentation for the groups.

1.4 Observe that the tiles can only fit in a quadrilled pattern. Analyze first what happens at the corners of each square, then complete the analysis by using the restrictions imposed by the asymmetric notches. You can also make a number of photocopies of the tiles and try tiling the plane with them.

Chapter 2

2.1 To prove the theorem of Menelaus, consider the composition of the homothety of center c' taking b to a with the homothety of center b' taking a to c. Find the center of this homothety.

To prove the theorem of Ceva, choose an appropriate affine frame and compute in these coordinates.

Just for fun, prove each theorem starting from the other by using the property of the complete quadrilateral (6.B).

2.2 Reduce the problem to the case when Y is zero-dimensional (i.e. a point). You can also reduce it to the case of the sphere ([B, 18.2]).

2.4 Apply the chain rule to $f \circ c$. To compute the curvature of conics, take the parametric representations $(a \cos t, b \sin t)$ for the ellipse, (t, t^2) for the parabola and $(a \operatorname{ch} t, b \operatorname{sh} t)$ for the hyperbola.

To prove existence and uniqueness of a curve with curvature given as a function of arclength, consider the differential equation

$$\vec{c}'''(\sigma) + K(\sigma)\vec{c}'(\sigma) = 0.$$

Chapter 3

3.1 Watch out for the case $p = 2$!

3.2 Observe that the barycentric coordinates of a point of the triangle are proportional to the areas of the three sub-triangles determined by this point.

For the quadrilateral, decompose it into two triangles in two different ways.

3.3 Consider the different types of points that can occur in the subdivision, and estimate the distances between them in all possible cases.

3.4 Take the vectorialization at a point, then change the point.

3.6 Remark that all partial derivatives of order p of f are homogeneous of degree $k - p$.

Chapter 4

4.2 Remark that the point whose projective coordinates are $(1, 1, \ldots, 1)$ belongs to all the charts.

4.3 Remark that the determinant of the automorphism $-\mathrm{Id}$ of $\mathbf{R}^{n+1}$ has the value $(-1)^{n+1}$.

4.4 Apply the duality relations for vector spaces that give the orthogonal complement of an intersection or sum of subspaces.

4.5 Enumerate the ordered p-tuples of free vectors.

4.6 Take a projective base that contains the four vertices of the first tetrahedron $\{a, b, c, d\}$. Consider the matrix of a homography taking a to a', b to b', c to c' and d to d'.

Chapter 5

5.2 Use the theorem of Desargues (cf. 5.D or [B, 5.4.3]) for the first part; use the first part to solve the second.

5.3 Use the theorem of Pappus (cf. 5.D or [B, 5.4.2]) for the case of three points, and Brianchon (cf. 16.C) for the case of a conic and a point. For the latter, you can also use a projective transformation to transform the figure into an affine conic whose center is the point being considered, and apply the existence of affine reflections through the diameters (see 15.C).

Chapter 6

6.1 Send u or v to infinity.

6.2 Differentiate the cross-ratio of the four solutions.

6.3 Use perspective (see 4.F).

6.4 Take appropriate projective coordinates to demonstrate the first relation. Then use the existence of a homography taking a projective base to another.

Chapter 7

7.1 Embed E into $\mathrm{Hom}_{\mathbf{R}}(\mathbf{C}, E)$ by associating with $x \in E$ the map $s: \mathbf{C} \to E$ such that $s(1) = x$ and $s(i) = 0$.

Chapter 8

8.1 Use a simultaneous orthogonal reduction of φ and ψ (cf. 13.B).

8.2 Use the intrinsic distance on the sphere $S(E)$ (see 18.B) and consider the distance between the midpoints of the arcs of great circles joining x to $f(x)$ and $f(x)$ to $f(f(x))$.

8.3 It boils down to finding the n-th roots of a complex number.

8.4 Consider the composition $\sigma_U \circ \sigma_T \circ \sigma_S$ of the reflections $\sigma_S, \sigma_T, \sigma_U$ through S, T, U, respectively (cf. 8.C).

8.5 Use the fact that the only automorphism of $\mathbf{R}$ is the identity.

8.6 Work with a convenient basis. Interpret the Gram determinants as areas (cf. 8.J).

Chapter 9

9.1 Consider a triple of non-collinear points such that the distance from one to the line passing through the other two is minimal in the set of all such distances.

9.2 When X is not a plane, compute analytically and find what type of the quadric is obtained.

9.3 Consider the maximum of the perimeter function over the topological product C^n of C with itself (n times). You must restrict yourself to an appropriate closed subset of the compact C^n!

9.4 For direct similarities, use the point $\langle a, b\rangle \cap \langle a', b'\rangle$. For inverse similarities, use the locus of the points such that the ratio of the distance to two fixed points is constant.

9.5 Start by constructing a square, two of whose vertices are on two sides of the triangle and one of whose sides is parallel to the third side of the triangle.

9.6 Use the formula $\rho^2 + 2\rho'^2 - \rho\rho''$, which gives the concavity of the curve $\rho(\vartheta)$ in polar coordinates.

Chapter 10

10.1 In part (iv), recall that the union of two lines is a conic, and that the two lines M, N are orthogonal if and only if their points at infinity ∞_M, ∞_N satisfy $[\infty_M, \infty_N, I, J] = -1$, where $\{I, J\}$ are the cyclical points (8.H).

10.2 Find the distance between the centers by using the triangle they form together with a vertex; use the formulas in 10.A. Don't give up! For the converse, use what you've just proved and a (delicate) uniqueness argument.

10.3 Use the fact that reflection through a line transforms an oriented angle between lines into its opposite (cf. 8.F).

Consider two lines that have the desired property, and find cocyclic points using 10.D.

10.4 There is an obvious candidate for the minimum!

10.5 Show that the diagonal AC passes through a well-defined point on the circle of diameter $[a, d]$.

10.6 Start by establishing a compass-only construction for the image of a given point by an inversion whose center and fixed circle are given. Then use this construction to solve the problem by carefully choosing the inversions.

10.7 Prove the chain of theorems by induction, based on the theorem of the six circles of Miguel: see 10.D.

10.8 To prove that we get all rational numbers in [0,1], you can show, by induction on l, that any point of [0,1] of the form i/l, where i and l are coprime, is the contact point between the x-axis and a Ford circle of radius $1/2l^2$.

Chapter 11

11.1 Find the common boundary and the partition induced on it.

11.2 Introduce a supporting line at a point in the closure of the set of extremal points.

11.3 Apply first problem 6.1 to show that there is a metric isomorphic to that of **R** on the open segment $]u, v[$. Then utilize perspective transformations (4.F; remember they preserve cross-ratios!).

11.4 Calculate the logarithmic derivative of P (expressed as a product of monomials); this gives the roots of P' as linear combinations of the roots of P, with positive coefficient. For the ellipse, use the second little theorem of Poncelet (17.B), and consider the angles as arguments of complex numbers.

Chapter 12

12.1 For the first construction, make liberal use of the properties of the inscribed and the central angles (10.D). For the second, study the conditions imposed by *folding* a piece of paper.

12.2 Remember that $G(P)$, the group of the polyhedron, acts simply transitively on the n-tuples of 12.C.

12.3 Compute the volume as the integral of the area of sections parallel to a plane perpendicular to D.

For the area you can use the first formula in 12.B, and consider the first terms of a series expansion in λ.

12.4 Use the theorem of Appolonius for the volume (15.D), and 10.B.

12.5 Express the convex set by means of its supporting function h, i.e. as the *envelope* of the lines $\cos t \cdot x + \sin t \cdot y = h(t)$ ("Euler equation", [B, 11.8, 12.5]), and the fact that for this function h the radius of curvature has the value $h(t) + h''(t)$.

Chapter 13

13.1 Use a non-singular completion (13.E).

13.2 Use the classification in dimension 1, which is given by $K/(K^*)^2$ (cf. 13.B).

13.3 Think of the prime numbers.

13.4 Utilize 13.F.

13.5 For an Artin space Art_{2s}, and an appropriate basis, study the maps f given by a matrix of the form

$$\begin{pmatrix} I & S \\ O & I \end{pmatrix}.$$

Find the conditions that such an S must satisfy.

Chapter 14

14.1 Associate with an orthonormal pair (x, y) of vectors in $\mathbf{R}^{n+1}$, which spans an oriented plane, the vector $x + iy$ in $\mathbf{C}^{n+1}$.

14.2 Apply the theorem given in 14.E on harmonically circumscribed quadrics.

14.3 Apply 14.E, 14.F and the preceding problem.

Chapter 15

15.1 Use polarity with respect to C and the properties of points at infinity. The area of the parabola can be obtained using a geometric series.

15.2 Think of the square!

15.3 Calculate the product of the roots of the equation $q(x + \lambda u) = 1$, where u is a unit vector. Then use the theorem of Appolonius (15.D).

15.4 Start as in the preceding problem, and use the fact that the trace of a matrix is invariant under conjugation. Finally, apply polarity relative to a sphere (10.B).

15.5 Find the coefficients of the equation of such a cone in an appropriate coordinate system.

15.6 Work backwards, finding formulas that give the coordinates of the intersection points $Q(u), Q(v), Q(w)$, where u, v, w are such that $Q(u)$ (resp. $Q(v), Q(w)$) is an ellipsoid (resp. a one- or two-sheeted hyperboloid). See 15.B.

Chapter 16

16.1 Take an equation $x^2 - yz = 0$, where $a = (1,0,0)$, $\beta = (0,1,0)$, $\gamma = (0,0,1)$.

16.3 Take an appropriate parametric representation (16.A).

16.4 Think of the complete quadrilateral (6.B).

16.5 Use the parametric representation (st, s^2, t^2) such that the two distinct fixed points (in the complexification, if necessary) are situated at $(0,1,0)$ and $(0,0,1)$. For the theorem of Poncelet on n-gons, consider the n-fold composition f^n of f with itself.

16.6 Treat first the case of conics tangent to four lines whose equation in an affine frame is particularly simple.

16.7 Don't forget to consider the tangents to the intersection points obtained.

Chapter 17

17.1 Use problem 16.5, or calculate analytically.

17.2 Part (i): a geometrical proof consists in first showing that two chords defined by four cocyclic points have same inclination relative to the axes of the ellipse (use 17.C); then passing to a parametrization by considering t on the circle $(a\cos t, a\sin t)$. Analytically, you can write down the fourth-degree equation giving the intersection points of the ellipse with a circle $x^2+y^2+2\alpha x+2\beta y+\gamma=0$. Then use the relations between the coefficients of this equation and its roots.

Part (ii): use the fact that the osculating circle intersects the ellipse at the contact point with multiplicity three.

Part (iii): for a given point, find the angle ϑ corresponding to the foot of each normal to the ellipse passing through this point.

17.3 Use problem 15.4 in the case $n=2$.

17.4 Choose carefully your coordinate system; once you obtain the equation of an ellipse with center at the origin, observe that the excentricity is a function of the quotient of the square of the trace by the determinant of the equation.

17.5 Parametrize by a well-chosen coordinate, and use the relations between the roots and the coefficients of the third-degree equation that gives the foot of the normals.

Chapter 18

18.1 The function of the levers is a physics question...

18.2 Use the fact that the stereographic projection preserves angles.

18.4 Use the Gram determinant of these $d+2$ points of $\mathbf{R}^{d+1}$.

18.5 Use the first terms of the expansion in $1/R$, around 0, of the sine and cosine functions.

18.6 Apply the fundamental formula of spherical trigonometry to the triangle formed by $m(t), n(t)$ and the intersection of C and D. Differentiate this formula with respect to t to find a relation between the velocities of the axes.

18.7 Consider the plane containing the centers of $\Sigma, \Sigma', \Sigma''$ and suppose that the sections of the three spheres by this plane are each tangent to two circles that do not intersect. Use inversion to transform these two circles into concentric circles (10.D). Now find two one-parameter families of spheres whose envelope is the torus. Apply also the properties of the circles of Villarceau (18.D) to find four families of circles on the Dupin cyclids. Study their angles.

Chapter 19

19.1 Lift the equilateral sets in projective space to the sphere (watch out for the two different possible kinds of triangles, 19.A). Also bear in mind that in all likelihood regular polyhedra have something to do with this problem.

19.2 Cut the quadrilateral in two and apply the formulas in 19.C.

19.3 Use a method analogous to that of the Gram determinant (8.J), but adapted to the quadratic form that defines hyperbolic space (19.B).

19.4 Assume the solution known and start working from the center of the polygon.

Chapter 20

20.2 Write down the cartesian equation of a torus. Find how to pass from the cartesian equation to the equation in the space of spheres.

20.3 Work in the vector space $\mathbf{R}^3$; think of the homography that takes the three points on the line to the centers of the spheres they describe.

Solutions

Chapter 1

First exercise in 1.H

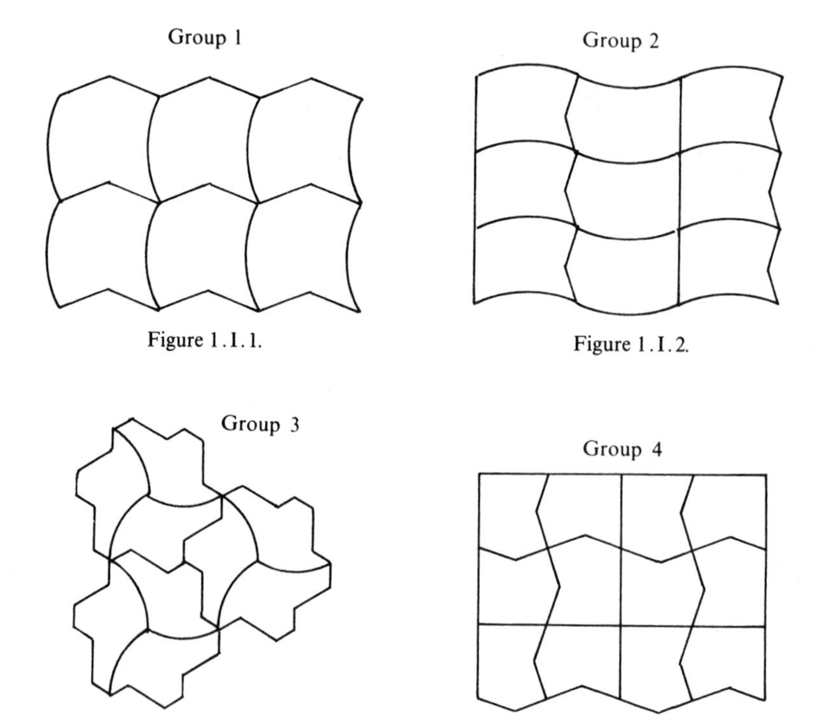

Figure 1.I.1.

Figure 1.I.2.

Figure 1.I.3.

Figure 1.I.4.

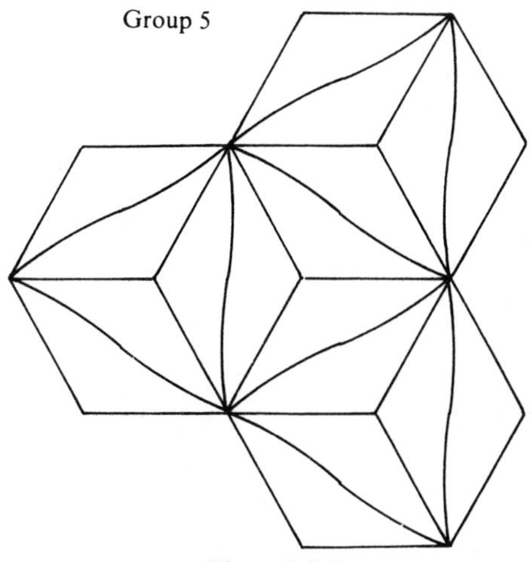

Figure 1.I.5.

Second exercise in 1.H

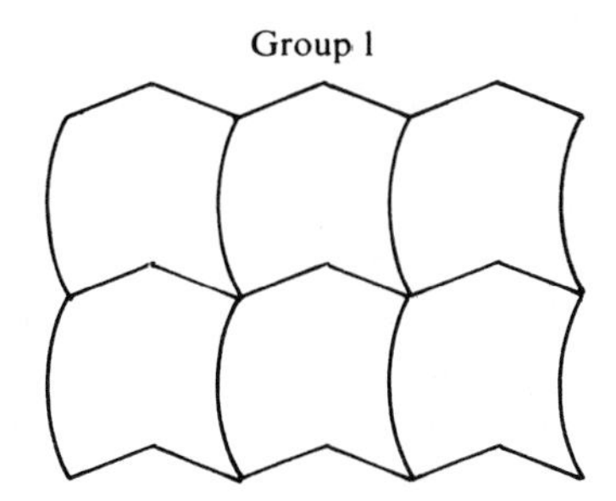

Figure 1.II.6.

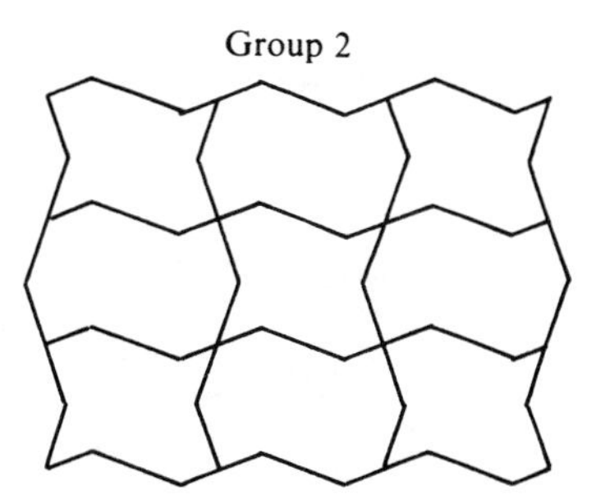

Figure 1.II.7.

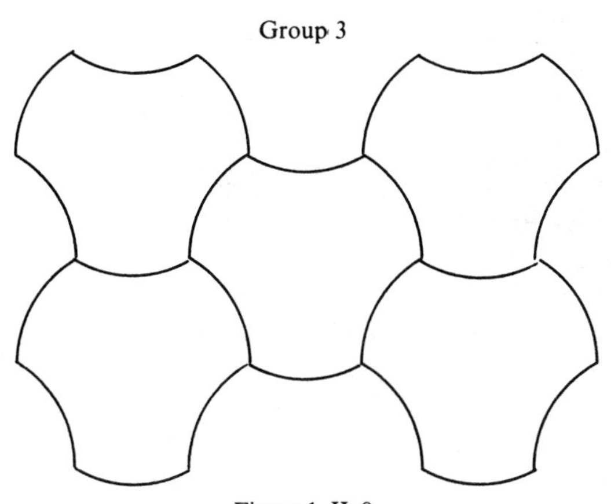

Figure 1.II.8.

Group 4

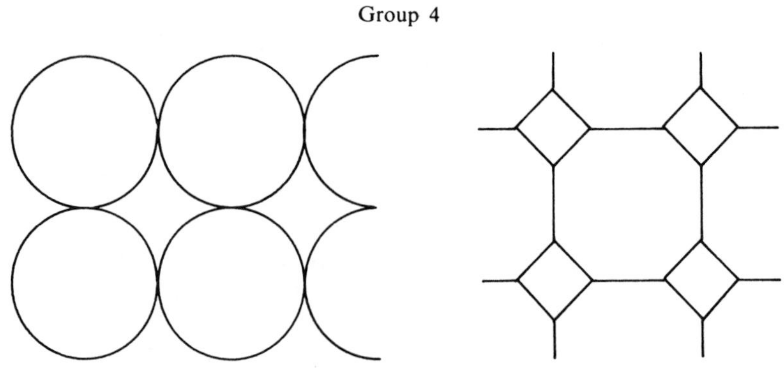

Figure 1.II.9.

Group 5

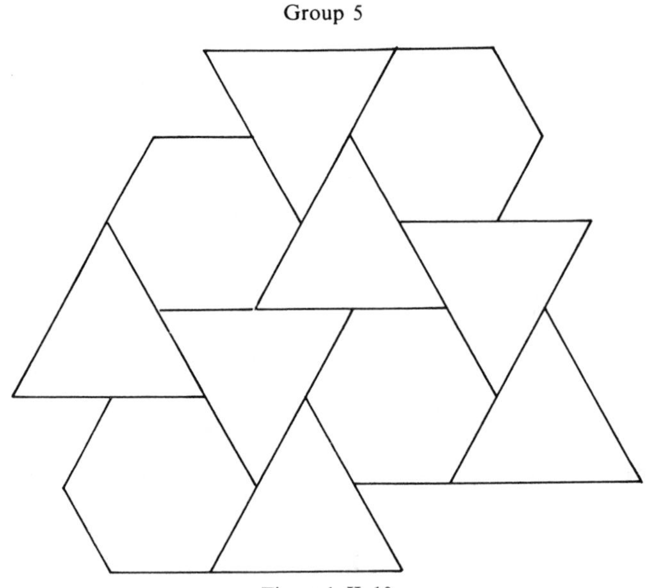

Figure 1.II.10.

1.1 The answer is "yes" in all three cases, as shown by the figures below. In all three cases the reflections through the midpoint of each side play an essential role. In the case of a quadrilateral, they are compatible as generators of (an index-two supergroup of) a lattice of translations because the midpoints of the sides of a quadrilateral always form a parallelogram; see 3.B or [B, 3.4.10].

In fact an even simpler way to solve the problem is by remarking, for the case of triangles first, that the figure formed by a triangle and its reflection through the midpoint of one side is a parallelogram. It is obvious that parallelograms tile the plane. In the case of a quadrilateral, it is enough to cut it up into two triangles and apply the remark above.

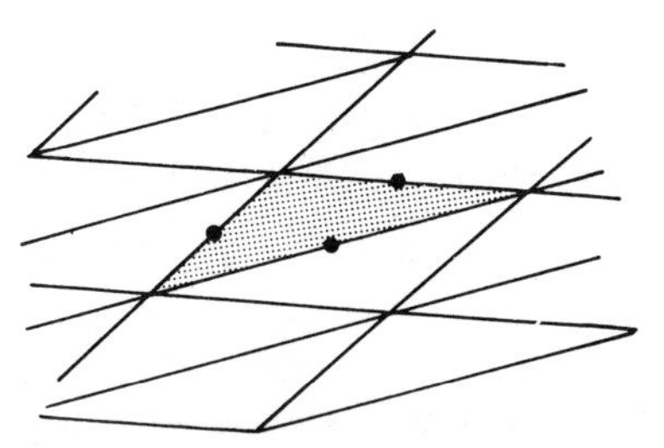

Figure 1.1.1.

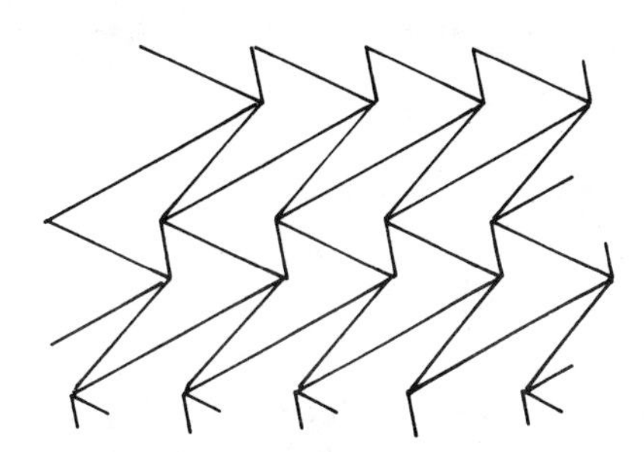

Figure 1.1.2.

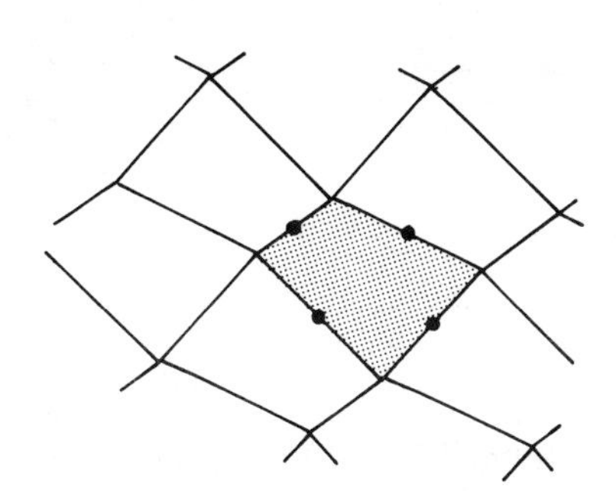

Figure 1.1.3.

1.2 We have two axioms to verify:

a) $\bigcup_{g \in G} g(P) = E$.

The only problem is showing the inclusion $E \subset \bigcup_{g \in G} g(P)$. For a fixed $x \in E$, let B be the ball of center x and radius $d(x, a)+1$. Since $\{g(a): g \in G\}$ is a discrete set, its intersection with B is finite. This proves the existence of an element $g \in G$ such that for any $h \in G$ we have $d(x, g(a)) \leq d(x, h(a))$. Since g is an isometry, we can apply g^{-1} to all these points while preserving the relation between the distances:

$$\forall h \in G \qquad d\left(g^{-1}(x), a\right) \leq d\left(g^{-1}(x), g^{-1} \circ h(a)\right),$$

and since G is a group this also means that

$$\forall k \in G \qquad d\left(g^{-1}(x), a\right) \leq d\left(g^{-1}(x), k(a)\right).$$

This shows that $g^{-1}(x) \in P$, or again that $x \in g(P)$.

b) $g(\mathring{P}) \cap h(\mathring{P}) \neq \phi \Rightarrow g(P) = h(P)$.

Since g is an isometry we can write

$$\begin{aligned} g(P) &= \left\{g(x): d(x, a) < d(x, k(a)), \forall k \in G \backslash \mathrm{Id}\right\} \\ &= \left\{y: d(y, g(a)) < d(y, k(a)), \forall k \in Gg\right\}. \end{aligned}$$

Hence

$$y \in g(\mathring{P}) \cap h(\mathring{P}) \begin{cases} d(y, g(a)) < d(y, k(a)), \forall k \in G \backslash g \\ d(y, h(a)) < d(y, k(a)), \forall k \in G \backslash h \end{cases} g(a) = h(a).$$

Since $g(P) = \{y : d(y, g(a)) \leq d(y, k(a)), \forall k \in G\}$, the preceding result implies $g(P) = h(P)$.

EXAMPLE. For a lattice group, we have the following figure:

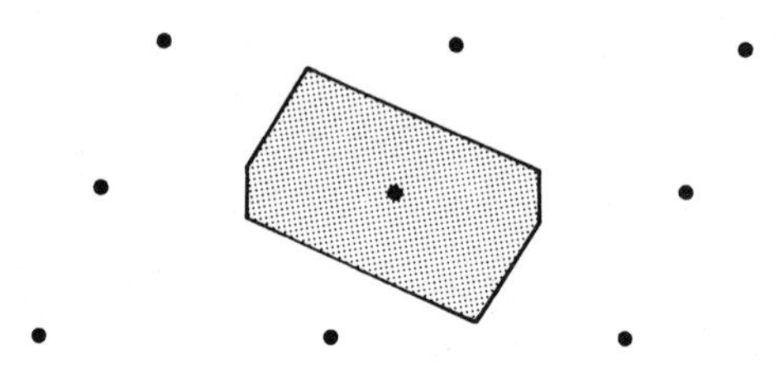

Figure 1.2.1.

REMARK. The form and size of the tile depend critically on the choice of a. The figure below illustrates this fact, where the shaded areas are the fundamental tiles corresponding to three different choices of a for the group generated by a $2\pi/3$ rotation around a point m and the reflection through a line D which does not contain m.

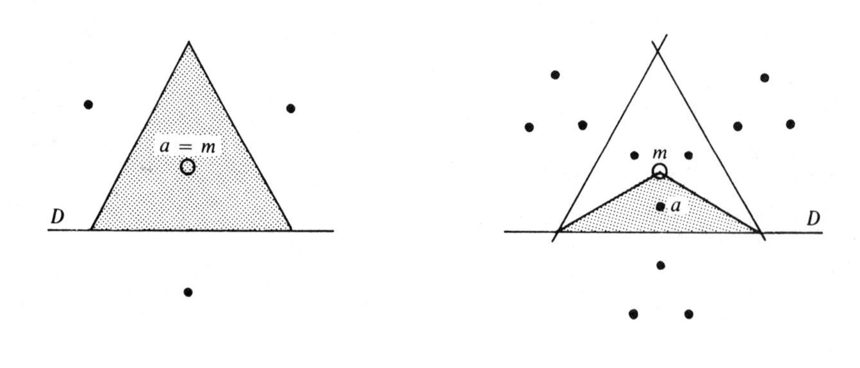

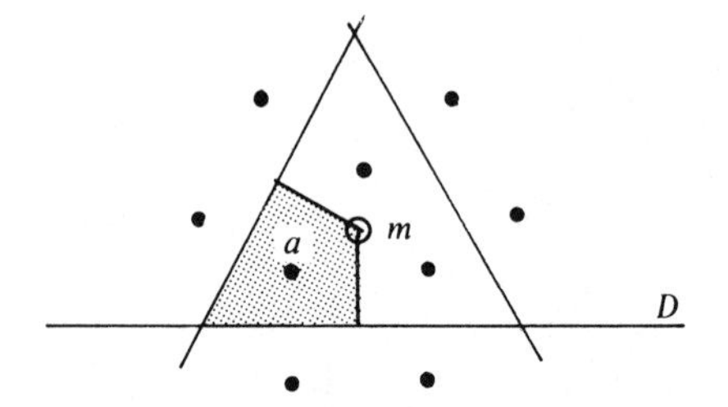

Figure 1.2.2.

1.3

1.3.1 Groups of Proper Motions: Stabilizers and Orbits

In general, the group G has a normal subgroup Γ consisting of the translations of G, and the quotient is a finite group; we choose representatives $g_1, \ldots, g_p$ for the classes in the quotient. If a is a point in the plane, the orbit Ga of a is the union of the orbits of the points $g_i a$ under the translation subgroup Γ, so the orbits all look more or less the same, namely, like a finite number of isometric lattices. The only important difference is the number of such lattices, which decreases when the stabilizer of a is not the identity.

In fact, the quotient group G/Γ acts on the set X of lattices; the action is transitive, and the stabilizer of the lattice containing a (in G/Γ) is the image in G/Γ of the stabilizer of a in G. So the number of lattices is equal to $\#(G/\Gamma)/\#G_a$. □

Now for the examples:

Group 1: $G = \Gamma$, so all stabilizers are trivial, and all orbits are lattices.

Group 2: $\#(G/\Gamma) = 2$. Only the vertices and the midpoints of the shortest sides of the fundamental tile have a non-trivial stabilizer, which contains a 180-degree rotation. An orbit contains in general two lattices, but there are two special ones which contain only one.

Figure 1.3.1.

Group 3: $\#(G/\Gamma) = 3$. Only the vertices of the tile have a nontrivial stabilizer, generated by a rotation of order three; there are four exceptional orbits.

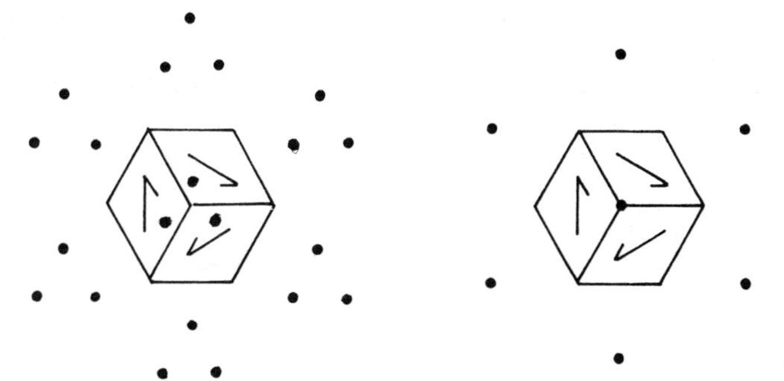

Figure 1.3.2.

Group 4: $\#(G/\Gamma)=4$. Only the vertices of the tile have a non-trivial stabilizer, generated by a rotation of order four; there are four exceptional orbits.

Group 5: $\#(G/\Gamma)=6$. The stabilizer of the vertex with an obtuse angle has order 3, and the stabilizers of the other two vertices have order 6. Thus in general the orbits have six lattices, and there are three exceptional orbits with 2, 1 and 3 lattices, respectively.

1.3.2 Groups of Proper Motions: Structure and Presentation

Group 1: We will show that G has the following presentation: 2 generators, namely the translations t_u and t_v, and a single relation $t_u t_v = t_v t_u$. In fact, since t_u and t_v commute, any other relation reduces to $t_u^n t_v^m = 1$, which implies $n\vec{u} + m\vec{v} = 0$, whence $n = m = 0$ since $\vec{u}$ and $\vec{v}$ are linearly independent.

The other groups are generated by two translations t_u and t_v and a rotation of order 2, 3, 4 or 6, respectively. In each case this rotation r induces, by conjugation, an automorphism of the group generated by the translations: if t is the translation by a vector $n\vec{u} + m\vec{v}$, $n, m \in \mathbf{Z}$, then $t' = rtr^{-1}$ is the translation by the vector $\vec{r}(n\vec{u} + m\vec{v})$, where $\vec{r}$ is the linear map tangent to r. Then G is isomorphic to the semi-direct product of the cyclical group generated by r and the free abelian group generated by the translations, i.e. there exists a bijection from G onto the product $\langle r \rangle \times \langle t_u, t_v \rangle$, by means of which the law of composition of G is expressed as

$$(r^a, t_u^n t_v^m)(r^b, t_u^p t_v^q) = (r^{a+b}, t_u^c t_v^d),$$

where $c\vec{u} + d\vec{v} = \vec{r}(m\vec{u} + n\vec{v}) + (p\vec{u} + q\vec{v})$. We will show that the group G is defined exactly by the relations $t_u t_v = t_v t_u$, $r^a = 1 (a = 2, 3, 4, 6)$, $rt_u r^{-1} = t_{r(u)}$, $rt_v r^{-1} = t_{r(v)}$. For the sake of clarity, we take the case $a = 2$: since r is a 180-degree rotation, $r^2 = 1$ and $r(\vec{u}) = -\vec{u}$, $r(\vec{v}) = -\vec{v}$. Suppose there is a relation

$$r^{a_1} t_u^{b_1} t_v^{c_1} \ldots r^{a_k} t_u^{b_k} t_v^{c_k} = 1$$

in the group G, where b_i, c_i are integers (possibly zero), and a_i is 0 or 1. Since the linear map tangent to the product is $\vec{r}^{a_1 + \cdots + a_k}$, the number of non-zero a_i is even. Because of the relation $t_u t_v = t_v t_u$, we can rearrange so that all the a_i are 1; then, using $rt_u r = t_{-u}$ and $rt_v r = t_{-v}$, we get

$$r^{a_1} t_u^{b_1} t_v^{c_1} r^{a_2} = t_u^{-b_1} t_v^{-c_1},$$

and the given relation reduces to a relation between translations. But we have shown above that any relation in the group of translations reduces to $t_u t_v = t_v t_u$.

Group 3: Since r is a rotation of order 3, we get $\vec{r}(\vec{u}) = -\vec{u} + \vec{v}$, $\vec{r}(\vec{v}) = -\vec{u}$, so G is defined by the generators t_u, t_v, r and the relations $t_u t_v = t_v t_u$, $r^3 = 1$, $rt_u r^{-1} = t_u^{-1}$ and $rt_v r^{-1} = t_u^{-1}$.

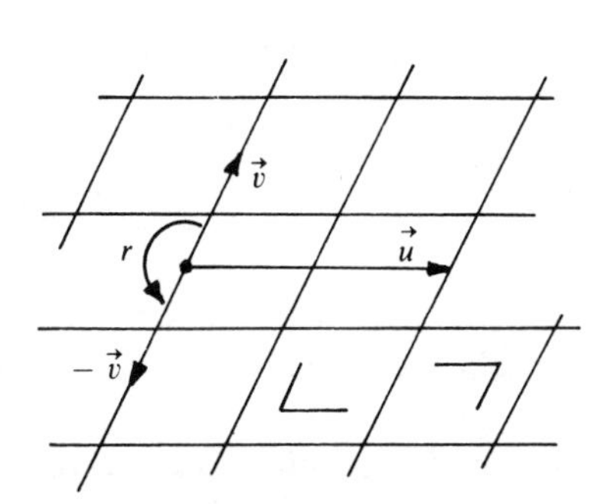

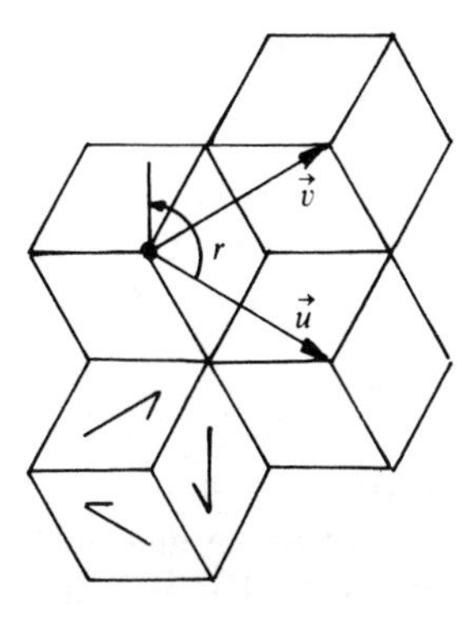

Figure 1.3.3.

1.4 *Note*. For this problem, you are allowed to use translations, rotations and reflections. If we wanted to use proper motions only, we would have to add the mirror images of tiles IV and V.

Step 1. We start by studying the corners of the quadrilled pattern that underlies any tiling (assuming one exists) using these tile shapes. Each corner must be filled exactly once. Now, from the point of view of the corners, there are only two kinds of tile: III, which we will *denote* by X, and all the others, to be *denoted* by 0. We will show that (perhaps after a 90-degree rotation) all the X's occur in alternating columns, and they occupy every other square of the columns in which they occur.

Only X can fill a corner, and then it forces all tiles around it to be 0's:

$$\begin{matrix} 0 & 0 & 0 \\ 0 & X & 0 \\ 0 & 0 & 0 \end{matrix}$$

In the next column, there must be an X to occupy the corners left empty by the 0's; up to a reflection, it must be in either position a or b:

$$\begin{matrix} 0 & 0 & 0 & b \\ 0 & X & 0 & a \\ 0 & 0 & 0 & \end{matrix}$$

(i) Suppose X occupies position b:

$$\begin{matrix} & & 0 & 0 & 0 \\ 0 & 0 & 0 & X & 0 \\ 0 & X & 0 & 0 & 0 \\ 0 & 0 & 0 & c & \end{matrix}$$

The upper left corner of square c must be occupied, so there's another X there:

$$\begin{matrix} & & 0 & 0 & 0 \\ 0 & 0 & 0 & X & 0 \\ 0 & X & 0 & 0 & 0 \\ 0 & 0 & 0 & X & 0 \\ & d & 0 & 0 & 0 \end{matrix}$$

By the same reasoning this requires an X in square d. Iterating we see that the squares X occupy alternating positions in two columns between which there is a column of 0's.

(ii) Now suppose there is an X in position a. We want to consider the position of the next X to its right. So we skip a column each time we encounter a tile X in position a. One of two things must occur:

— either we encounter only X's in position a, in which case we have shown that the X's occupy every other square of a row. This is the general case we want to establish, up to a 90-degree rotation.
— or we end up by finding an X in position b, in which case the reasoning above shows that at that place there are two columns with X's in alternating positions. It is easy to see that if this is the case all other X's must be arranged in columns:

							0	0	0
0	0	0	0	0	0	0	0	X	0
X	0	X	0	X	0	X	0	0	0
0	0	0	0	0	0	0	0	X	0
		X		X	0	X	0	0	0
					0	0	0	X	0
		X		X	0	X	0	0	0
					0	0	0	X	0
		X		X	0	X	0	0	0

This proves our assertion. □

Step 2. We place ourselves at a column of X's, i.e. one with tiles of type III. Either above or below us there will be an asymmetrical notch; we start moving in that direction. We will encounter a 0 tile, then another tile of type III. The 0 tile can only be of type I or IV, from the figure below (where the round notches can be either kind—symmetrical or asymmetrical):

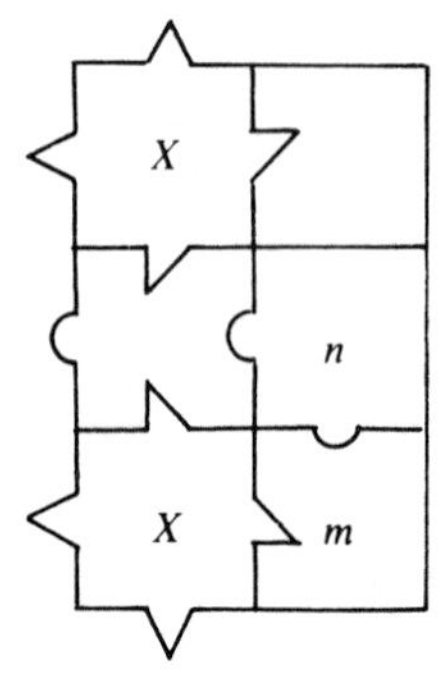

Figure 1.4.2.

Tile m can only be of type I, IV or V; in all cases it must have a concave notch at the top, as shown. Tile n, which has two convex notches, cannot be of type III (because it is next to a III), so it must be of type VI.

The reasoning in step 1 shows that there must be a tile of type III in position a or b; but, because of the notches, position b is excluded. So we obtain the configuration shown below, which will be *called* X_3.

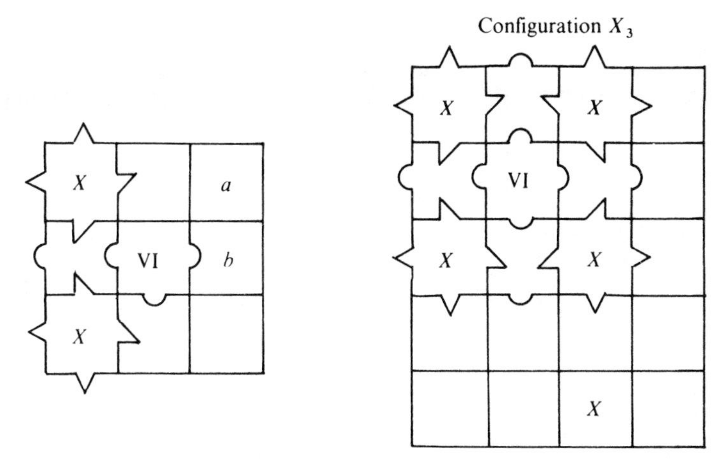

Figure 1.4.3.

Step 3. Notice that every tile X belongs to a unique configuration X_3, determined by the direction of its asymmetrical notches. Such is in particular the case of the tile X located at the bottom right of figure 1.4.3.

By an argument analogous to that of step 1, we show that the X_3's are arranged in columns, separated by one layer of tiles of type I, II, IV or V.

Now we look more closely at one X_3. It looks like figure 1.4.4 below:

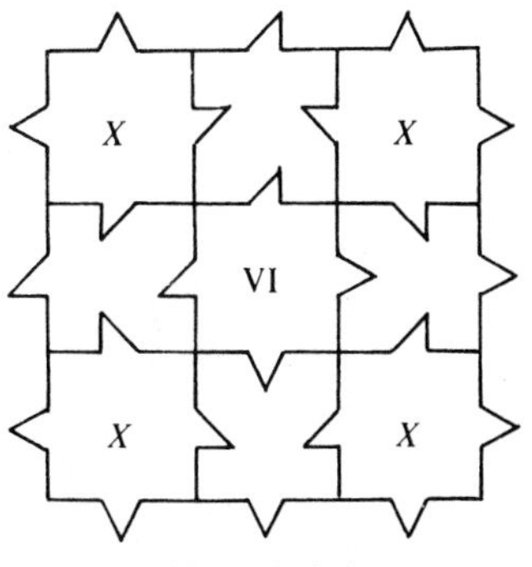

Figure 1.4.4.

where only the notches at the center of a face can be asymmetrical. The types of such central notches (the only ones we'll care about from now on) are the same as those of the central tile of type VI in this X_3. We move up or down, in the direction of an asymmetrical notch. The central tile of the right-hand X_3 in figure 1.4.5 below

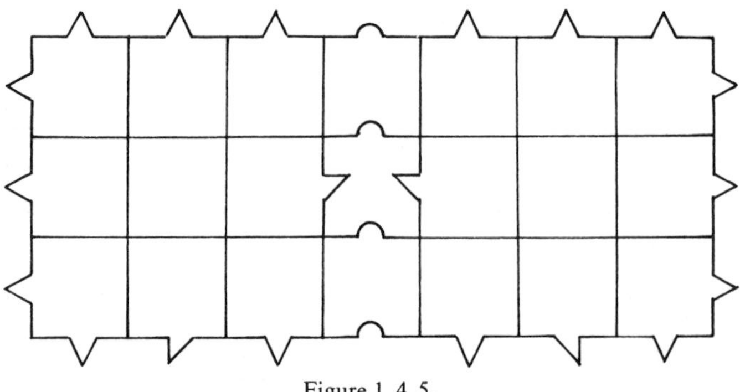

Figure 1.4.5.

can only be of type VI, since it has two convex notches and cannot be of type III. A similar reasoning to the one in step 2 shows that four X_3's must necessarily form a larger structure, which will be *called* X_7, as in figure 1.4.6. In an X_7 configuration, all the notches around the outer edges are symmetrical, except possibly those at the middle of each edge. The latter are the same as the corresponding notches in the central tile of type VI.

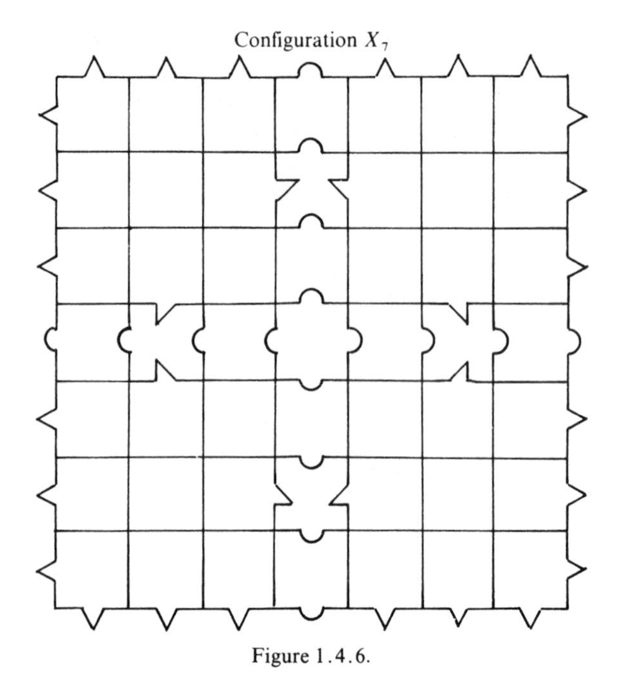

Figure 1.4.6.

Step 4. The same procedure allows us to obtain a configuration X_{15} by grouping four X_7 around a tile of type VI; the structure of this configuration is imposed by the central tile. By induction we form the configurations X_{2^n-1}, which are squares of side $2^n - 1$, all of whose notches point out and are symmetric, except for those located at the middle of each outer edge.

An increasing nested sequence of X_i must cover either the plane, or a half-plane, or a quadrant. In the last two cases, we build a new sequence starting from a tile of type III which is not in the first one, and keep doing this until we have covered the whole plane. Now we have either one sequence of X_i, or two sequences covering a half-plane each, or three sequences covering a half-plane and two quadrants, or four, each covering one fourth of the plane.

Step 5. Periodicity.

Every translation that leaves the tiling invariant must map a tile of type III into another of the same type and with the same orientation. We now consider the X_3 to which these two tiles belong. The fact that the orientation of the two tiles of type III is the same and the uniqueness in the construction of X_3 guarantee that this translation must send one X_3 into the other; and they must be oriented in the same way, since the translation must leave invariant the whole plane. In other words, the position of the original tile III inside the X_3, X_7, etc. determined by it is preserved by the translation. This means every translation takes an X_{2^n-1} into another; but then it must translate by a distance equal to at least 2^n times the side of a tile (unless it is the identity), which cannot be true for all n. Thus we have shown that every tiling with the six tiles given is necessarily non-periodic.

REMARK. The above solution was based on a tiling that was assumed to exist, and we determined its structure, showed it could exist and was not periodical. We could also adopt the converse point of view and start constructing tilings. Each time we form a X_{2^n-1} we have four choices for the orientation of the tile of type VI at the center, so after n steps we get 2^n possibilities. Thus there is an uncountable number of ways to tile the plane with the six given tiles. □

Chapter 2

2.1 To prove the theorem of Menelaus, the idea is to use the group of dilatations, more precisely the composition of homotheties. We will show first the following

Assertion: If $f = H_{z,\lambda}$ (resp. $g = H_{y,\mu}$) is the homothety of center x and ratio λ (resp. center y and ratio μ), and if $\lambda\mu \neq 1$, then the composition $g \circ f$ is a homothety of ratio $\lambda\mu$ whose center z belongs to the line $\langle x, y \rangle$.

We already know that $g \circ f$ is a homothety (2.C or [B, 2.3.3.12]). By definition, the three points z, x, $g(f(x))$ are collinear; but then so are x, $g(x)$, y. Now $g(f(x)) = g(x)$, so x, $g(x)$, y and z are all on the same line. □

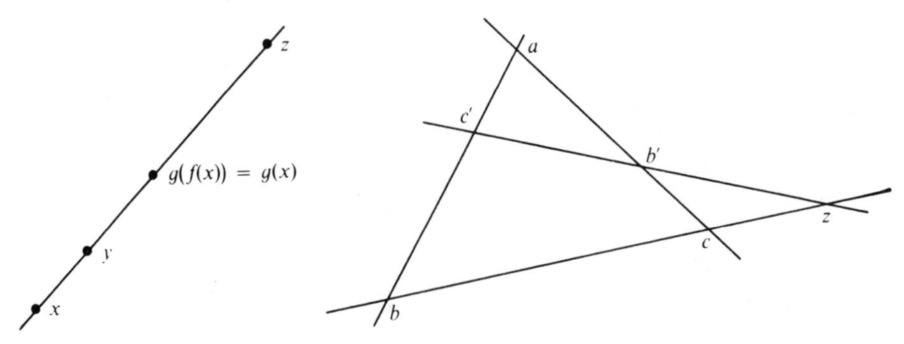

Figure 2.1.1. Figure 2.1.2.

Now we apply this fact to the problem in question by taking $f = H_{c',\lambda'}$ and $g = H_{b',\mu'}$, where $\lambda' = -\overrightarrow{c'a}/\overrightarrow{c'b}$ and $\mu' = -\overrightarrow{b'c}/\overrightarrow{b'a}$. By construction we have $f(b) = a$ and $f(a) = c$, so $(g \circ f)(b) = c$. So the center z of the homothety is on the line $\langle b, c \rangle$, but also on the line $\langle b', c' \rangle$ because of the result above. Thus we have

$$z = \langle b, c \rangle \cap \langle b', c' \rangle.$$

On the other hand, since $(g \circ f)(b) = c$, we have the relation $\lambda'^{-1}\mu'^{-1} = -\overrightarrow{zb}/\overrightarrow{zc}$. Summing up, a', b', c' are collinear if and only if

$$\frac{\overrightarrow{c'a}}{\overrightarrow{c'b}} \frac{\overrightarrow{b'c}}{\overrightarrow{b'a}} \frac{\overrightarrow{a'b}}{\overrightarrow{a'c}} = 1. \qquad \square$$

Observe that the case $\lambda'\mu' = 1$ cannot occur, because we cannot have $\overrightarrow{a'b}/\overrightarrow{a'c} = 1$.

The theorem of Ceva will be proved analytically. Take $\{a, \overrightarrow{ab}, \overrightarrow{ac}\}$ as an affine frame, wherein the coordinates of a, b, c are by definition given by $(0,0)$, $(1,0)$ and $(0,1)$ respectively. Put $c' = (\gamma, 0)$, $b' = (0, \beta)$, $a' = (u, 1-u)$, the latter notation coming from the fact that the equation of $\langle b, c \rangle$ is

$$x + y - 1 = 0.$$

The equation of the line $\langle b, b' \rangle$ is $x + y/\beta = 1$, and that of $\langle c, c' \rangle$ is $x/\alpha + y = 1$. So the coordinates of $x = \langle b, b' \rangle \cap \langle c, c' \rangle$ are

$$\left(\frac{\alpha(1-\beta)}{\alpha\beta - 1}, \frac{\beta(1-\alpha)}{\alpha\beta - 1} \right).$$

The points a, a', x will be collinear if and only if we have $[\alpha(1-\beta)/\beta(1-\alpha)] = u/(1-u)$.

On the other hand, we have

$$\frac{\overrightarrow{c'b}}{\overrightarrow{c'a}} = \frac{1-\alpha}{\alpha}, \quad \frac{\overrightarrow{b'a}}{\overrightarrow{b'c}} = -\frac{1-\beta}{\beta}, \quad \frac{\overrightarrow{a'c}}{\overrightarrow{a'b}} = -\frac{1-\gamma}{\gamma}. \qquad \square$$

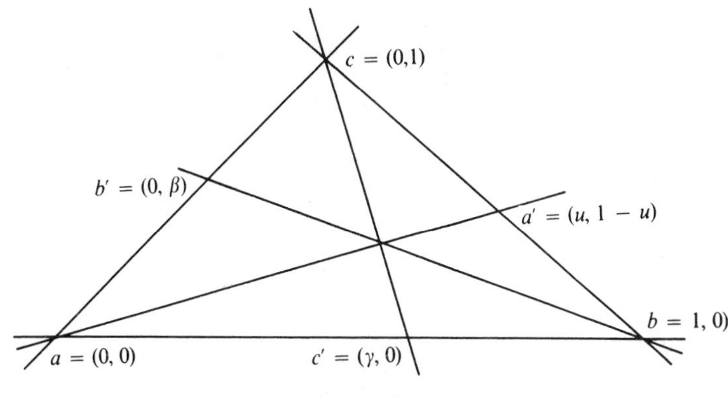

Figure 2.1.3.

There are many different proofs of these two famous and easily stated theorems. The theory of barycenters (chapter 3) furnishes very short ones:

Menelaus: write $b' = \beta c + (1-\beta)a$ and $c' = \gamma a + (1-\gamma)b$. The intersection point $a' = \langle b', c' \rangle \cap \langle b, c \rangle$ can be obtained by finding a barycenter of b' and c' where the variable a does not appear. It must be then

$$a' = \frac{\beta\gamma c - (1-\beta)(1-\gamma)b}{\beta\gamma - (1-\beta)(1-\gamma)}.$$

But then we have $\overrightarrow{a'b}/\overrightarrow{a'c} = \beta\gamma/[(1-\beta)(1-\gamma)]$, whereas

$$\frac{\overrightarrow{c'a}}{\overrightarrow{c'b}} = \frac{1-\gamma}{-\gamma}, \quad \frac{\overrightarrow{b'c}}{\overrightarrow{b'a}} = \frac{1-\beta}{-\beta}. \qquad \square$$

We leave the barycentric proof of Ceva to the reader; to conclude, we will show how to pass from one theorem to the other. This is best done by using the "complete quadrilateral" or the "polar line of a point relative to two lines", mentioned in 6.B or [B, 6.5.7]. In the figure below, the property is that the four points a', a'', b, c are in harmonic division, i.e. $\overrightarrow{a'b}/\overrightarrow{a'c} = -\overrightarrow{a'b}/\overrightarrow{a'c}$. This proves each theorem can be deduced from the other. $\square$

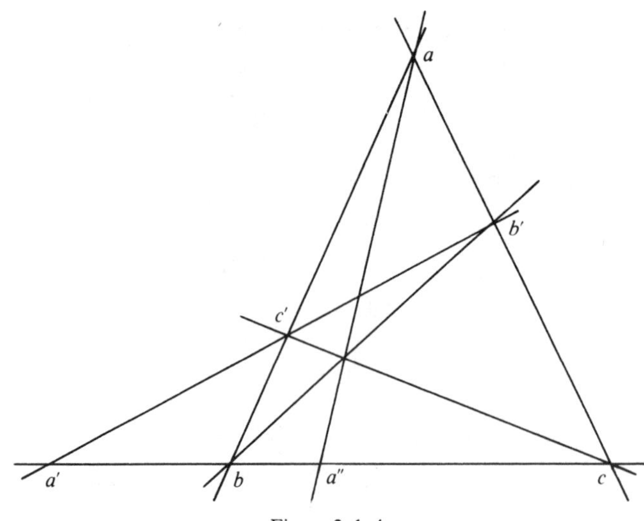

Figure 2.1.4.

2.2 We start by establishing some notations that will be in effect throughout the exercise. We choose a vectorialization of X at a point 0 of Y, and a complementary space Z; we denote projection onto Z and parallel to Y by p.

a) *$X \setminus Y$ is not connected when Y is a hyperplane.*

Here $\dim Z = 1$. Then $Z \setminus 0$ is composed of two half-lines Z^+ and Z^-. It is enough to observe that $X \setminus Y = p^{-1}(Z^+) \cup p^{-1}(Z^-)$, and the two sets are open and disjoint.

b) *$X \setminus Y$ is (path-)connected when* $\operatorname{codim} Y = \dim Z > 1$.

Let x_1, x_2 be two points of $X \setminus Y$, and let $z_i = p(x_i)$ be their projections on Z. Since the segments $x_1 z_1$ and $z_2 x_2$ are contained in $X \setminus Y$, it is enough to show that we can connect z_1 to z_2 inside $Z \setminus 0$. When z_1 and z_2 are linearly independent, the segment $z_1 z_2$ does not pass through 0, and we're done. On the other hand, if z_1 and z_2 are linearly dependent, the hypothesis $\dim Z > 1$ implies we can take a point z_3 which is not on the line supporting z_1 and z_2; we then connect z_1 to z_2 by the segments $z_1 z_3$ and $z_3 z_2$, neither of which contains 0.

c) *Each component of $X \setminus Y$ is simply connected when Y is a hyperplane.*

Let $x(t)$ be a loop (i.e. $x(0) = x(1)$) contained in $X \setminus Y$. Putting $z(t) = p(x(t))$, the projection of the loop onto Z, we can deform $x(t)$ into $z(t)$ by the homotopy $(1-s)x(t) + sz(t)$, which remains inside $X \setminus Y$. Now $z(t)$ can be deformed into a point by the homotopy $(1-s)z(t) + sy(0)$, which does not cross 0.

d) $X\setminus Y$ *is not simply connected when* $\operatorname{codim} Y = \dim Z = 2$.

We identify Z with $\mathbf{C}$, the complex plane, and we show that the loop $e^{2i\pi t}$ is not homotopic to a point. In fact, if there existed a homotopy $x(s,t)$ with $x(0,t) = e^{2i\pi t}$ and $x(1,t) = a \in X\setminus Y$, there would also be a homotopy *on the circle* and doing the same thing (just take $z(s,t) = p(x(x,t))/|p(x(s,t))|$). Now it is well known that such a homotopy does not exist. (One way to prove it is to define the argument of $z(s,t)$ by continuity, starting from $\arg z(0,0) = 0$; the function $\arg z(s,1) - \arg z(s,0)$ is continuous in s, its range is $2\pi\mathbf{Z}$, and it has value 2π at 0 and 0 at 1 – contradiction.)

e) $X\setminus Y$ *is simply connected when* $\operatorname{codim} Y = \dim Z > 2$.

Let $x(t)$ be a loop in $X\setminus Y$. The homotopy $(1-s)x(t) + sz(t)$ transforms this loop into its projection $z(t) = p(x(t))$. We then endow Z with a Euclidean structure, and transform $z(t)$ into a loop $y(t) = z(t)/\|z(t)\|$, which lies on the sphere S^d, $d = \dim Z - 1$ (use the homotopy $(1-s)z(t) + sy(t)$). The result now follows from the classical fact that S^d is simply connected.

To prove this fact, we start by covering S^d by the two sets $S^d \setminus y(0)$ and $S^d \setminus -y(0)$, which are open. This implies that for any t there exists a neighborhood V_t such that $y(V_t)$ is entirely contained in one of the two sets. Using compactness of the interval $[0,1]$, we find a partition $0 < t_1 < t_2 < \cdots < t_m < 1$ of the interval, so that $y([t_i, t_{i+1}])$ is contained in either open set. If the image of any subinterval contains the point $-y(0)$, we deform the loop in that subinterval so as to avoid that point. This is done by applying the stereographic projection π with the pole $y(0)$ (see 18.A); we choose a point w such that the half-lines $w\pi(y(t_i))$ and $w\pi(y(t_{i+1}))$ don't pass through $\pi(-y(0))$, then we take a vector hyperplane transversal to the two lines. Now we replace the path by its projection onto the two half-lines, taken parallel to the chosen hyperplane, and finally we return to the sphere by using the inverse stereographic projection π^{-1}.

All this shows that we can assume that $y([0,1])$ does not contain the point $-y(0)$. We now use stereographic projection based at $-y(0)$ to pass to $\mathbf{R}^d$, which is certainly simply connected. This completes the proof. □

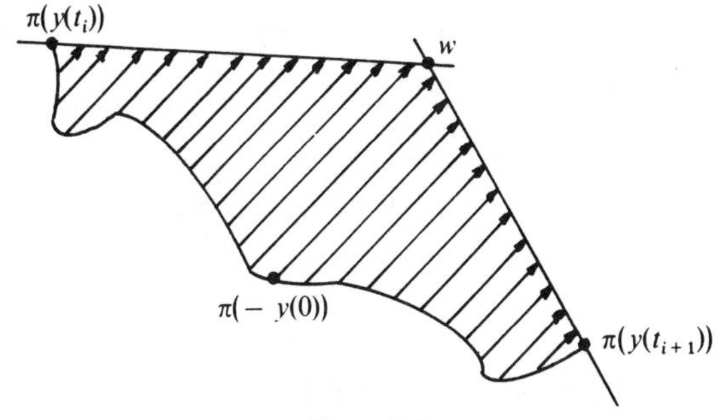

Figure 2.2.

2.3 Since $\mu(H)=0$ we have $\int_X=\int_{X'}+\int_{X''}$, so that $\mu(K)=\mu(K')+\mu(K'')$ and

$$\overrightarrow{a\,\mathrm{cent}(K)}=(\mu(K))^{-1}\left(\int_{x\in X}\chi_{K'}(x)\overrightarrow{ax}\,\mathrm{d}\mu+\int_{x\in X}\chi_{K''}(x)\overrightarrow{ax}\,\mathrm{d}\mu\right)$$

$$=\frac{\mu(K')}{\mu(K)}\overrightarrow{a\,\mathrm{cent}(K')}+\frac{\mu(K'')}{\mu(K)}\overrightarrow{a\,\mathrm{cent}(K'')}.$$

The interior of a compact convex set is the intersection of all open half-spaces $\mathring{X}'$ such that $X'\supset K$; thus it suffices to show that $K\subset X'$ and $\mathring{K}\neq\phi$ imply $\mathrm{cent}(K)\in\mathring{X}'$. Let $a\in H$ and $m\in\mathring{K}$. For every $x\in X$ we call $f(x)$ the real number such that $f(x)\overrightarrow{am}$ is the projection of x on the line am, parallel to H. If $x\in K$ then $f(x)\geq 0$, and if $x\in\mathring{K}$ then $f(x)>0$. The projection of $\overrightarrow{a\,\mathrm{cent}(K)}$ on the line am is equal to $\int_{x\in X}\chi_K(x)f(x)\,\mathrm{d}\mu\cdot\overrightarrow{am}$. Since f is non-negative everywhere and strictly positive on a set of non-zero measure, the integral is strictly positive, implying that $\mathrm{cent}(K)$ is inside $\mathring{X}$. □

2.4 We will study each of the three problems in turn:

a) Invariance of the equiaffine length and curvature

First remark that the expression given for the length does not depend on the chosen parametrization; if $j:[\alpha,\beta]\to[a,b]$ is a monotonic function of class C^2, we have

$$\int_\alpha^\beta\left[\det\left((\overrightarrow{c\circ j})'(\tau),(\overrightarrow{c\circ j})''(\tau)\right)\right]^{1/3}\mathrm{d}\tau$$

$$=\int_\alpha^\beta\left[\det\left(j'(\tau)\vec{c}'\circ j(\tau),\,j''(\tau)\vec{c}'\circ j(\tau)+(j'(\tau))^2\vec{c}''\circ j(\tau)\right)\right]^{1/3}\mathrm{d}\tau$$

$$=\int_\alpha^\beta\left[\det(\vec{c}'\circ j(\tau),\vec{c}''\circ j(\tau))\right]^{1/3}j'(\tau)\,\mathrm{d}\tau$$

$$=\int_a^b\left[\det(\vec{c}'(t),\vec{c}''(t))\right]^{1/3}\mathrm{d}t,\quad\text{by putting}\quad j(\tau)=t.$$

Now let f be such that $\det\vec{f}=1$. Then

$$\int_\alpha^\beta\left[\det\left((\overrightarrow{c\circ j})'(t),(\overrightarrow{c\circ j})''(t)\right)\right]^{1/3}\mathrm{d}t=\int_\alpha^\beta\left[\det(\vec{f}\circ\vec{c}'(t),\vec{f}\circ\vec{c}''(t))\right]^{1/3}\mathrm{d}t.$$

We associate with a pair of vectors $(\vec{x},\vec{y})$ the linear map given by the matrix whose columns are $\vec{x}$, $\vec{y}$. Then we can write $(\vec{f}(\vec{x}),\vec{f}(\vec{y}))=\vec{f}\circ(\vec{x},\vec{y})$. Hence

$$\int_\alpha^\beta\left[\det\left((\overrightarrow{f\circ c})'(t),(\overrightarrow{f\circ c})''(t)\right)\right]^{1/3}\mathrm{d}t$$

$$=\int_\alpha^\beta\left[\det\vec{f}\right]^{1/3}\left[\det(\vec{c}'(t),\vec{c}''(t))\right]^{1/3}\mathrm{d}t,$$

which implies the invariance property, in view of the fact that $\det\vec{f}=1$.

If $\det(\vec{c}'(t), \vec{c}''(t)) \neq 0$ this determinant has constant sign, and the function $\sigma(t) = \int_a^t [\det(\vec{c}'(\tau), \vec{c}''(\tau))]^{1/3} d\tau$ is monotonic and has non-zero derivative everywhere; this proves that the curve can be parametrized by its equiaffine length.

To show that the equiaffine curvature is invariant, we proceed as above:

$$K = \det\left((\overrightarrow{f \circ c})'', (\overrightarrow{f \circ c})'''\right) = \det(\vec{f} \circ \vec{c}'', \vec{f} \circ \vec{c}''') = \det \vec{f} \cdot \det(\vec{c}'', \vec{c}'''),$$

whence invariance under mappings f of determinant 1. □

REMARK. The sign of the equiaffine arclength depends on the orientation of the curve; the sign of the curvature does not.

b) Equiaffine length and curvature of conics

Any ellipse can be parametrized by $x = a\cos t$, $y = b\sin t$ in some affine frame (see 17.A). In this frame

$$\sigma(t) = \int_0^t (ab)^{1/3} d\tau = (ab)^{1/3} t;$$

thus the total length is $2\pi(ab)^{1/3}$. In the new parametrization, we have $x = a\cos(\sigma(ab)^{-1/3})$ and $y = b\sin(\sigma(ab)^{-1/3})$; using calculus we get $K = (ab)^{-2/3}$ (the equiaffine curvature of an ellipse is constant).

For a parabola, choose an affine frame in which it can be written as $y = x^2/2p$ (see 17.A); the length is $\sigma(x) = \int_0^x dt/p = x/p$, which gives the new parametrization $x = p\sigma$ and $y = p\sigma^2/2$. Finally, the curvature is $K = 0$ since the third derivatives are zero (the equiaffine curvature of a parabola is zero).

A hyperbola takes the form $x = a\,\mathrm{ch}\,t$, $y = b\,\mathrm{sh}\,t$ in some affine frame; after establishing the length $\sigma(t) = \int_0^t -(ab)^{1/3} d\tau = -(ab)^{1/3} t$, we obtain the new parametrization

$$x = a\,\mathrm{ch}\left(-\sigma(ab)^{-1/3}\right) \quad \text{and} \quad y = b\,\mathrm{sh}\left(-\sigma(ab)^{-1/3}\right),$$

which shows that the curvature has value $K = -(ab)^{-2/3}$ (the equiaffine curvature of a hyperbola is again a constant). □

c) Intrinsic equation

We have established in *a*) that any curve of class C^3 verifying the condition $\det(\vec{c}'(t), \vec{c}''(t)) \neq 0$ for all t can be parametrized by its equiaffine length, and satisfies an equation $K = K(\sigma)$, where K is a C^0 function of the equiaffine length. All these considerations hold modulo $\mathrm{SA}(X)$, which means that if $f \in \mathrm{SA}(X)$, the composition of f with that curve satisfies the same intrinsic equation $K = K(\sigma)$. We shall now examine the following converse: given a function K of class C^0, there exists a curve of class C^3 whose equiaffine curvature satisfies the equation $K = K(\sigma)$ as a function of its equiaffine length, and moreover this curve is unique modulo $\mathrm{SA}(X)$.

Let $(m, \vec{u}, \vec{v})$ be a plane frame, where $(\vec{u}, \vec{v})$ is the reference basis; the thoerem of Cauchy-Lipschitz guarantees the existence and uniqueness of a solution for the problem

$$\begin{cases} \vec{c}'''(\sigma)+K(\sigma)\vec{c}'(\sigma)=0 \\ c'(0)=m, \quad \vec{c}'(0)=\vec{u}, \quad \vec{c}''(0)=\vec{v}. \end{cases} \tag{I}$$

Any solution $\vec{c}$ for (I) satisfies the desired conditions, as follows:

— σ is the equiaffine arclength, since $\det(\vec{c}'(0),\vec{c}''(0))=1$, and the derivative of the function $\det(\vec{c}'(\sigma),\vec{c}''(\sigma))$ is zero (from equations (I));
— $K(\sigma)$ is indeed the equiaffine curvature of $\vec{c}$, because, again from (I), we get $\det(\vec{c}''(\sigma),\vec{c}'''(\sigma))=-K(\sigma)\det(\vec{c}''(\sigma),\vec{c}'(\sigma))=K(\sigma)$.

For uniqueness (modulo $\mathrm{SA}(X)$), suppose that $k(\sigma)$ is a curve that verifies the intrinsic equation; then there is an $f\in\mathrm{SA}(X)$ such that $f\circ k(0)=m$, $\vec{f}\circ\vec{k}'(0)=\vec{u}$, $\vec{f}\circ\vec{k}''(0)=v'$. This is because σ is the equiaffine length, which means $\det(\vec{k}'(0),\vec{k}''(0))=1$, and $\mathrm{SA}(X)$ is transitive on affine frames of determinant one. From the fact that $\det(\vec{f}\circ\vec{k}'(\sigma),\vec{f}\circ\vec{k}''(\sigma))=1$, we obtain by differentiation that $\det(\vec{f}\circ\vec{k}'(\sigma),\vec{f}\circ\vec{k}'''(\sigma))=0$, whence it follows that $\vec{f}\circ\vec{k}'(\sigma)$ and $\vec{f}\circ\vec{k}'''(\sigma)$ are collinear; the coefficient of proportionality can be found in the intrinsic equation itself, since the latter can be written as

$$K(\sigma)=\det(\vec{f}\circ\vec{k}''(\sigma),\vec{f}\circ\vec{k}'''(\sigma)),$$

i.e. $f\circ k$ satisfies the equation $\vec{f}\circ\vec{k}'''(\sigma)+K(\sigma)\vec{f}\circ\vec{k}'(\sigma)=0$. Checking this against the definition of f, we see that the function $f\circ k$ must be a solution for equation (I). Now the uniqueness guaranteed by the theorem of Cauchy-Lipschitz shows that $f\circ k=c$, which concludes our proof. □

REMARK. This result allows us to state a converse for *b*): The only curves whose equiaffine curvature is constant are the conics (ellipses if the curvature is positive, parabolas if it is zero, hyperbolas if negative). In particular we notice that any parabola can be obtained from any other by an affine transformation.

Chapter 3

3.1 Let m be the centroid of the points $x_{1,i}$, so that $\sum_{i=1}^{p}\overrightarrow{mx_{1,i}}=\vec{0}$. We can then write $\overrightarrow{mx_{2,i}}=(1/p-1)\sum_{j\neq i}\overrightarrow{mx_{1,j}}=-(1/p-1)\overrightarrow{mx_{1,i}}$. Now observe that m is again the centroid of points $x_{2,i}$ (since $\sum_{i=1}^{p}\overrightarrow{mx_{2,i}}=-(1/p-1)\sum_{i=1}^{p}\overrightarrow{mx_{1,i}}=\vec{0}$), so it is natural to use induction: $\overrightarrow{mx_{k,i}}=(1/p-1)\sum_{j\neq i}\overrightarrow{mx_{k-1,j}}=-(1/p-1)\overrightarrow{mx_{k-1,i}}$, and with the induction hypothesis we get $\overrightarrow{mx_{k,i}}=-(-1/p-1)^k\overrightarrow{mx_{1,i}}$. Next we verify that m is still the centroid of points $x_{k,i}$ by writing $\sum_{i=1}^{p}\overrightarrow{mx_{k,i}}=-(-1/p-1)^k\sum_{i=1}^{p}\overrightarrow{mx_{1,i}}=\vec{0}$. We distinguish two cases:

— if $p=2$, the sequences do not converge (unless the two points coincide), since at each step they jump from one of the original points to the other;
— if $p>2$, $\lim_{k\to\infty}\overrightarrow{mx_{k,i}}=\vec{0}$, so for every i the sequence $x_{k,i}$ tends towards m. □

3.2 We fix a positively oriented orthonormal basis of the Euclidean plane, and consider the three vertices A, B, C of the triangle in that order. We denote by det the determinant of pairs of vectors in this basis.

The centroid G of the object described in the problem is the barycenter of the weighted points (A', a), (B', b) and (C', c), where A' is the midpoint of BC and a is its length, and so on. We can write, for example, $\overrightarrow{A'G} = (b\overrightarrow{A'B'} + c\overrightarrow{A'C'})/(a+b+c)$, so that

$$\det\left(\overrightarrow{A'G}, \overrightarrow{A'B'}\right) = \frac{c}{a+b+c} \det\left(\overrightarrow{A'C'}, \overrightarrow{A'B'}\right).$$

Now $|\det(\overrightarrow{A'G}, \overrightarrow{A'B'})| = \mathrm{d}(A', B')\,\mathrm{d}(G, A'B') = c\,\mathrm{d}(G, A'B')$, whence

$$\mathrm{d}(G, A'B') = \frac{\left|\det\left(\overrightarrow{A'C'}, \overrightarrow{A'B'}\right)\right|}{a+b+c}$$

where $\det|(\overrightarrow{A'C'}, \overrightarrow{A'B'})|$ is the area of the triangle A', B', C'. The distances from G to the sides of the triangle A', B', C' are equal, which means G is the center of the circle inscribed in the triangle A', B', C'.

The point G is constructed by drawing the sides of triangle A', B', C', then two of the inner bisectors of this triangle. Their intersection point is G. □

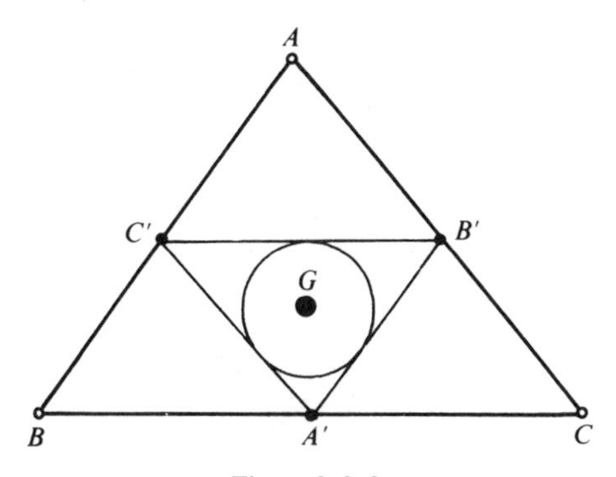

Figure 3.2.2.

Now let's consider a homogeneous plate in the shape of a quadrilateral with vertices A, B, C, D. Its centroid G is the barycenter of the centroids G_1, G_2 of the triangles ABC and ADC, considered with their respective masses; in particular, G, G_1 and G_2 are collinear. Analogously, the point G and the centroids G_3, G_4 of the triangles ABD and DBC are collinear.

Now the centroid of a triangular plate of vertices u, v, w coincides with the barycenter of its vertices, given by $(u+v+w)/3$. To see this, choose a frame $(O, \vec{i}, \vec{j})$ for the plane, and calculate the centroid of the homogeneous triangular plate of vertices O, $I = O + \vec{i}$, $J = O + \vec{j}$. Its coordinates are

$$x_g = \frac{1}{1/2}\int_0^1 dx \int_0^{1-x} x\,dy = \frac{1}{1/2}\int_0^1 (x - x^2)\,dx = 2\left(\frac{1}{2} - \frac{1}{3}\right) = \frac{1}{3},$$

$$y_g = 2\int_0^1 dx \int_0^{1-x} y\,dy = \int_0^1 (1-x)^2\,dx = \frac{1}{3}.$$

So for this triangle our assertion holds. But the triangle u, v, w is the image of the triangle O, I, J under an affine transformation, and affine transformations preserve both centroids of plates and barycenters, so the centroid of the plate uvw is indeed $(u + v + w)/3$.

To construct the point G, it is enough to draw the two diagonals AC and BD of the quadrilateral, then locate their midpoints as well as the midpoints of the sides and finally draw eight medians to obtain the four barycenters G_1, G_2, G_3, G_4. Then G is the intersection of the lines G_1G_2 and G_3G_4.

We remark that this construction resorts only to the affine notions of midpoint and of a line joining two points; it includes no metrical notions, unlike the case of the triangle made of wire. This is due to the fact that affine transformations preserve the centroid of plates, but not the centroid of systems of bars. □

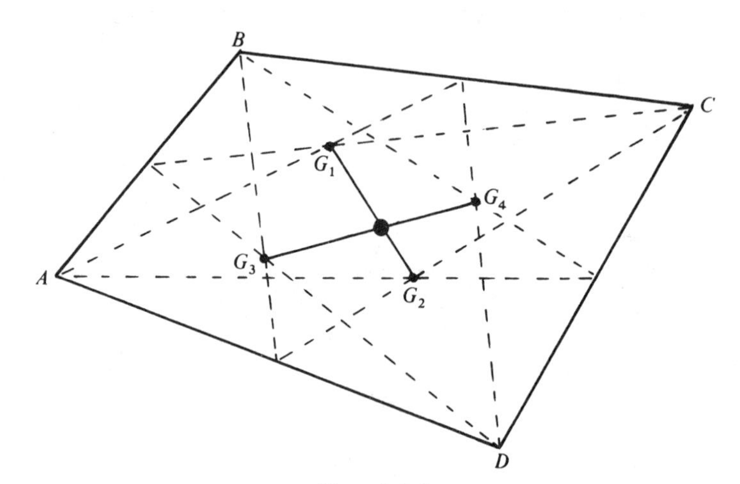

Figure 3.2.3.

3.3 Let C be a simplex in the subdivision. It contains three kinds of points: 1) the points x_i of Σ; 2) the points obtained as a subdivision of an $(n-1)$-dimensional simplex of Σ; 3) the barycenter of the points of Σ. We will study the distance between two points of C, considering the possible cases:

— two old points. This is impossible, since each subsimplex contains only one old point. In fact, if there were two, they would be part of a single one-dimensional simplex contained in Σ, and the latter would have been subdivided.

— one point of type 2) and one of type 1) or 2). The two would be part of a single $(n-1)$-dimensional simplex, and such a simplex arises by subdivision of an $(n-1)$-dimensional simplex of Σ. Since every $(n-1)$-dimensional simplex of Σ has diameter less than or equal to d, any subdivision has diameter $\leq (n-1)d/n$, by the induction hypothesis. The induction step is completed by observing that $(n-1)/n \leq n/(n+1)$.

— a point of type 1) and the point of type 3). We have to find the distance from x_i to $(x_0 + \cdots + x_n)/(n+1)$. This distance is equal to the norm

$$\left\| \overrightarrow{x_i \frac{x_0 + \cdots + x_n}{n+1}} \right\| = \left\| \frac{\overrightarrow{x_i x_0} + \cdots + \overrightarrow{x_i x_{i-1}} + \overrightarrow{x_i x_{i+1}} + \cdots + \overrightarrow{x_i x_n}}{n+1} \right\|$$

$$\leq \frac{\|\overrightarrow{x_i x_0}\| + \cdots + \|\overrightarrow{x_i x_{i-1}}\| + \|\overrightarrow{x_i x_{i+1}}\| + \cdots + \|\overrightarrow{x_i x_n}\|}{n+1};$$

but $\|x_i x_j\| \leq d$ since $x_i, x_j \in \Sigma$, which implies the distance is $\leq nd/(n+1)$.

— a point of type 2) and the point of type 3). A point y of type 2) is the barycenter of points of type 1) with weights ≥ 0, i.e.

$$y = \sum \lambda_i x_i, \quad \text{with} \quad \sum \lambda_i = 1;$$

so that $\|\overrightarrow{Oy}\| = \|O\Sigma\vec{\lambda}_i x_i\|$, where $O = (x_0 + \cdots + x_n)/(n+1)$ is the equibarycenter of the points of Σ. Then $\|\overrightarrow{Oy}\| = \|\Sigma\lambda_i\overrightarrow{Ox_i}\| \leq \Sigma\lambda_i\|\overrightarrow{Ox_i}\|$. But, according to the previous paragraph, $x_i \in \Sigma$ implies $\|\overrightarrow{Ox_i}\| \leq d'$, for $d' = nd/(n'+1)$, so that $\|\overrightarrow{Oy}\| \leq \Sigma\lambda_i d' = d'$.

This takes care of the last case. We have proved the induction step, so the diameter is indeed $\leq nd/(n+1)$.

All it takes to show that the diameter of successive simplices tends toward zero is understanding what the assertion means. Call C_p a simplex obtained by p barycentric subdivisions (and there are a great many of them!). Then the diameter of C_p is bounded above by the number $(n/n+1)^p d$, which approaches 0 when p approaches ∞. This means we can make a simplex as small as we like by subdividing it enough times. □

3.4 It is clear that we should define a polynomial function over X as a polynomial function over X_a, the vectorialization of X at a point a. The only problem is showing that this definition does not depend on the choice of a. Denote by $\mathscr{P}_k(E; W)$ the space of polynomials of degree less than or equal to k over E and with values in W, and let $f \in \mathscr{P}_k(X_a; W)$. By definition, there is a map $\varphi: X_a^k \to W$ such that $f = \varphi \circ \Delta_k$, where Δ_k is the diagonal map.

Now for any point b of X there exists a natural bijection $X_a \to X_b$ taking u to $u + \overrightarrow{ba}$, and preserving the underlying structure of X. We can thus associate to f the map $g\colon X_b \to W$ given by $g(u) = f(u + \overrightarrow{ab})$; for this map we have

$$\begin{aligned} g(u) &= f(u + \overrightarrow{ab}) \\ &= \varphi(u + \overrightarrow{ab}, u + \overrightarrow{ab}, \dots, u + \overrightarrow{ab}) \\ &= \varphi(u, u, \dots, u) + k\varphi(\overrightarrow{ab}, u, \dots, u) + \cdots \\ &\quad + C_k^p \varphi(\underbrace{\overrightarrow{ab}, \overrightarrow{ab}, \dots, \overrightarrow{ab}}_{p\ times}, u, \dots, u) + \cdots + \varphi(\overrightarrow{ab}, \overrightarrow{ab}, \dots, \overrightarrow{ab}). \end{aligned}$$

But the function $(\overrightarrow{ab}, \overrightarrow{ab}, \dots, \overrightarrow{ab}, \underbrace{\cdot, \cdot, \dots, \cdot}_{h\ times})$ is h-linear, whence $(\overrightarrow{ab}, \overrightarrow{ab}, \dots, \overrightarrow{ab}, u, \dots, u)$ is a homogeneous polynomial of degree h, and the sum is a polynomial of degree less than or equal to k. This shows that $g \in \mathscr{P}_k(X_b; W)$, and we identify f with g. We define a polynomial over X as an equivalence class of polynomials over X_a, $a \in X$, obtained in this way. □

REMARK. The argument above would not work for homogeneous polynomials, as can be easily seen from the calculation; this means a homogeneous polynomial over X_a is not transformed into a homogeneous polynomial over X_b; and there is no way to define homogeneous polynomials over an affine space.

3.5 Since φ is symmetric, we can expand an expression like $f(v_{i_1} + \cdots + v_{i_j})$ by using the multinomial formula. We obtain

$$\begin{aligned} (*) &= \frac{1}{k!} \sum_{j=1}^{k} (-1)^{k-j} \sum_{1 \le i_1 < \cdots < i_j \le k} f(v_{i_1} + \cdots + v_{i_j}) \\ &= \frac{1}{k!} \sum_{j=1} \sum_{1 \le i_1 < \cdots < i_j \le k} \sum_{l_1 + \cdots + l_j = k} \frac{(-1)^{k-j} k!}{l_1! \dots l_j!} v_{i_1}^{l_1} \dots v_{i_j}^{l_j}, \end{aligned}$$

where we have used the handy notation $v_1^{l_1} \dots v_j^{l_j} = \varphi(v_1, v_1, \dots, v_j, v_j)$, each vector v_i appearing l_i times.

To prove that $(*)$ is equal to $\varphi(v_1, \dots, v_k)$, we count how many times each monomial $v_1^{m_1} \dots v_k^{m_k}$ $(m_1 + \cdots + m_k = k)$ appears in the sum $(*)$, and with what coefficient.

The equality $v_{i_1}^{l_1} \dots v_{i_j}^{l_j} = v_1^{m_1} \dots v_k^{m_k}$ implies that:

i) The number N of non-zero terms in the sequence $m_1, \dots, m_k$ is equal to the number of non-zero terms in the sequence $l_1, \dots, l_j$, whence
ii) $j \ge N$;
iii) the non-zero terms in the sequence $l_1, \dots, l_j$ are well-determined; they are the non-zero m_i;
iv) giving the sequence $i_1 < \cdots < i_j$ determines the sequence $l_1, \dots, l_j$;
v) the sequence $i_1, \dots, i_j$ is determined by the indices h such that $m_{i_h} = 0$.

As a consequence of the above, for fixed $i_1, \dots, i_j$, there is at most one term in $v_1^{m_1} \dots v_k^{m_k}$ in the sum $\sum_{l_1 + \cdots + l_j = k}$; for fixed j, there are exactly as many sequences $i_1, \dots, i_j$ that give rise to a term in $v_1^{m_1} \dots v_k^{m_k}$ as there are ways to choose $j - N$ terms among the $k - N$ non-zero terms of the sequence m_i, and there are $C_{k-N}^{j-N}=(k-N)!/((j-N)!(k-j)!)$ such ways. The coefficient of $v_1^{m_1}\dots v_k^{m_k}$ in the double sum $\Sigma_{1\le i_1<\cdots<i_j\le k}\Sigma_{l_1+\cdots+l_j=k}$ is thus equal to $(-1)^{k-j}C_{k-N}^{j+N}k!/(m_1!\dots m_k!)$, and the coefficient of $v_1^{m_1}\dots v_k^{m_k}$ in the sum (*) is equal to

$$\begin{aligned}\frac{1}{k!}\sum_{j=N}^{k}(-1)^{k-j}C_{k-N}^{j-N}\frac{k!}{m_1!\dots m_k!}&=\frac{1}{m_1!\dots m_k!}\sum_{j=0}^{k-N}(-1)^{k-N-j}C_{k-N}^{j}\\&=\frac{1}{m_1!\dots m_k!}(1-1)^{k-N}\\&=0\quad\text{if}\quad N\neq k,\quad\text{and}\\&=\frac{1}{m_1!\dots m_k!}\quad\text{if}\quad N=k;\end{aligned}$$

but in the latter case the condition $m_1+\cdots+m_k$ implies $m_1=\cdots=m_k=1$. Thus, there is only one non-zero term, equal to $v_1\dots v_k$ in our shorthand notation, or $\varphi(v_1,\dots,v_k)$ in the original notation. □

3.6 We differentiate the identity $f(\lambda x)=\lambda^k f(x)$ with respect to λ. By the chain rule, we get $f'(\lambda x)(x)=k\lambda^{k-1}f(x)$, and the Euler identity is obtained by putting $\lambda=1$.

Differentiating again with respect to λ, we get

$$f''(\lambda x)\cdot(x,x)=k(k-1)\lambda^{k-2}f(x),$$

and so on up to the p-th derivative which will be (if it exists):

$$f^{(p)}(\lambda x)(x,\dots,x)_{p\text{ times}}=k(k-1)\dots(k-p+1)\lambda^{k-p}f(x),$$

and we obtain, by putting $\lambda=1$:

$$f^{(p)}(x)\cdot(x,\dots,x)_{p\text{ times}}=k(k-1)\dots(k-p+1)f(x),\quad\text{for any } x\in\lambda.$$

Now we differentiate the identity $f(\lambda x)=\lambda^k f(x)$ with respect to x. For every vector h, we obtain $f'(\lambda x)(\lambda h)=\lambda^k f'(x)\cdot h$, or $f'(\lambda x)\cdot h=\lambda^{k-1}f'(x)\cdot h$, i.e. the derivative in the direction h is homogeneous of degree $k-1$. By differentiating p times (if possible), we conclude that for any $h_1,\dots,h_p\in X$ and any $\lambda\in\mathbf{R}$ we have $f^{(p)}(\lambda x)\cdot(h_1,\dots,h_p)=\lambda^{k-p}f^{(p)}(x)\cdot(h_1,\dots,h_p)$. □

Remark that if f is of class C^0 and homogeneous of degree 0, then $f(x)=f(0)$ for any $x\in X$, and f is constant. This is the first step of a proof by induction. Assume we have shown that the only functions of class C^{k-1} that are homogeneous of degree $k-1$ are the homogeneous polynomials of degree $k-1$. Now take $f: X\to\mathbf{R}$ of class C^k and homogeneous of degree k. Then, for any $h\in X$, the function $x\mapsto f'(x)\cdot h$ is of class C^{k-1} and homogeneous of degree $k-1$. Thus we can write $f'(x)\cdot h=\varphi_h(x,\dots,x)_{k-1\text{ times}}$, where φ_h is a symmetric $(k-1)$-linear form over X, which is uniquely determined by h. By uniqueness, we conclude that the map $h\circ\varphi_h$ is linear. This implies that the map $\varphi: X\times X^{k-1}\to\mathbf{R}$ given by $\varphi(h,x_1,\dots,x_{k-1})=\varphi_h(x_1,\dots,x_{k-1})$ is k-linear. Now write $f(x)$ as

$$f(x) = [f(tx)]_{t=0}^{t=1} = \int_0^1 \frac{\mathrm{d}}{\mathrm{d}t}(f(tx))\,\mathrm{d}t = \int_0^1 f'(tx)\cdot x\,\mathrm{d}t$$

$$\int_0^1 \varphi(x, tx, \dots, tx)\,\mathrm{d}t = \int_0^1 t^{k-1}\varphi(x, x, \dots, x)\,\mathrm{d}t = \frac{1}{k}\varphi(x, \dots, x).$$

If we *denote* by $S\varphi$ the symmetrization of φ, given by $S\varphi(x_1, \dots, x_k) = (1/k!)\Sigma_{\sigma \in \mathfrak{S}_k}\varphi(x_{\sigma(1)}, \dots, x_{\sigma(k)})$, we get that for any $x \in X$, $f(x) = (1/k)S\varphi(x, \dots, x)$. Since $S\varphi$ is a symmetric k-linear form over X, we conclude that f is a homogeneous polynomial of degree k.

Chapter 4

4.1 Recall (referring to [B, 4.6.16] if necessary) that $P^2(\mathbf{Z}_2)$ can be represented by the following triangle:

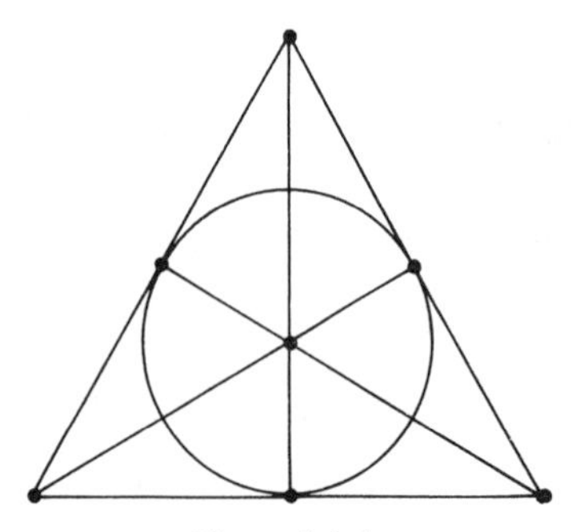

Figure 4.1.1.

The points are represented by the three vertices, the three midpoints of the sides, and the barycenter of the triangle. The lines are represented by the curves that join the three points they contain; they are the three sides, the medians and the circle inscribed in the (equilateral) triangle.

To represent $P^3(\mathbf{Z}_2)$, we proceed in the same way: the role of points is assigned to the vertices of a tetrahedron, that of lines to curves joining the three points contained in them, that of planes to surfaces containing not only seven points, but also the seven curves that represent the seven lines in the plane. We found it confusing to depict all this in a single figure, so we start by enumerating the elements of the configuration, and then proceed to the description of the figures.

The fifteen points will be: the four vertices, the six midpoints of the edges, the four barycenters of the faces, and the barycenter of the tetrahedron. The lines will be accounted for as follows: the edges of the tetrahedron (six), the medians (four times three) and inscribed circles (four) of each face, the four segments joining a vertex to the barycenter of the opposite face and the three segments joining the midpoints of two opposite edges (these seven lines pass through the barycenter of the tetrahedron), and finally six ellipses connecting the barycenter of two faces and the center of the opposite edge; this gives the representation of the thirty-five lines of $P^3(\mathbf{Z}_2)$. The fifteen planes are divided into four types: the faces (four), the bisector planes of each pair of faces (six), cones having as vertex a vertex of the tetrahedron and as directrix the circle inscribed in the opposite face (four), and finally the surface formed by the union of four cones whose vertex is the center of the tetrahedron and whose directrices are again the circles inscribed in the faces.

In order to show how the elements are related among themselves, we have first drawn, for each type of plane, the lines that are contained in it (figure 4.1.2). Then we have the picture from the "dual" point of view; figure 4.1.3 shows, for each type of point, the pencil of lines passing through this point. Finally, we draw a number of "self-dual" pictures, showing, for each type of line, the pencil of planes containing this line (figure 4.1.4).

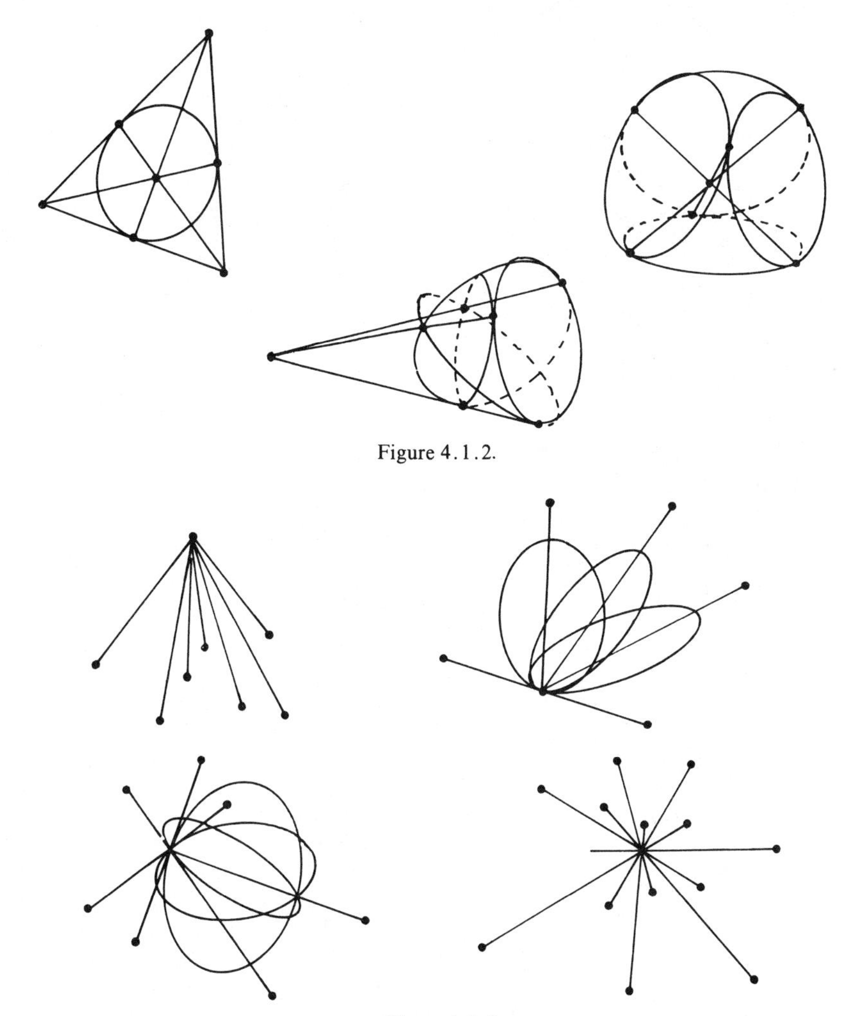

Figure 4.1.2.

Figure 4.1.3.

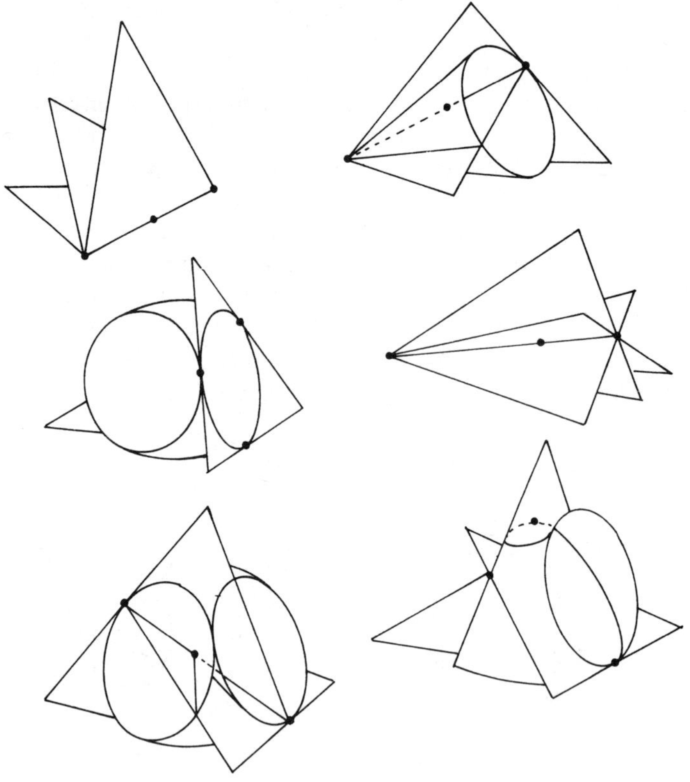

Figure 4.1.4.

4.2 An easy calculation shows that the Jacobian of the transition map

$$\pi_j \circ \pi_i^{-1} : (v_1, \ldots, v_j)$$

$$\mapsto \left(\frac{v_1}{v_{j-1}}, \ldots, \frac{v_{i-1}}{v_{j-1}}, \frac{1}{v_{j-1}}, \frac{v_i}{v_{j-1}}, \ldots, \frac{v_{j-2}}{v_{j-1}}, \frac{v_j}{v_{j-1}}, \ldots, \frac{v_n}{v_{j-1}} \right)$$

is given by

$$J(\pi_j \circ \pi_i^{-1}) = \det \begin{bmatrix} 1/v_{j-1} & & & & & & & \vdots \\ & \ddots & & & & & & \vdots \\ & & 1/v_{j-1} & 0 & 0 & & & \vdots \\ & & & 0 & 1/v_{j-1} & & & \vdots \\ & & & & 0 & \ddots & & \vdots \\ & & & & & 0 & 1/v_{j-1} & \vdots \\ -v_1/v_{j-1}^2 & \cdots & \cdots & \cdots & \cdots & \cdots & -v_{j-2}/v_{j-1}^2 & \cdots \; -v_n/v_{j-1}^2 \\ & & & & & & & 1/v_{j-1} \\ & & & & & & & \ddots \\ \cdots & \cdots & \cdots & \cdots & \cdots & \cdots & \cdots & 1/v_{j-1} \end{bmatrix}$$

Now this determinant is easy to compute, since there is only one way to choose one non-zero element in each row and column; the result is

$$J=(-1)^{j-i-1}\left(\frac{1}{v_{j-1}}\right)^{n-1}\times\left(\frac{-1}{v_{j-1}^2}\right)$$

$$=(-1)^{j-i}\left(\frac{1}{v_{j-1}}\right)^{n+1}$$

Now we will try to orient $P^n(\mathbf{R})$. Remark that the point $p(1,1,\ldots,1)$ belongs to all the charts $P(\mathbf{R}^{n+1})\backslash P(H_i)$. In order to orient $P^n(\mathbf{R})$, it is sufficient to choose an orientation at each point, and to require that this orientation be changed according to the sign of the Jacobian when we pass from one chart to another. We choose an orientation at $I=(1,\ldots,1)$; for instance, the positive orientation $(+1)$ in the chart π_0. Then the orientation at I will be $(-1)^i$ in the chart π_i. If $P_n(\mathbf{R})$ is orientable, the orientation at an arbitrary point will be $(-1)^i$ in the chart π_i, since it must be compatible with the orientation of $\mathbf{R}^n$, and π_i is a homeomorphism from $P^n(\mathbf{R})\backslash P(H_i)$ onto $\mathbf{R}^n$. Then $P^n(\mathbf{R})$ is oriented if and only if the orientations that we have just described do not contradict one another at the same point. Take an arbitrary point with homogeneous coordinates $(v_0, v_1,\ldots, v_n)$. Its orientation is $(-1)^i$ according to the chart π_i and $(-1)^j$ according to the chart π_j. To pass from one chart to the other we use a map whose Jacobian is $(-1)^{j-i}(1/v_{j-1})^{n+1}$. The chosen orientation will be consistent if and only if $(1/v_{j-1})^{n+1}$ is positive for any point, which is the case if and only if $n+1$ is even, or, alternatively, n is odd.

□

4.3 $\mathrm{GP}(P^n(\mathbf{R}))=\mathrm{GP}(\mathbf{R}^{n+1})=\mathrm{GL}(\mathbf{R}^{n+1})/\mathbf{R}^*\,\mathrm{Id}$. But $\mathrm{GL}(\mathbf{R}^{n+1})$ has two connected components, one containing the maps with determinant 1, the other those with determinant -1. Taking the quotient by $\mathbf{R}^*\,\mathrm{Id}$, which has determinant $(-1)^{n+1}$, preserves the two components if n is odd and lumps them together if n is even.

In order to orient projective space, we can work as follows: choose a projective base anywhere, and ascribe to it an orientation α. Any other base can be obtained from the first by an element of $\mathrm{GP}(P^n(\mathbf{R}))$, and it will be ascribed orientation α if the homography it defines belongs to the component of the identity in $\mathrm{GP}(P^n(\mathbf{R}))$. This way of ascribing orientations to bases is continuous, since two bases that are close to one another correspond to elements of $\mathrm{GP}(P^n(\mathbf{R}))$ that are also close in the topology of this group.

If $\mathrm{GP}(\mathbf{R}^{n+1})$ has two components, there are two classes of orientations, and $P^n(\mathbf{R})$ is orientable. If n is even, $\mathrm{GP}(\mathbf{R}^{n+1})$ is connected, and there is only one class of orientations (i.e. all bases have the same orientation); this means $P^n(\mathbf{R})$ is not orientable. □

4.4 We work in the vector space E; for each i, let K_i be the vector hyperplane of E that corresponds to the H_i. Each K_i is the orthogonal space of a non-zero linear form k_i under the natural pairing, and the intersection of the K_i is the orthogonal complement of the subspace of E^* spanned by these linear forms, whence

$$\dim\left(\bigcap_i K_i\right) = n+1-\dim\left(\left\langle \bigcup_i k_i \right\rangle_{E^*}\right).$$

Passing to the corresponding projective spaces, we can write

$$\dim\left(\bigcap_i H_i\right)+1 = n+1-\dim\left(\left\langle \bigcup_i k_i \right\rangle\right)-1,$$

or again

$$\dim\left(\bigcap_i H_i\right)+\dim\left(\left\langle \bigcup_i H_i \right\rangle\right) = n-1. \qquad \square$$

4.5 Enumerating the p-dimensional subspaces of $P(E)$ is the same as enumerating the $(p+1)$-dimensional vector subspaces of E, or those of K^{n+1}, since choosing a basis for E gives an isomorphism between K^{n+1} and E. We start by enumerating the (ordered) $(p+1)$-tuples of linearly independent vectors in K^{n+1}. The only condition on the first vector is that it be different from zero, so there are $k^{n+1}-1$ possible choices; the second vector should not belong to the line spanned by the first, whence $k^{n+1}-k$ choices; the $(q+1)$-th vector should not belong to the vector space spanned by the first q, which is isomorphic to K^q, whence $k^{n+1}-k^q$ choices. The number we want is then

$$(k^{n+1}-1)(k^{n+1}-k)\dots(k^{n+1}-k^p).$$

Every space possesses a base, so the number of $(p+1)$-dimensional subspaces of K^{n+1} is equal to the total number of ordered $(p+1)$-tuples of linearly independent vectors in K^{n+1}, divided by the number of $(p+1)$-tuples that span the same subspace; the latter is equal to the number of (ordered) bases for a space of dimension $(p+1)$, i.e. the number of (ordered) $(p+1)$-tuples of linearly independent vectors in K^{p+1}. But we've just found this number for K^{n+1}, so the same formula works for K^{p+1}, replacing n by p. Thus the number of p-dimensional subspaces of $P(E)$ is equal to

$$\frac{(k^{n+1}-1)(k^{n+1}-k)\dots(k^{n+1}-k^p)}{(k^{p+1}-1)(k^{p+1}-k)\dots(k^{p+1}-k^p)}.$$

We use the isomorphism $\mathrm{GP}(E)\simeq \mathrm{GL}(E)/K^*\mathrm{Id}_E$ to count the elements of $\mathrm{GL}(E)$, or again the bases of E, and this too is given by the formula for the $(p+1)$-tuples, this time replacing p by n. We finally divide by the cardinality of K^*, and we obtain the following formula:

$$\begin{aligned}\#\mathrm{GP}(E) &= \frac{(k^{n+1}-1)(k^{n+1}-k)\dots(k^{n+1}-k^n)}{(k-1)}\\ &= (k^{n+1}-1)(k^{n+1}-k)\dots(k^{n+1}-k^{n-1})k^n. \qquad \square\end{aligned}$$

4.6 Fix a 4-dimensional vector space E over a commutative field K, and a tetrahedron, i.e. four non-coplanar points a_1, a_2, a_3, a_4, in $P(E)$. In order to perform calculations, we complete this tetrahedron into a projective base by fixing a point a_0 which is not in any of the faces of the tetrahedron. We shall see that, if a tetrahedron b_1, b_2, b_3, b_4 is in "Möbius position" relative to the a_i, then one of the homographies taking a_i to b_i is expressed as a skew-symmetric matrix in the base $a_0, \ldots, a_4$.

1) Necessary condition

There is an infinite number of homographies taking each a_i to b_i; we choose the one taking the projective base $a_0, \ldots, a_4$ to the base $a_0, b_1, \ldots, b_4$. (For the calculation we assume a_0 is not in any of the faces of the tetrahedron b_i, and we will also assume that no b_i is in an edge of the tetrahedron a_i.) All the matrices associated with this homography f are proportional among themselves; we choose one and call it X. Now we can express the condition of being in "Möbius position": the point b_i is in the face a_j, a_k, a_l if and only if $x_{ii} = 0$, and, consequently, $\{b_i\}$ is in Möbius position relative to $\{a_i\}$ if and only if the matrices X and X^{-1} have only zeros in the diagonal. The additional hypothesis that b_i is not in an edge of the tetrahedron a_i means that the elements of X off the diagonal are non-zero. The l-th diagonal element of X^{-1} is proportional to the minor

$$\begin{vmatrix} 0 & x_{ij} & x_{ik} \\ x_{ji} & 0 & x_{jk} \\ x_{ki} & x_{kj} & 0 \end{vmatrix},$$

which is equal to $x_{ji}x_{ik}x_{kj} + x_{ij}x_{jk}x_{ki}$; the Möbius condition implies $x_{ll} = 0$, or

$$\frac{x_{ij}x_{jk}x_{ki}}{x_{ji}x_{kj}x_{ik}}.$$

It is convenient to put $y_{ij} = -x_{ij}/x_{ji}$, so that

$$y_{ij}y_{jk}y_{ki} = 1 \quad \text{and} \quad y_{ij}y_{ji} = 1. \tag{*}$$

Now the twelve numbers y_{ij}, $i \neq j$, satisfy the relations (*) above if and only if there are four numbers z_i, different from zero, such that $y_{ij} = z_j/z_i$; in fact, given the z_i, it is easily seen that the z_j/z_i satisfy (*), and, conversely, for numbers y_{ij} satisfying (*), we can put $z_1 = 1$, $z_i = y_{1i}$ for $i \neq 1$, and we can see that $y_{i1} = y_{1i}^{-1} = z_1/z_i$, and, for $i \neq j \neq 1$, $y_{ij} = y_{i1}y_{1j} = z_j/z_i$ (the family z_i is determined, up to a common factor, by the y_{ij}). Let Z be the diagonal matrix whose coefficients are $z_1, \ldots, z_4$; we put $X' = XZ$, and this matrix defines a homography of $P(E)$ taking each a_i to b_i. It is also skew-symmetric, since $x'_{ij} = z_i x_{ij}$, $x'_{ji} = z_j x_{ji}$, implying that

$$x'_{ij}x'^{-1}_{ji} = (z_i/z_j)(x_{ij}/x_{ji}) = -y_{ij}y_{ji} = -1. \qquad \square$$

2) Sufficient condition

Conversely, suppose we are given an invertible skew-symmetric matrix X'. It defines, in the projective base $a_0,\ldots,a_4$, a homography f which sends $a_1,\ldots,a_4$ to a tetrahedron $b_1,\ldots,b_4$, where $b_i \in \langle a_j, a_k, a_l\rangle$. The homography f^{-1} has the same property, since its diagonal elements are zero; this shows that $\{a_i\}$ and $\{b_i\}$ are Möbius tetrahedra. Thus the Möbius tetrahedra associated with a given tetrahedron depend on five parameters belonging to the field K: six defining the skew-symmetric matrix, minus one since two proportional matrices give rise to the same homography. Here is a matrix that gives a solution for any field (even a non-commutative one):

$$\begin{pmatrix} 0 & 1 & -1 & 1 \\ -1 & 0 & 1 & -1 \\ 1 & -1 & 0 & 1 \\ -1 & 1 & 1 & 0 \end{pmatrix} \qquad \square$$

3) Another solution

Fix in $P(E)$ a tetrahedron $a_1,\ldots,a_4$, and choose the face a_2, a_3, a_4 as the plane at infinity (see 5.D). The unknown tetrahedron $b_1,\ldots,b_4$ has one vertex, b_1, at infinity; this vertex can be represented as a vector $\overrightarrow{b_1}$ (up to a constant), and so can the points at infinity $\overrightarrow{a_2},\overrightarrow{a_3},\overrightarrow{a_4}$. Since $\overrightarrow{a_2}$ is in the plane $\overrightarrow{b_1}, b_3, b_4$, we can write $\overrightarrow{a_2}=\overrightarrow{b_1}+u\overrightarrow{b_3b_4}$, $u \in K$, up to a constant. In the same way, the conditions on $\overrightarrow{a_3}$ and $\overrightarrow{a_4}$ allow us to write $\overrightarrow{a_3}=\overrightarrow{b_1}+v\overrightarrow{b_2b_4}$, and $\overrightarrow{a_4}=\overrightarrow{b_1}+w\overrightarrow{b_2b_3}$. The condition that a_1 is on the face b_2, b_3, b_4 is expressed by writing $a_1 = xb_2 + yb_3 + zb_4$, where $x, y, z \in K^*$, and $x+y+z=1$. All that remains now is to write down the conditions on b_2, b_3 and b_4. The first point is on the plane at infinity defined by $a_1,\overrightarrow{a_3},\overrightarrow{a_4}$, so $\overrightarrow{a_1b_2}$ is a linear combination of $\overrightarrow{a_3}$ and $\overrightarrow{a_4}$, namely

$$\overrightarrow{a_1b_2}=s\overrightarrow{a_3}+t\overrightarrow{a_4}=(s+t)\overrightarrow{b_1}+sv\overrightarrow{b_2b_4}+tw\overrightarrow{b_2b_3};$$

but the point $\overrightarrow{b_1}$ is not parallel to the plane b_2, b_3, b_4, so that $s+t=0$, and

$$\overrightarrow{a_1b_2}=s\left(v\overrightarrow{b_2b_4}-w\overrightarrow{b_2b_3}\right).$$

Now

$$\overrightarrow{a_1b_2}=(x-1)b_2+yb_3+zb_4=y\overrightarrow{b_2b_3}+z\overrightarrow{b_2b_4},$$

whence the relation $wz=vy$. Analogously, the conditions on b_3 and b_4 lead to the relations $vy=ux$ and $ux=wz$. Conversely, given six numbers x, y, z, u, v, w that satisfy the relations $ux=vy=wz$, and a tetrahedron $b_1,\ldots,b_4$ such that $\overrightarrow{b_1}$ is at infinity, the formulas above give four points $a_1,\ldots,a_4$ which form a Möbius tetrahedron relative to $\{b_i\}$. $\square$

Chapter 5

5.1 Since the proof was obtained by sending points α, β to infinity, we draw here what happens when we send c to infinity, then when we send both c and c'.

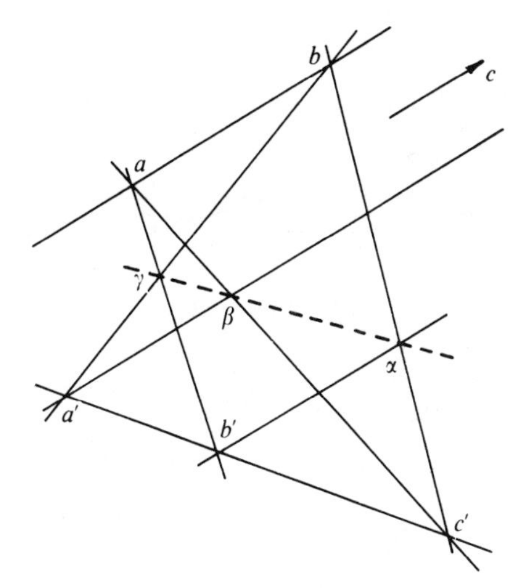

Figure 5.1.1.

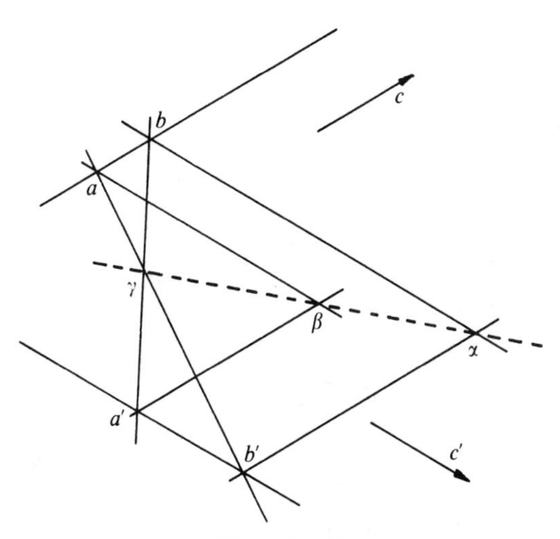

Figure 5.1.2.

5.2 Given two lines D and D' and a point a, we draw two lines starting at a and intersecting D and D' at points b and c, respectively. Then we choose a point b' in D distinct from b, and we draw from b' two lines T and T', parallel to $\langle a, b\rangle$ and $\langle b, c\rangle$, putting $c' = D' \cap T'$. We draw the parallel T'' to $\langle a, c\rangle$ passing through c'; the intersection a' of T and T'' allows us to draw $\langle a, a'\rangle$, which passes through the common point to D and D'. □

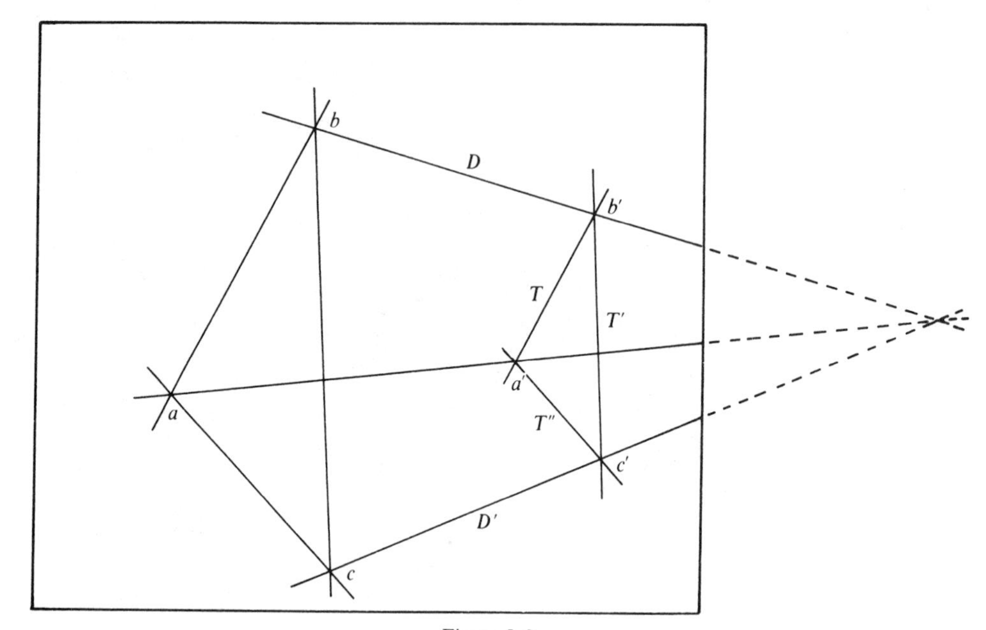

Figure 5.2.

The second case is an application of the first: given two points a and b, we draw, little by little, two lines starting at b and passing not too far from a, say at a distance of less than one fourth of the length of the ruler. Then the construction above allows us to locate a third point on the line $\langle a, b\rangle$ that lies close to a. All we have to do now is draw the segment obtained. □

5.3 We divide this exercise into three parts:

a) Existence of the points b_i

We want to prove a topological property, so we'll have to introduce the topology of $\mathbf{R}^2$ in one way or another; since all the norms are equivalent, we choose the Euclidean norm, which allows us to work with angles.

The only two obstacles in defining the points b_i are the following: if one of the points b_i is outside A, we cannot define $d_j(b_i)$ anymore; and if the two lines that define b_i are parallel, there is no b_i since we're working in the affine plane. We have to show that if b is close enough to a these two obstacles do not occur.

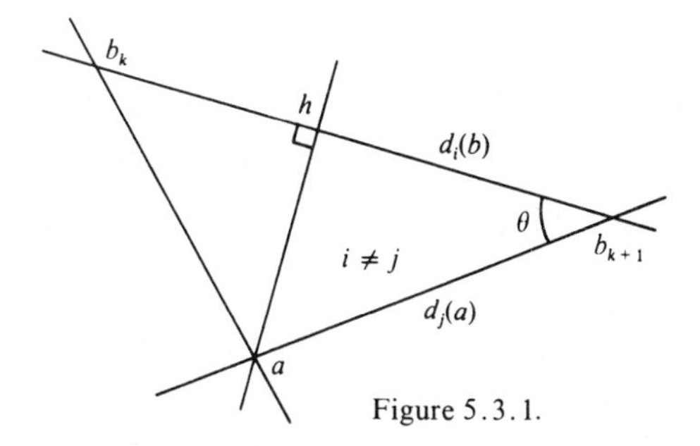

Figure 5.3.1.

Consider the three lines defined by the point a. They form three angles (i.e. non-oriented angles between non-oriented lines); we call α the smallest of the three. Now choose ε so that $b \in B(a, \varepsilon)$ implies $b \in A$ and $\overline{d_i(a)d_i(b)} < \alpha/2$ for all i (this is done by using the continuity of the d_i). If $b_k \in B(a, \varepsilon)$, then, since $\vartheta > \alpha/2$, we have $ab_{k+1} = ah/\sin\vartheta < ab_k/\sin(\alpha/2)$. So if we take b inside the ball $B(a, \varepsilon\sin^6(\alpha/2))$, we can certainly define the points b_i by induction, because by the formula above these points will each be inside $B(a, \varepsilon)$, which ensures that they are inside A and that the two lines which define b_{i+1} are not parallel, since they form an angle of at least $\alpha/2$. □

b) Complement of three lines

Given points a and b, we will construct b_1, b_2, b_3, b_4 and b_5. We consider the lines $\langle b_1, b_2\rangle$, containing p_1, and $\langle b_3, b_4\rangle$, containing p_3; next we remark that $b = \langle p_1, b_3\rangle \cap \langle p_3, b_1\rangle$, $b_5 = \langle p_1, b_4\rangle \cap \langle p_3, b_2\rangle$, $p_2 = \langle b_1, b_4\rangle \cap \langle b_2, b_3\rangle$.

The theorem of Pappus (cf. 5.D or [B, 5.4.2]) says that these three points are collinear, hence that $\langle p_2, p_5\rangle$ intersects $\langle p_1, a\rangle$ at the point $b_6 = b$.

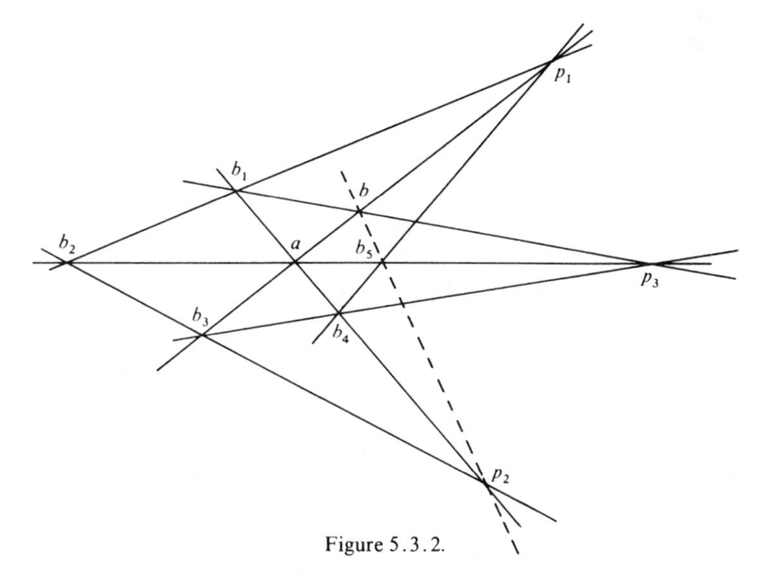

Figure 5.3.2.

c) Web defined by a conic and a point

Given points a and b, we will construct b_1, b_2, b_3, b_4 and b_5. We consider the hexagon $bb_1b_2b_5b_4b_3(b)$ (in this order!), and remark that this hexagon circumscribes conic C. The theorem of Brianchon (16.C) says that its three diagonals meet in a point; since $\langle b_4, b_1\rangle$ and $\langle b_3, b_2\rangle$ intersect at p, the third diagonal $\langle b_5, p\rangle$ must also pass through p, showing that the line $\langle b_4, p\rangle$ intersects the line $\langle a, b\rangle$ at b, i.e. $b = b_6$. □

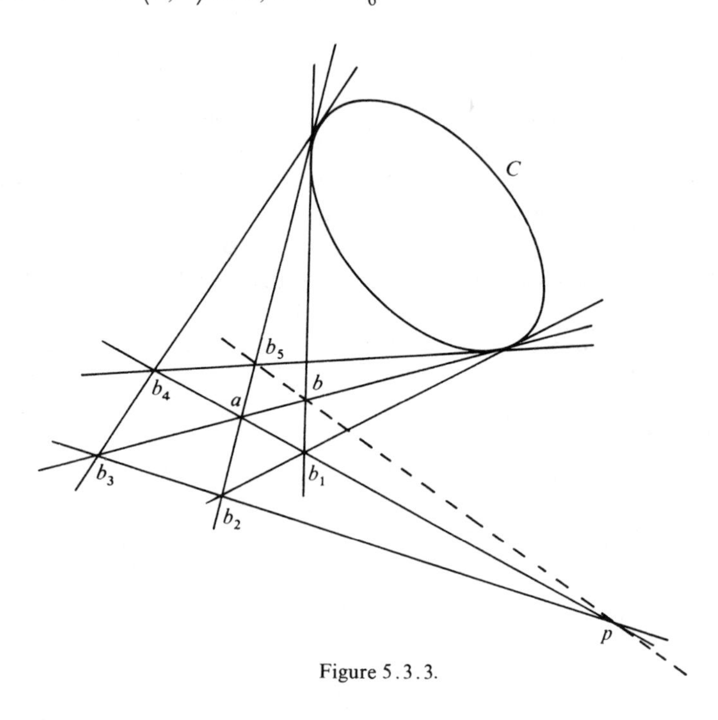

Figure 5.3.3.

REMARK. The two cases above are degenerate cases of the following situation: we are given a plane algebraic curve C of third class (i.e. such that through each point in the plane there are at most three tangents to C; such curves are dual to the cubics, the algebraic curves of degree 3). Then the associated web is hexagonal (and, in fact, the converse is also true: every hexagonal web is obtained in this way). The frontispiece of this book is an example of such a web.

Chapter 6

6.1 The idea is to send v to infinity and apply formula (1) in 6.A. In fact, if $v = \infty_D$, the point at infinity in the line under consideration, we get

$$[x, y, u, \infty_D] = \frac{\overrightarrow{xu}}{\overrightarrow{yu}}, \quad [y, z, u, \infty_D] = \frac{\overrightarrow{yu}}{\overrightarrow{zu}}, \quad [z, x, u, \infty_D] = \frac{\overrightarrow{zu}}{\overrightarrow{xu}},$$

whence the desired result follows:

$$\frac{\overrightarrow{xu}}{\overrightarrow{yu}} \cdot \frac{\overrightarrow{yu}}{\overrightarrow{zu}} \cdot \frac{\overrightarrow{zu}}{\overrightarrow{xu}} = 1. \qquad \square$$

6.2 Recall that the cross-ratio of the y_i can be written as

$$k = \frac{(y_3 - y_1)(y_4 - y_2)}{(y_3 - y_2)(y_4 - y_1)},$$

with the following conventions: $k = \infty$ if the denominator is zero and the numerator is not, and $k = -1$ if both are zero (we assume the four points y_i don't all coincide).

Let's first take care of the particular cases: if for some t_0 we have $y_3(t_0) = y_2(t_0)$, the theorem of Cauchy-Lipschitz implies that for all t in the interval of definition of the functions we have $y_3(t) = y_2(t)$. This shows that the cross-ratio $[y_i(t)]$ is independent of t in the cases when $(y_3 - y_2)(y_4 - y_1)$ vanishes.

Suppose now that the denominator does not vanish; then we can differentiate k with respect to t, and we obtain:

$$k' = \frac{\begin{array}{r}(ay_3^2 + by_3 + c - ay_1^2 - by_1 - c)(y_4 - y_2)\\ +(y_3 - y_1)(ay_4^2 + by_4 + c - ay_2^2 - by_2 - c)\end{array}}{(y_3 - y_2)(y_4 - y_1)}$$

$$-(y_3 - y_1)(y_4 - y_2)\frac{\begin{array}{r}(ay_3^2 + by_3 - ay_2^2 - by_2)(y_4 - y_1)\\ +(y_3 - y_2)(ay_4^2 + by_4 - ay_1^2 - by_1)\end{array}}{(y_3 - y_2)^2(y_4 - y_1)^2}$$

$$k' = \frac{(ay_3 + ay_1 + b)(y_3 - y_1)(y_4 - y_2) + (ay_4 + ay_2 + b)(y_3 - y_1)(y_4 - y_2)}{(y_3 - y_2)(y_4 - y_1)}$$

$$-(y_3 - y_1)(y_4 - y_2)\frac{\begin{array}{r}(ay_3 + ay_2 + b)(y_3 - y_2)(y_4 - y_1)\\ +(ay_4 + ay_1 + b)(y_3 - y_2)(y_4 - y_1)\end{array}}{(y_3 - y_2)^2(y_4 - y_1)^2}$$

$$k' = k(ay_3 + ay_1 + b + ay_4 + ay_2 + b - ay_3 - ay_2 - b - ay_4 - ay_1 - b) = 0.$$

This last equation implies that the cross-ratio $k(t) = [y_i(t)]$ is independent of t. $\square$

6.3 Denote by $(s_i)_{i=1,\ldots,4}$ the vertices of T, by $(H_i)_{i=1,\ldots,4}$ its faces, and by $(a_i)_{i=1,\ldots,4}$ the intersections of the line D with the faces. Let H be an arbitrary plane containing D but not s_4; the perspective of center s_4 from H onto H_4 transforms line D into the line $D_4 = H_4 \cap \langle D, s_4\rangle$, while the points a_1, a_2 and a_3 are transformed into the points b_1, b_2, b_3 whose configuration is indicated in the figure. Further, write $D_i = H_4 \cap \langle D, s_i\rangle$ for $i = 1,2,3$ (as has already been done for $i=4$), and draw these points, remarking that $D_i = \langle a_4, s_i \rangle$. Now consider in H_4 the perspective with center s_1, from D_4 onto $\langle s_2, s_3 \rangle$, which implies the following equalities:

$$\begin{aligned}[a_i] &= [b_1, b_2, b_3, a_4] = [b_1, s_3, s_2, c] = [D_4, D_3, D_2, D_1] \\ &= [D_1, D_2, D_3, D_4],\end{aligned}$$

whence $[a_i] = [\langle D, s_i \rangle]$. □

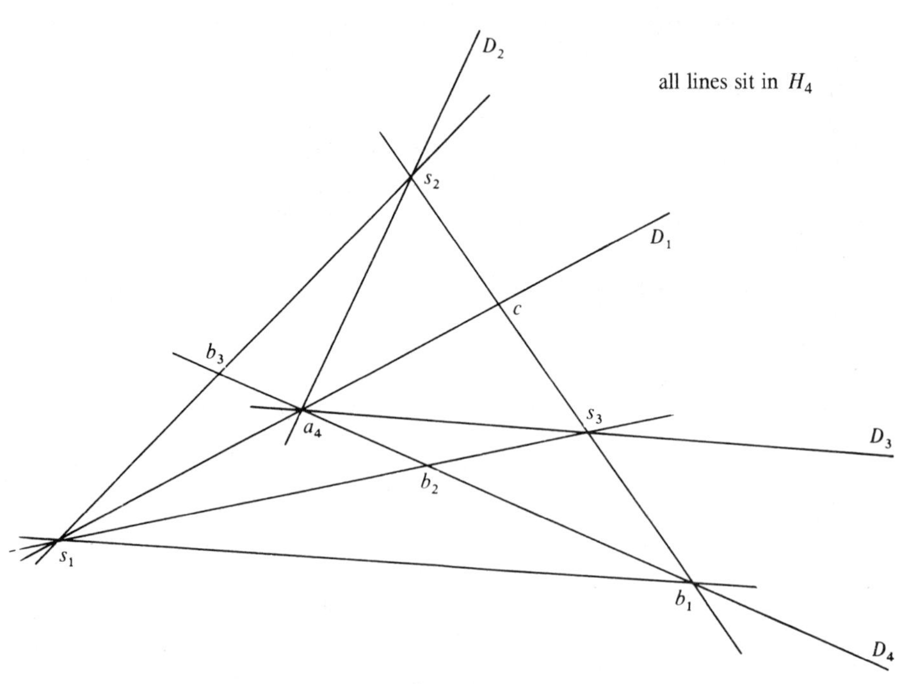

Figure 6.3.

6.4 Since the $(a_i)_{i=1,\ldots,4}$ form a projective base, we can write, in the corresponding system of homogeneous coordinates, $a_1 = (1,0,0)$, $a_2 = (0,1,0)$, $a_3 = (0,0,1)$, $a_4 = (1,1,1)$ and $a_5 = (\alpha, \beta, \gamma)$. Then we compute the cross-ratios, noting that since a_5 does not belong to any of the lines $\langle a_1, a_2 \rangle, \langle a_2, a_3 \rangle$, $\langle a_3, a_1 \rangle$, we have $\alpha\beta\gamma \neq 0$. We obtain:

$$\begin{aligned}[d_{12}, d_{13}, d_{14}, d_{15}] &= [(1,0),(0,1),(1,1),(\beta,\gamma)] = \beta/\gamma; \\ [d_{23}, d_{21}, d_{24}, d_{25}] &= [(0,1),(1,0),(1,1),(\alpha,\gamma)] = \gamma/\alpha; \\ [d_{31}, d_{32}, d_{34}, d_{35}] &= [(1,0),(0,1),(1,1),(\alpha,\beta)] = \alpha/\beta.\end{aligned}$$

Taking the product of the three cross-ratios we verify that it is equal to one.

This shows that giving three cross-ratios is excessive if we want to determine the point a_5 relative to the projective base $(a_i)_{i=1,\ldots,4}$, which is reasonable, since in a two-dimensional space two "coordinates" should be enough. The next question is intended to make this idea precise, by leaving aside the condition $a_4 \notin \langle a_1, a_2 \rangle \cup \langle a_2, a_3 \rangle \cup \langle a_3, a_1 \rangle$.

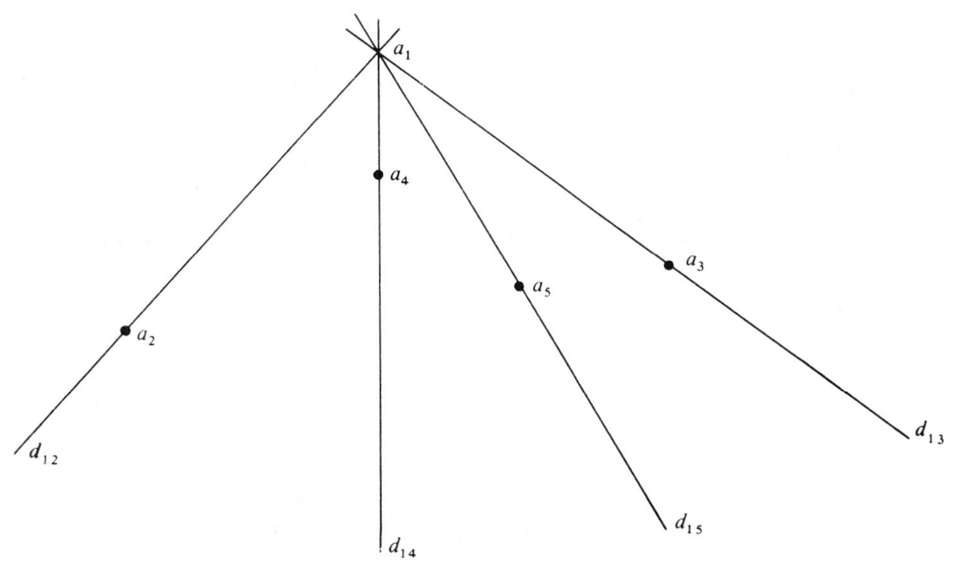

Figure 6.4.

If there is a homography taking the points $(a_i)_{i=1,\ldots,5}$ into $(a_i')_{i=1,\ldots,5}$ it is clear that any projective quantity calculated for the (a_i) will be the same if calculated for the (a_i'). We are interested in the converse: assuming that $[d_{ij}, d_{ik}, d_{i4}, d_{i5}] = [d_{ij}', d_{ik}', d_{i4}', d_{i5}']$ and that

$$\left[d_{ji}, d_{jk}, d_{j4}, d_{j5}\right] = \left[d_{ji}', d_{jk}', d_{j4}', d_{j5}'\right],$$

we will try to show there exists a homography taking the points (a_i) into (a_i').

There is a unique homography f taking the $(a_i)_{i=1,\ldots,4}$ into $(a_i')_{i=1,\ldots,4}$, since they are both projective bases; so we have to show that the hypothesis implies $a_5' = f(a_5)$. Denote by d_i'' the lines $\langle a_i', f(a_5)\rangle$ and by d_j'' the lines $\langle a_j', f(a_5)\rangle$. Then we can write

$$\left[d_{ij}, d_{ik}, d_{i4}, d_{i5}\right] = \left[d_{ij}', d_{ik}', d_{i4}', d_i''\right]$$

and

$$\left[d_{ji}, d_{jk}, d_{j4}, d_{j5}\right] = \left[d_{ji}', d_{jk}', d_{j4}', d_j''\right],$$

since f is a homography; but this implies that $d_i'' = \langle a_i', a_5'\rangle$ and $d_j'' = \langle a_j', a_5'\rangle$. But since a_i, a_j and a_5 are not collinear, we have $d_i'' \neq d_j''$ (again because f is an isomorphism); this means the two lines intersect in a single point $a_5' = f(a_5)$. □

Generalizations. Let $(a_i)_{i=1,\ldots,p+3}$ be $p+3$ points in a p-dimensional projective space, so that the first $p+2$ points form a projective base. For $\{i, j\} \subset$

$\{1, \ldots, p+1\}$, we denote by I_{ij} the set $\{1, \ldots, p+1\}$ minus the elements, i and j. The four hyperplanes $H_i = \langle (a_k)_{k \in I_{ij}}, a_j \rangle$, $H_j = \langle (a_k)_{k \in I_{ij}}, a_i \rangle$, $K_{ij} = \langle (a_k)_{k \in I_{ij}}, a_{p+2} \rangle$ and $L_{ij} = \langle (a_k)_{k \in I_{ij}}, a_{p+3} \rangle$ are in the same pencil, and their cross-ratio will be denoted by $b_{ij} = [H_j, H_i, K_{ij}, L_{ij}]$.

Imposing the same conditions on a_{p+3} as we did on a_5 in the case of the plane (the reader can figure out explicitly what they are), we shall show that the cross ratios b_{ij} satisfy certain relations. To do that, we work in the projective plane determined by a_i, a_j and a_k, where i, j, k are three distinct indices in the set $\{1, \ldots, p+1\}$. In addition to the points a_i, a_j and a_k, we consider the points α_{ijk} and β_{ijk}, whose i-th, j-th and k-th homogeneous coordinates are the same as the corresponding coordinates of a_{p+2} and a_{p+3}, respectively, and whose other coordinates are zero. Applying the result we obtained in dimension 2, we obtain the relations $b_{ij} b_{jk} b_{ki} = 1$. To investigate a necessary and sufficient condition for the existence of a homography taking the points $(a_i)_{i=1,\ldots,p+3}$ into the points $(a_i')_{i=1,\ldots,p+3}$, we proceed exactly as in the case $p = 2$.

We start by remarking that if $p+1$ points span a p-dimensional projective space, then the intersection of the $p+1$ hyperplanes spanned by each p of these points is empty (this can be proved by projectivizing the result that if $p+1$ vectors span a $(p+1)$-dimensional vector space, then they are linearly independent). Let's first apply this remark to the hyperplanes H_i: we have $\cap_{i=1}^{p+1} H_i \neq \phi$, so there must be a $j \in \{1, \ldots, p+1\}$ such that $a_{p+3} \notin H_j$. For this value of j, observe that a_{p+3} and H_j span the whole space, so we can again apply the remark to the L_{ij} and H_j: since $(\cap_{i \neq j} L_{ij}) \cap H_j = \phi$, we get $\cap_{i \neq j} L_{ij} = \{a_{p+3}\}$, since if there were another point in this intersection there would be a whole line, and the intersection of the line with H_j cannot be empty.

Putting all this together, we can show, using the same method as in the case $p = 2$, that a necessary and sufficient condition for the existence of a homography taking the points $(a_i)_{i=1,\ldots,p+3}$ into $(a_i')_{i=1,\ldots,p+3}$ is that the p equalities $b_{ij} = b_{ij}'$ hold where j is as above and $i = 1, \ldots, j-1, j+1, \ldots, p+1$. The modifications in the proof are left to the reader. □

REMARK. This exercise shows that we can introduce in projective geometry a system of "coordinates" more geometric than the homogeneous coordinates, in the same way that we introduce barycentric coordinates in affine geometry.

6.5 Interchanging the fixed points a and b transforms $[a, b, m, f(m)]$ in its inverse; this shows that the pair $\{k, 1/k\}$ depends only on f.

According to 6.D, we have $k = \lambda/\mu$, where λ and μ are the eigenvalues associated with f, i.e. the solutions of

$$\begin{vmatrix} \alpha - x & \beta \\ \gamma & \delta - x \end{vmatrix} = 0 \Leftrightarrow x^2 - (\alpha + \delta)x + \alpha\delta - \gamma\beta = 0. \tag{1}$$

But k and $1/k$ are roots of the equation

$$(X-k)\left(X-\frac{1}{k}\right)=0 \Leftrightarrow X^2-\left(k+\frac{1}{k}\right)X+1=0,$$

so it is enough to know

$$k+\frac{1}{k}=\frac{\lambda}{\mu}+\frac{\mu}{\lambda},$$

where we know that λ and μ are solutions of (1).

This expression is symmetric in λ and μ, so it must be possible to write it as a function of the elementary symmetric polynomials, which we know from (1). In other words, it's a waste of time to solve (1). We can use just the coefficients:

$$\begin{aligned}\frac{\lambda}{\mu}+\frac{\mu}{\lambda}&=\frac{\lambda^2+\mu^2}{\lambda\mu}\\&=\frac{(\lambda+\mu)^2-2\lambda\mu}{\lambda\mu}\\&=\frac{(\alpha+\delta)^2-2(\alpha\delta-\gamma\beta)}{\alpha\delta-\gamma\beta}\\&=\frac{\alpha^2+2\beta\gamma+\delta^2}{\alpha\delta-\beta\gamma},\end{aligned}$$

so that k and $1/k$ are roots of

$$X^2-\frac{\alpha^2+2\beta\gamma+\delta^2}{\alpha\delta-\beta\gamma}X+1=0 \Leftrightarrow (\alpha\delta-\beta\gamma)X^2 -(\alpha^2+2\beta\gamma+\delta^2)X+(\alpha\delta-\beta\gamma)=0. \qquad \square$$

6.6 Remark that, even with the normalization $\alpha\delta-\beta\gamma=1$, there are still two matrices for each homography (one the negative of the other). Consequently, the number $t=\operatorname{trace} f$ is only defined up to its sign. So it's rather $t^2=(\alpha+\delta)^2$ that must be considered as an invariant of f. This number does not depend on the projective base of the complex projective line that we are using to express f in.

a) Since the homographies being considered have two distinct fixed points, the equation

$$k^2-(t^2-2)k+1=0 \tag{I}$$

never has 1 as a root; this means $t^2\neq 4$. The numbers k and t^2 are connected by the relation $t^2=2+k+1/k$.

If the homography f is elliptic, i.e. if $|k|=1$, then $k+1/k$ is real and lies in the interval $[-2,2[$, so that $t^2\in[0,4[$.

Conversely, if $t^2 \in [0,4[$, there exists a real number ϑ such that $t^2 - 2 = 2\cos\vartheta$; then the complex number $k = e^{i\vartheta}$ has absolute value 1 and is root of (I), showing that f is elliptic.

If the homography f is hyperbolic, i.e. if k is positive real, then $k + 1/k$ is real and greater than 2, so t^2 is real and greater than 4.

Conversely, if $t^2 \in]4, +\infty[$, there exists a real number ϑ such that $t^2 - 2 = \operatorname{ch}\vartheta$; then the positive real number $k = e^{\vartheta}$ is a root of (I), and f is hyperbolic.

The only remaining case is when t^2 is not positive real, which must be equivalent to f being loxodromic. This is the case when $t = a + d$ is not real.

b) Behavior under iteration.

If the homography f has two distinct fixed points a and b, they will also be fixed for any iterate f^n; consequently, f^n falls into one of the classes above (elliptic, hyperbolic or loxodromic) if and only if f^n is not the identity. So we must put aside the set U_n of "roots of unity" in GP (i.e. the set of homographies f such that $f^n = id$).

In a projective base $\{a, b, m\}$, where m is arbitrary, the matrix of f can be written $\begin{pmatrix} \lambda & 0 \\ 0 & \mu \end{pmatrix}$; the matrix of f^n can then be written $\begin{pmatrix} \lambda^n & 0 \\ 0 & \mu^n \end{pmatrix}$. Thus we have $\lambda\mu = 1$, and

$$k_n = k(f^n) = k(f)^n, \quad t(f) = \lambda + \mu, \quad \text{and} \quad t_n = t(f^n) = \lambda^n + \mu^n.$$

If $f \in U_n$, then $k(f)^n = 1$, so $|k| = 1$, so all the elements of U_n are elliptic. Clearly, $|k^n| = 1$ if and only if $|k| = 1$; consequently, f^n is elliptic if and only if f is elliptic but does not belong to U_n.

If f is hyperbolic, then k is positive real; the same is true of k^n, hence f^n is also hyperbolic. Conversely, if f^n is hyperbolic, we have $k^n = \lambda^{2n} \in \mathbf{R}_+$; we write $\lambda = ru$ where $r > 0$ and $|u| = 1$. We necessarily have $u^{2n} = 1$, so the homography u whose matrix in the base $\{a, b, m\}$ is $\begin{pmatrix} u & 0 \\ 0 & \bar{u} \end{pmatrix}$ lies in U_n, whereas the homography r whose matrix is $\begin{pmatrix} r & 0 \\ 0 & r^{-1} \end{pmatrix}$ is hyperbolic, and has the same fixed points as u.

We conclude that if f^n is hyperbolic, f is either hyperbolic or loxodromic, and in the latter case can be written as the product of a hyperbolic homography r and a homography u in U_n, where r and u have the same fixed points.

Finally, if f is loxodromic, the same holds for f^n, unless f is a product $f = r_0 u$, where r is hyperbolic, $u \in U_n$, and r and u have the same fixed points.

c) Loxodromes.

Loxodromes, or rhumb lines in navigation, are defined in 18.2; they can be characterized as follows: every loxodrome, expressed in an appropriate projective base, is a logarithmic spiral.

If f is a homography with fixed points a and b, its matrix, in a base $\{a, b, m\}$, is diagonal; this means that, modulo a homographic change of coordinates, f can be written as $z \mapsto f(z) = kz$, where $k \neq 0, 1$. Writing $k = e^v$, we see that the iterates $f^n(z)$ are located on the logarithmic spiral $t \mapsto e^{tv}z$. If f is elliptic, this curve is a circle (rhumb line due west); if f is hyperbolic, it is a straight line (rhumb line due north).

Chapter 7

7.1 We study the space $E^C = \{s : \mathbf{C} \to E, s \text{ is } \mathbf{R}\text{-linear}\}$. First we embed E in E^C by the map $x \mapsto \{1 \mapsto x, i \mapsto 0\}$. This is well-defined because a linear map is uniquely determined by its value on the elements of a basis; here we take the **R**-basis $\{1, i = \sqrt{-1}\}$ for **C**.

The conjugation σ on E^C will be defined by

$$\sigma(s) = \sigma(s(1), s(i)) = (s(1), -s(i)).$$

Extending an **R**-linear map $f: E \to E'$ to the complexifications E^C, E'^C is very easy; the complexification f^C is simply defined by

$$f^C(s) = f \circ s.$$

The functoriality relation $(g \circ f)^C = g^C \circ f^C$ is trivial. The relation $\sigma \circ f^C = f^C \circ \sigma$ is also easy:

$$\begin{aligned}\sigma(f^C(s)) &= \sigma(f \circ s) = \sigma(f(s(1)), f(s(i))) = (f(s(1)), -f(s(i))) \\ &= f((s(1), -s(i))) = f(\sigma(s)).\end{aligned}$$

□

Chapter 8

8.1 The geometrical idea here is that the axes of the ellipsoid $\{x : \psi(x) = 1\}$ in the Euclidean space E, φ must be invariant under G, which implies reducibility, unless the length of all the axes is the same, in which case $\psi = k\varphi$.

We know (cf. 13.B or [B, 13.5]) that there is a basis $\{e_i\}$ of E which is orthonormal for both φ and ψ. The proof of this fundamental result consists in introducing the endomorphism f of E defined by

$$P(f(x), y) = Q(x, y) \quad \text{for every } x, y \in E,$$

where P (resp. Q) denotes the polar form of φ (resp. ψ). This relation uniquely defines f because φ is non-degenerate (cf. 13.C or [B, 13.2]). One shows then that all the eigenvalues of f are real; let them be (λ_i). We have to show that they are all equal if G is irreducible. But $G \subset O(E, \varphi) \cap O(E, \psi)$, hence G leaves both φ and ψ invariant, and consequently also P and Q. This implies that f commutes with every element of G, which is intuitively obvious, but can nevertheless be explicitly verified:

$$\begin{aligned}P(f(g(x)), y) &= Q(g(x), y) = Q(x, g^{-1}(y)) = P(f(x), g^{-1}(y)) \\ &= P(g(f(x)), y)\end{aligned}$$

for every y, whence $f(g(x)) = g(f(x))$ for every $g \in G$.

Now let V be an eigenspace of f, with eigenvalue λ; since $g \circ f = f \circ g$, we have $g(V) = V$ for every $g \in G$, whence $V = E$ since G is irreducible. □

8.2 The best thing to work with is the intrinsic metric of the unit sphere $S(E)$ (see 18.B or [B, 18.4]). Let $f \in O(E)$; the function $z \mapsto \overline{zf(z)}$ is defined and continuous on the compact set $S(E)$, so it achieves its minimum at one point x at least. If $f(x) = x$ or $f(x) = -x$ we're done since then the line Ox is invariant under f.

Otherwise the distance $\overline{xf(x)}$ is non-zero and less than π; in particular, $(x, f(x))$ are joined by a unique segment (an arc of great circle). Let y be the midpoint of this segment. Since f induces an isometry from $S(E)$ onto itself, the image $f(y)$ is the midpoint of the unique segment joining $f(x)$ to $f(f(x)) = f^2(x)$.

We claim now that $x, f(x), f^2(x)$ belong to the same plane (through the origin). Otherwise $y, f(x)$ and $f(y)$ would not be in the same plane either; therefore, they would form a spherical triangle, and we would have (cf. 18.B)

$$\begin{aligned}\overline{yf(y)} < \overline{yf(x)} + \overline{f(x)f(y)} &= \tfrac{1}{2}\overline{(xf(x))} + \tfrac{1}{2}\overline{(f(x)f^2(x))}\\ &= \tfrac{1}{2}\overline{(xf(x))} + \tfrac{1}{2}\overline{(xf(x))} = \overline{xf(x)},\end{aligned}$$

which is a contradiction.

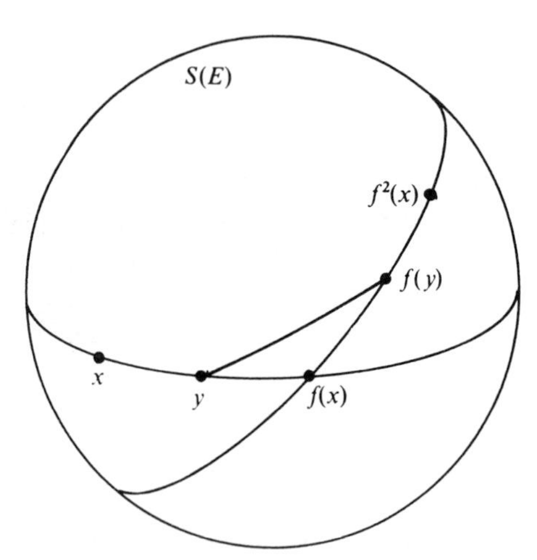

Figure 8.2.

8.3 Solving the equation $nx = a$ in $\widetilde{\mathfrak{U}}(E)$ is the same as solving $f^n = g$ in $O^+(E)$, where g is a rotation by an angle a. We introduce the following notation:

$$g = \begin{pmatrix} b & -c \\ c & b \end{pmatrix}, \quad f = \begin{pmatrix} d & -e \\ e & d \end{pmatrix},$$

where $b^2+c^2=d^2+e^2=1$; then, for $i=\sqrt{-1}$, we introduce the matrices $P=\begin{pmatrix}1 & 1\\ i & -i\end{pmatrix}$ and its inverse $P^{-1}=1/2\begin{pmatrix}1 & -i\\ 1 & i\end{pmatrix}$. We have the equations:

$$\begin{pmatrix} b-ci & 0\\ 0 & b+ci\end{pmatrix}=P^{-1}gP \quad\text{and}\quad \begin{pmatrix} d-ei & 0\\ 0 & d+ei\end{pmatrix}=P^{-1}fP;$$

this means that if $f^n=g$, then $(d-ei)^n=(b-ci)$ and $(d+ei)^n=b+ci$, as can be seen by taking $P^{-1}fP$ to the n-th power. Conversely, if

$$(d-ei)^n=(b-ci) \quad\text{and}\quad (d+ei)^n=b+ci$$

(by conjugation, one of the equations is enough), we get $f^n=g$ because

$$g=P\begin{pmatrix} b-ci & 0\\ 0 & b+ci\end{pmatrix}P^{-1} \quad\text{and}\quad f=P\begin{pmatrix} d-ei & 0\\ 0 & d+ei\end{pmatrix}P^{-1}.$$

Thus we have shown that $f^n=g$ is equivalent to $(z_f)^n=z_g$, where we have put

$$z_f=d+ei \quad\text{and}\quad z_g=b+ci$$

(observe that the condition $b^2+c^2=d^2+e^2=1$ implies that z_f and z_g have absolute value 1). The desired result follows now very simply from the fact that every complex number of absolute value one has exactly n distinct n-th roots, and they have absolute value one. □

EXAMPLES.

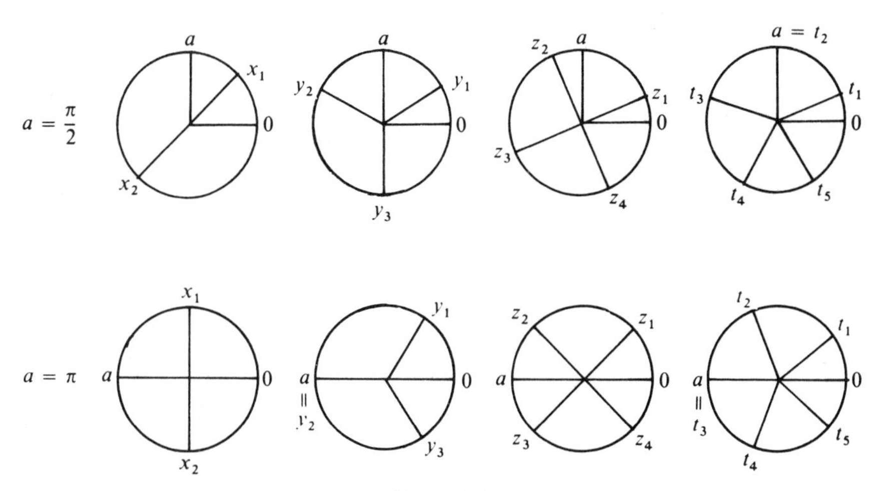

Figure 8.3.

8.4 The idea is to consider the reflections $\sigma_S, \sigma_T, \sigma_U$ through the axes S, T, U, respectively (cf. 8.C or [B, 8.2.9]). They are elements of $O^+(E)$ and, by assumption, satisfy the properties $\sigma_S(A)=B$, $\sigma_T(B)=C$ and $\sigma_U(C)=A$. Set

$f = \sigma_U \circ \sigma_T \circ \sigma_S$; then $f(A) = A$. Since $f \in O^+(E)$, it must be either the identity or a rotation around a well-defined axis (cf. 8.E or [B, 8.4.7.1]).

If f is not the identity, the problem has exactly one solution: A is the axis of f, and we have $B = \sigma_S(A)$ and $C = \sigma_T(B)$. If f is the identity, we can take an arbitrary line for A, and make $B = \sigma_S(A)$ and $C = \sigma_T(B)$ as before. What is the condition on S, T, U for the composition $f = \sigma_U \circ \sigma_T \circ \sigma_S$ to be the identity?

It is easy to see that the product $\sigma_T \circ \sigma_U$ of two reflections through lines is a rotation around the line perpendicular to both S and T, by an angle double that between S and T. If f is the identity, we must have

$$\sigma_T \circ \sigma_S = \sigma_U,$$

which implies that U must be perpendicular to S and T, but also that the angle between S and T is $\pi/2$, so that the angle of σ_U can be π. Hence, S, T and U must be pairwise orthogonal, and this condition is also sufficient for f to be the identity. □

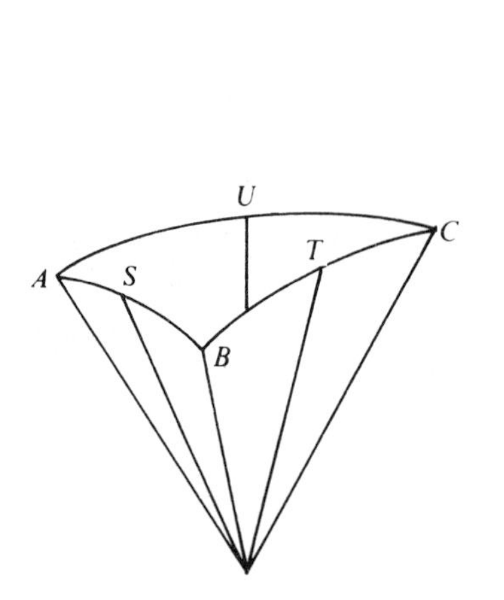

Figure 8.4.1.

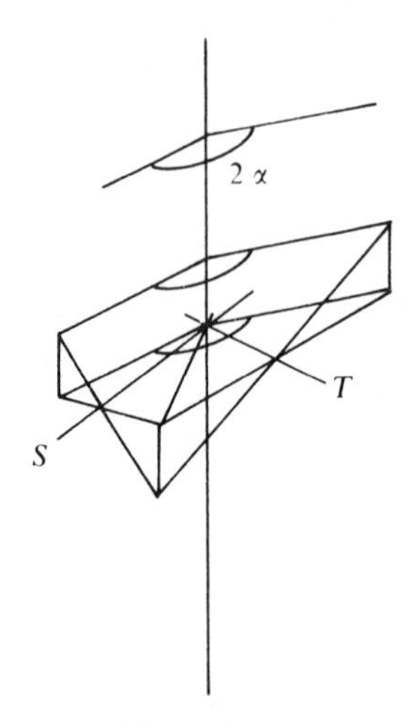

Figure 8.4.2.

Considering now *half-lines* instead of lines, we start working as before. We find f and its axis Δ; the half-line A must necessarily be the half-line determined by Δ and making an angle less than or equal to $\pi/2$ with S. But then it can happen that $B = \sigma_S(A)$ is a half-line that makes an angle greater than $\pi/2$ with T, in which case the problem has no solution.

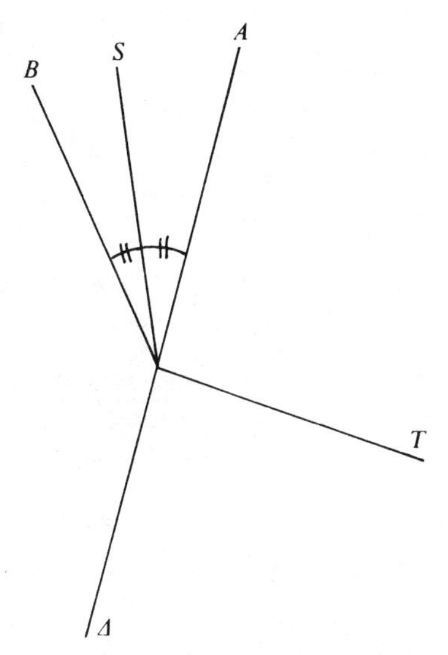

Figure 8.4.3.

Generalizations. We can generalize the preceding result by considering n given lines $(S_i)_{i=1,\ldots,n}$, and looking for n lines $(A_i)_{i=1,\ldots,n}$ such that the line S_1 is the bisector of A_1 and A_2, the line S_2 is the bisector of A_2 and A_3, and so on up to S_n, the bisector of A_n and A_1. The same technique can be carried over without any changes; we find there is a solution, and it is unique unless the composition $\sigma_{S_n} \circ \cdots \circ \sigma_{S_1}$ is the identity instead of a proper rotation.

If we try to generalize to Euclidean vector spaces of dimension $k > 3$, on the other hand, it may happen that the composition $\sigma_U \circ \sigma_T \circ \sigma_S$ leaves no line invariant (this will be the case "generically", cf. 8.E or [B, 8.4.7.3]), and then the problem will have no solution. □

REMARK. The idea of introducing $\sigma_U \circ \sigma_T \circ \sigma_U$ is very fruitful in geometry, and can be used in many analogous problems. You can start by applying it to the following problem: find, in an affine space, a polygon for which midpoints of the sides are given. Then try the following two applications of this idea: the problem of Castillon—finding a polygon inscribed in a given circle and whose sides pass through prescribed points (cf. [B, 10.11.4, 16.3.10.3]), and the great theorem of Poncelet about polygons inscribed in one conic and circumscribed around another, for the case of two bitangent conics (cf. problem 16.5).

8.5 Recall first that **R** possesses only one automorphism, the identity. In fact, if φ is an automorphism of **R**, we have $\varphi(1)=1$, and $\varphi(p/q)=p/q$ for every pair of integers (p,q), $q \neq 0$. The result follows from the density of **Q** in **R** and the fact that φ is increasing, this because if $a \geq 0$ we have

$$\varphi(a) = \left(\varphi(\sqrt{a})\right)^2 \geq 0.$$

Now let φ denote an automorphism of $\mathbf{H}$; since $\mathbf{R}$ is the center of $\mathbf{H}$, we have $\varphi(\mathbf{R}) \subset \mathbf{R}$ and $\varphi^{-1}\mathbf{R} \subset \mathbf{R}$, which implies that $\varphi(\mathbf{R}) = \mathbf{R}$, hence $\varphi|_{\mathbf{R}} = \mathrm{Id}_{\mathbf{R}}$. Now remark that the square roots of -1 are the pure quaternions of norm 1; denote their set by S^2. The images of i, j, k will be in S^2, and must form a positively oriented orthonormal basis for $\mathbf{R}^3$, since $\varphi(i)\varphi(j) = \varphi(k)$. We conclude by writing $a = b + ci + dj + ek$:

$$\begin{aligned}\varphi(a) &= \varphi(b) + \varphi(c)\varphi(i) + \varphi(d)\varphi(j) + \varphi(e)\varphi(k)\\ &= b + c\varphi(i) + d\varphi(j) + e\varphi(k);\end{aligned}$$

since $(\varphi(i), \varphi(j), \varphi(k))$ is a positively oriented orthonormal basis, the map $ci + dj + ek \mapsto c\varphi(i) + d\varphi(j) + e\varphi(k)$ is a rotation in $\mathbf{R}^3$. Thus we have shown that $\varphi(a) = \mathscr{R}(a) + \rho(\mathscr{P}(a))$, where $\rho \in O^3(3)$ depends only on φ. □

8.6 Formula (1)

- If b and c are collinear, the formula is obviously verified; both sides vanish.
- If b and c are linearly independent, the three vectors b, $b \times c$ and $b \times (b \times c)$ form an orthogonal basis for $\mathbf{R}^3$, from which we can get an orthonormal basis by normalizing. We assume b has coordinates $(b, 0, 0)$, $c = (c_1, c_2, 0)$ and $a = (a_1, a_2, a_3)$. We will calculate the coordinates of the two sides of (1) in this basis.

Since the basis is orthonormal, we can apply the explicit formulas for the vector product:

$$b \times c = \left| \begin{matrix} 0 \\ 0 \\ bc_2 \end{matrix} \right. \qquad a \times (b \times c) = \left| \begin{matrix} a_2 bc_2 \\ -a_1 bc_2; \\ 0 \end{matrix} \right.$$

on the other hand,

$$\begin{aligned}(a|c)b - (a|b)c &= (a_1c_1 + a_2c_2) \left| \begin{matrix} b \\ 0 \\ 0 \end{matrix} \right. - a_1 b \left| \begin{matrix} c_1 \\ c_2 \\ 0 \end{matrix} \right. \\ &= \left| \begin{matrix} a_2 bc_2 \\ -a_1 bc_2, \\ 0 \end{matrix} \right.\end{aligned}$$

concluding the proof of formula (1). □

Note. This was not an elegant solution; writing down coordinates in a basis and calculating with them somehow jars our sense of mathematical ethics. But in this case we could hardly have avoided it. It is easy enough, by using the definition of vector products by means of mixed products, plus the linearity of the latter, to prove (1) up to a multiplicative factor; but we have to remember that the mixed product is defined by its natural properties only up to this same multiplicative factor, and to define it completely we have to postulate that the mixed product of an orthonormal basis is 1. So orthonormal bases necessarily come into the picture. Now from the moment we resort to them, we might as well write down the vectors in coordinates; contriving a way not to make coordinates appear explicitly is a bit hypocritical...

By using (1) we get:

$$(a\times b)\times(a\times c)=(a\times b|c)a-0=(a,b,c)a$$

and

$$\begin{aligned}(a\times b,a\times c,b\times c)&=((a\times b)\times(a\times c)|b\times c)\\&=(a,b,c)(a|b\times c)\\&=(a,b,c)^2.\end{aligned}$$

That $\mathbf{R}^3$ is a Lie algebra is a simple matter; it is obviously an algebra since $\times$ is distributive relative to addition (because of linearity), so all we have to verify is Jacobi's identity. Using (1):

$$\begin{aligned}&a\times(b\times c)+b\times(c\times a)+c\times(a\times b)\\&\qquad=(a|c)b-(a|b)c+(b|a)c-(b|c)a-(c|b)a+(c|a)b=0.\end{aligned}$$

As for formula (2), we can start using coordinates without compunction. Choose an orthonormal basis, and write

$$a=\begin{vmatrix}a_1\\a_2,\\a_3\end{vmatrix}\quad b=\begin{vmatrix}b_1\\b_2\\b_3\end{vmatrix}$$

in this basis. Then

$$a\times b=\begin{vmatrix}a_2b_3-b_2a_3\\a_3b_1-b_3a_1,\\a_1b_2-b_1a_2\end{vmatrix}$$

whence

$$\|a\times b\|^2=(a_2b_3-b_2a_3)^2+(a_3b_1-b_3a_1)^2+(a_1b_2-b_1a_2)^2.$$

On the other hand,

$$p(a)=\begin{vmatrix}0\\a_2\\a_3\end{vmatrix}\quad p(b)=\begin{vmatrix}0\\b_2\\b_3\end{vmatrix}$$

$$\begin{aligned}\mathrm{Gram}(p(a),p(b))&=\begin{vmatrix}a_2^2+a_3^2 & a_2b_2+a_3b_3\\a_2b_2+a_3b_3 & b_2^2+b_3^2\end{vmatrix}\\&=(a_2^2+a_3^2)(b_2^2+b_3^2)-(a_2b_2+a_3b_3)^2\\&=a_2^2b_2^2+a_3^2b_2^2+b_3^2a_2^2+a_3^2b_3^2-a_2^2b_2^2-2a_2a_3b_2b_3-a_3^2b_3^2\\&=a_2^2b_3^2-2a_2a_3b_2b_3+a_3^2b_2^2\\&=(a_2b_3-a_3b_2)^2.\end{aligned}$$

The other two terms are obtained in an analogous way. □

Interpretation. The square of the length of $a \times b$, which is also equal to the volume of the parallelepiped built on the vectors $\{0, a, b, a \times b\}$ (since $|a \times b, a \times b| = (a, b, a \times b)$), is equal to the sum of the surfaces of the three parallelograms obtained as projections of the parallelogram $\{0, a, b, a+b\}$ onto the three coordinate planes. Choosing the basis conveniently, we can make this parallelogram lie in a coordinate plane, showing that

$$\|a \times b\|^2 = \mathrm{vol}(0, a, b, a \times b). \qquad \square$$

Note. The fact that we are equating a distance, an area and a volume shows that these notions are not intrinsic, i.e. they require an arbitrary choice of a unit length. The world around us is Euclidean only up to a multiplicative factor. This means we couldn't have an equality like the above in physics, where the formulas must be invariant under multiplication of the metric by a constant; this explains the well-known homogeneity of physics formulas, which does not have to hold here.

Equation $x \times a = b$

Uniqueness: $x \times a = x' \times a \Leftrightarrow (x - x') \times a = 0$

$$\Leftrightarrow x' = x + \lambda a.$$

Existence: we evidently must have $(a|b) = 0$, otherwise there can be no solution. So suppose this condition is satisfied, and put $x = \alpha(a \times b)$; then

$$x \times a = \alpha(a \times b) \times a = -\alpha a \times (a \times b) = -\alpha\left[(a|b)a - a^2 b\right]$$

by (1); but $(a|b) = 0$ by assumption, so $x \times a = \alpha a^2 b$, and choosing $\alpha = 1/a^2$ gives a solution. To sum up:

— if a is not perpendicular to b, there is no solution;
— if a is perpendicular to b, the set of solutions is an affine line whose direction is a and which passes through $(1/a^2) a \times b$. $\square$

Chapter 9

9.1 We endow the plane with a Euclidean structure, and prove the theorem by contradiction. Suppose there is a triple (i, j, k) such that $x_i \notin \langle x_j, x_k \rangle$. Consider the distances $d(x_k, \langle x_h, x_l \rangle)$ for all triples (k, h, l) such that $x_k \notin \langle x_h, x_l \rangle$, and choose three points $a = x_i$, $b = x_j$ and $c = x_k$ such that this distance is minimal for the set of all such triples.

By assumption, the line $\langle b, c \rangle$ contains a third point d among the (x_i). Hence at least two of the three points b, c, d are located on the same side of the perpendicular D to the line $\langle b, c \rangle$ and passing through a. Call O the intersection of D and $\langle b, c \rangle$; we can assume the points are arranged as in the figure below. But the distance from c to $\langle a, d \rangle$ is strictly less than the distance from O to $\langle a, d \rangle$, which is also less than the distance $\overline{aO}$ from a to $\langle b, d \rangle$, contradicting the minimality assumption. $\square$

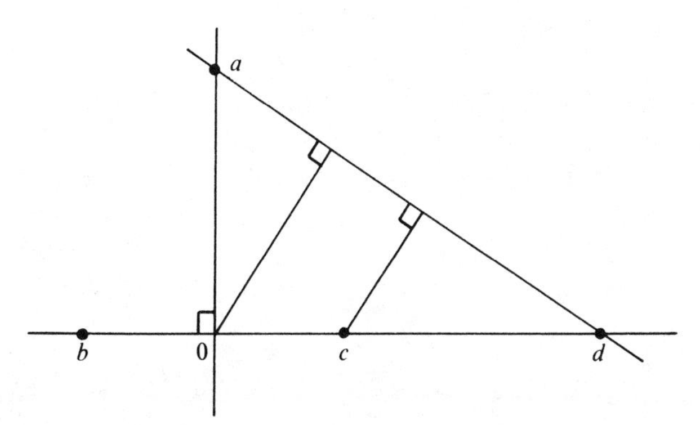

Figure 9.1.

Note. This theorem can be refined: if the n points $(x_i)_{i=1,\dots,n}$ are not all collinear, then there are at least $3n/7$ lines that contain only two among the n points.

9.2 The equation $d(x, D) = d(x, D')$ is equivalent to

$$(d(x, D))^2 - (d(x, D'))^2 = 0;$$

thus the set we want is a quadric.

Case 1: *X is a plane.*

We start by recalling that the bisectors of two lines D, D' are the two lines T, T' which satisfy $\widehat{DT} = \widehat{TD'}$, where the angles are oriented angles between non-oriented lines (i.e. elements of $\mathfrak{U}(E)$; cf. 8.F). Recall also that, if we have previously oriented the space, such an angle is uniquely determined by its sine, i.e. the number $\sin \widehat{DD'} = b \in [-1,1]$, where b is such that one of the two rotations taking D into D' can be written $\begin{pmatrix} a & -b \\ b & a \end{pmatrix}$ with a positive.

We *denote* by p the intersection of the two lines, and by m and m' the orthogonal projections of a point x on D and D'. The sine of the angle $\widehat{\langle p, m\rangle\langle p, x\rangle}$ can be calculated thanks to the right triangle pmx:

$$\left|\sin \widehat{\langle p, m\rangle\langle p, x\rangle}\right| = mx/px.$$

So the desired points p satisfy

$$\left|\sin \widehat{\langle p, m\rangle\langle p, x\rangle}\right| = \left|\sin \widehat{\langle p, m'\rangle\langle p, x\rangle}\right|,$$

whence

$$\widehat{D\langle p, x\rangle} = \pm \widehat{\langle p, x\rangle D'}$$

since $D \neq D'$, the line $\langle p, x\rangle$ is one of the bisectors of D, D'. The set we're looking for is thus the conic formed by the two bisectors of D, D'. □

Case 2: *X is not a plane.*

Let H be the codimension-2 subspace orthogonal to both D and D'. If D and D' intersect, the set of points equidistant from D and D' is the union of the two hyperplanes $\langle T, H\rangle$ and $\langle T', H\rangle$ containing the bisectors T, T' of D, D'.

Now suppose that $D \cap D' = \phi$; we'll find the type (in the affine classification, which is essentially sufficient) of the quadric $d(x, D) = d(x, D')$. Let $p = H \cap D$, $p' = H \cap D'$, and choose unit vectors $\vec{e}_1$ in the direction of D, $\vec{e}_2$ in $\vec{e}_1^{\perp}$ and such that the direction of D' is $\alpha\vec{e}_1 + \beta\vec{e}_2$, then $\vec{e}_3$ such that $\overrightarrow{pp'} = \gamma\vec{e}_3$, and finally $\vec{e}_4, \dots, \vec{e}_n$ such that $(p, \vec{e}_3, \dots, \vec{e}_n)$ is an orthonormal frame for H. Then our quadric has the following equation in the orthonormal frame $(p, \vec{e}_1, \dots, \vec{e}_n)$:

$$x_2^2 - (\beta x_1 - \alpha x_2)^2 + 2\gamma x_3 - \gamma^2 = 0;$$

putting $x_1' = \beta x_1 - \alpha x_2$ and $x_3' = \gamma x_3 - \gamma^2/2$ (which is a valid change of variables since $\beta \neq 0$ from the fact that D and D' are not parallel, and $\gamma \neq 0$ from the fact that D and D' are not coplanar), we get the new equation $x_2^2 + x_1'^2 + 2x_3' = 0$.

In the notation of [B, 15.3.2], this quadric is of type III (1,1); in dimension 3, it's a hyperbolic paraboloid. □

9.3 The idea is to look for a polygon with maximal perimeter, and then to show that it must be a light polygon, using the first variation formula (cf. 9.G). We must leave out self-intersecting polygons, however, since for these the maximal perimeter is achieved (when n is even), by the diameter of the set taken n times. For a fixed integer n, we *denote* by $\Delta \subset C^n$ the subset of the product $C \times \cdots \times C$ (n times) formed by the n-tuples $(x_1, \dots, x_n)$ such that each x_i is on the closed arc from x_{i-1} to x_{i+1} (modulo n), along the curve. Since Δ is a closed subset of the compact set C^n, the perimeter function $p: \Delta \to \mathbf{R}$ defined by $p((x_1, \dots, x_n)) = \Sigma_{i=1}^n x_i x_{i+1}$ achieves its maximum for at least one n-tuple (a_i). We show that such an n-tuple forms a light polygon.

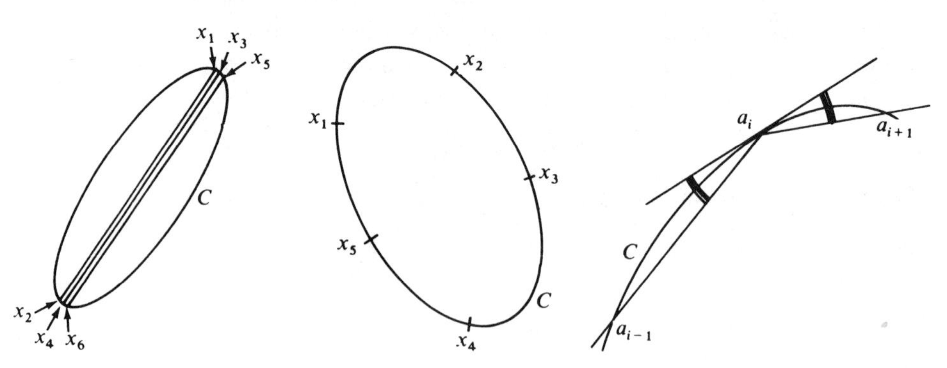

Figure 9.3.1. Figure 9.3.2. Figure 9.3.3.

The function $x \mapsto a_{i-1}x + xa_{i+1}$, defined on the arc of the curve from a_{i-1} to a_{i+1}, is maximal at $x = a_i$, so its derivative at this point is zero. By the first variation formula and because C is strictly convex, this shows that the angles between the tangent to C at a_i and each of the lines $\langle a_{i-1}, a_i\rangle$ and $\langle a_i, a_{i+1}\rangle$ are equal. □

See also [B, 17.6.5].

9.4 a) Construction using invariance of angles

We want to find the center ω^+ of an element $f \in \mathrm{Sim}^+(\mathbf{R}^2)$ such that $f(a) = a'$ and $f(b) = b'$.

Let's first take care of two trivial cases: when $a = a'$, the center is the point $a = a'$; and if $\langle a, b\rangle // \langle a', b'\rangle$, f is a homothety or a translation, and in the first case the center is the point $\langle a, a'\rangle \cap \langle b, b'\rangle$.

Now let $e = \langle a, b\rangle \cap \langle a', b'\rangle$; this point is well-defined since we've excluded the case when the two lines are parallel. We have $\widehat{\langle a, \omega^+\rangle\langle a, b\rangle} = \widehat{\langle a, \omega^+\rangle\langle a, e\rangle}$, and, since f preserves angles, we get

$$\widehat{\langle a, \omega^+\rangle\langle a, b\rangle} = \widehat{\langle a, \omega^+\rangle\langle a', b'\rangle}$$

(here we're dealing with oriented angles between non-oriented lines, i.e. elements of $\mathfrak{U}(E)$). Thus we can write

$$\widehat{\langle a, \omega^+\rangle\langle a, e\rangle} = \widehat{\langle a, \omega^+\rangle\langle a', e\rangle},$$

and, because of the property of angles inscribed in a circle (cf. 10.D or [B, 10.9.3]), this implies that a, a', ω^+ and e lie on the same circle; the same applies to b, b', ω^+ and e.

All the above justifies the following construction: after having found e, we draw the circle C_1 containing a, a' and e (if $a = e$, C_1 is the circle passing through a and a' and tangent to $\langle a, b\rangle$ at a). We also draw the circle C_2 passing through b, b' and e. The intersection points of the two circles are then e and ω^+.

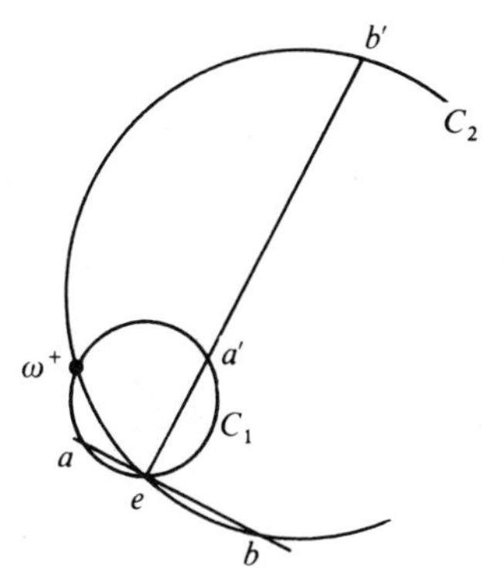

Figure 9.4.1.

REMARK. This kind of method is no good for finding the center ω^- of a similarity $g \in \mathrm{Sim}^-(\mathbf{R}^2)$ such that $g(a) = a'$ and $g(b) = b'$, since the set of points x such that $\widehat{\langle a, x\rangle\langle a, e\rangle} = -\widehat{\langle a', x\rangle\langle a', e\rangle}$ is a curve which cannot be drawn with the ruler and the compass.

b) Constructions using invariance of ratios of distances

We will assume here that the similarities we're looking for are not isometries, i.e. we have $ab \neq a'b'$ (this is not an important restriction since the problem is much simpler in the case of isometries). Given the points a, b, a' and b', the set $C_1 = \{x \in \mathbf{R}^2 : xa'/xa = a'b'/ab\}$ is a circle passing through the points

$$c_1 = \big(ab/(ab - a'b')\big)a' - \big(a'b'/(ab - a'b')\big)a$$

and

$$d_1 = \big(ab/(ab - a'b')\big)a' + \big(a'b'/(ab - a'b')\big)a$$

(cf. 9.F or [B, 9.7.6.5]). The set $C_2 = \{x \in \mathbf{R}^2 : xb'/xb = a'b'/ab\}$ is also a circle. The two intersection points of C_1 and C_2 are the centers ω^+ and ω^- of the similarities $f \in \mathrm{Sim}^+$ and $g \in \mathrm{Sim}^-$ taking (a, b) to (a', b'). The problem will be solved if we know how to draw C_1 and C_2. Here is the construction for C_1: find b'' such that

$$\langle a', b''\rangle // \langle a, b\rangle \quad \text{and} \quad a'b'' = a'b';$$

the point $c_1 = \langle a, a'\rangle \cap \langle b, b''\rangle$ is in C_1. Then find b''' such that b''' is in $\langle a, b\rangle$ and $\langle b', b'''\rangle // \langle a', b\rangle$; the point d_1 such that $d_1 \in \langle a, a'\rangle$ and $\langle b, d_1\rangle // \langle b''', a'\rangle$ is also in C_1. The construction of C_2 is analogous. □

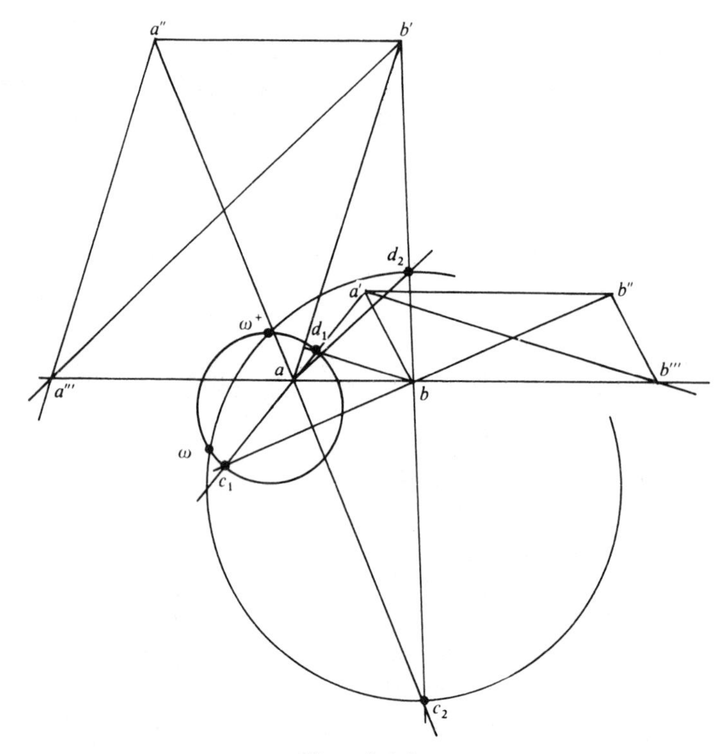

Figure 9.4.2.

c) Construction using the complex numbers

The analytical expression of similarities in terms of complex numbers is so simple that one is tempted to try to use them to find the centers of the similitudes taking (a, b) into (a', b'). The constructions obtained by using this point of view, however, are much more complicated and less geometrical than the preceding ones.

If we put $a=0$, $b=1$, $f(z)=\alpha z+\beta$ and $g(z)=\alpha\bar{z}+\beta$, we find that $a'=\beta$ and $b'=\alpha+\beta$. The center ω^+ is given by $\omega^+=a\omega^+ +\beta$, whence $\omega^+=\beta/(1-\alpha)$, whereas ω^- is given by

$$\omega^-=\alpha\overline{\omega}^-+\beta, \quad \text{whence} \quad \overline{\omega}^-=\bar{\alpha}\omega^-+\bar{\beta} \quad \text{and} \quad \omega^-=\alpha(\bar{\alpha}\omega^-+\bar{\beta})+\beta,$$

which finally gives $\omega^-=(\alpha\bar{\beta}+\beta)/(1-\alpha\bar{\alpha})$. So the problem boils down to the following: given the points 0, 1, β and $\alpha+\beta$, find the points $\beta/(1-\alpha)$ and $(\alpha\bar{\beta}+\beta)/(1-\alpha\bar{\alpha})$. We show the solution in the figures, without a step-by-step explanation. □

9.5 The detour is typical of this kind of question. We build an arbitrary square $\{a, b, c, d\}$ such that a lies on $\{x, y\}$ and b lies on $\{x, z\}$, and so that $\langle a, b\rangle$ is parallel to $\langle y, z\rangle$. Denoting by c', d' the intersection points of $\langle y, z\rangle$ with $\langle x, d\rangle$ and $\langle x, c\rangle$, respectively, the homothety of center x and ratio

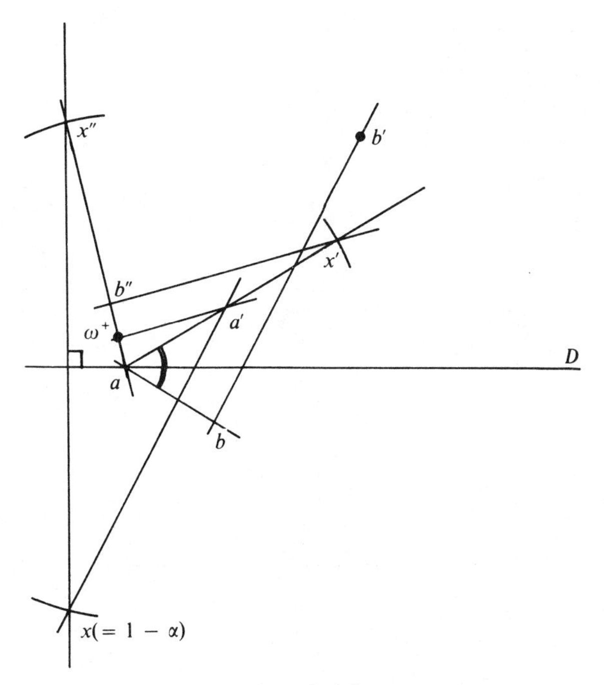

Figure 9.4.3.

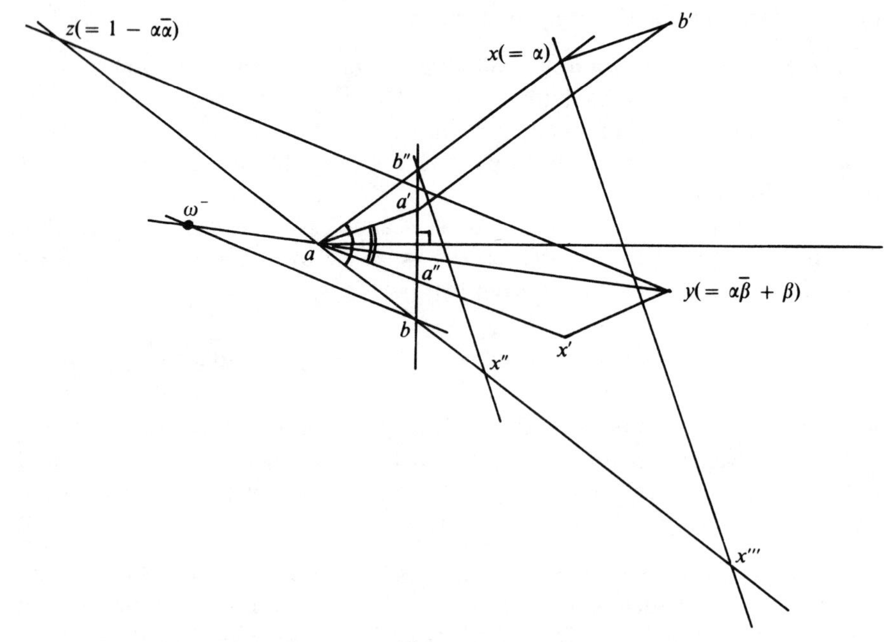

Figure 9.4.4.

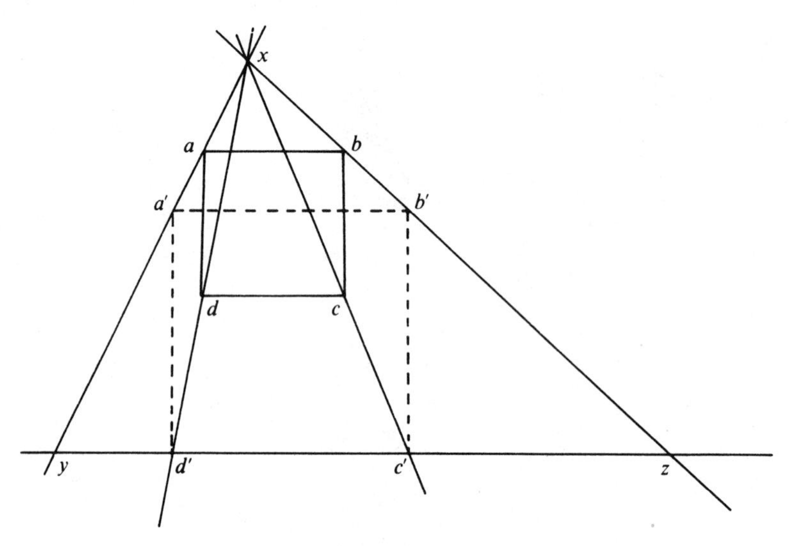

Figure 9.5.2.

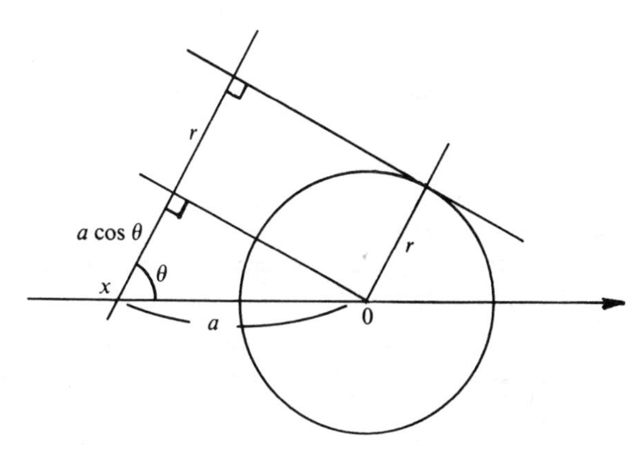

Figure 9.6.2.

$xd'/xd = xc'/xc$ will transform the square $\{a, b, c, d\}$ into a new square $\{a', b', c', d'\}$ satisfying the conditions of the problem. □

Compare with [B, 9.6.6].

9.6 Denoting by $a = xO$ the distance from the point x to the center O of the circle C of radius r, the equation of the polar line of x with respect to C is $\rho = a\cos\vartheta + r$, in polar coordinates with center at x and axis $\overrightarrow{xO}$.

The classical formula that gives the sign of the concavity of a curve $\rho = f(\vartheta)$ with respect to the pole is

$$f^2 + 2f'^2 - ff''.$$

Here we get the function $2a^2 + 3ar\cos\vartheta + r^2$. Thus our pedal curves will always be concave as seen from point x, except for such angles ϑ as satisfy $2a^2 + 3ar\cos\vartheta + r^2 < 0$. This cannot happen unless

$$\cos\vartheta = -\frac{2a^2 + r^2}{3ar} > -1.$$

But

$$\frac{2a^2 + r^2}{3ar} - 1 = \frac{r}{3a}\left(\frac{2a}{r} - 1\right)\left(\frac{a}{r} - 1\right),$$

so this case occurs only when $r/2 < a < r$, and ϑ lies in an interval $]-\pi - \vartheta_0, -\pi + \vartheta_0[$.

This is natural from an intuitive point of view. For $a = r$, the curve is a cardioid, which has a cusp, and for a very small, the pedal curve is almost a circle, so its concavity cannot change. □

Chapter 10

10.1 (i) We denote by a, b and c the vertices of T, and by a', b' and c' those of T'; since $a'bac$ is a parallelogram, we have $a'b = ca$ and $a'c = ba$. Repeating the argument for the other points, we prove that c is the midpoint of $a'b'$, b is the midpoint of $a'c'$ and a is the midpoint of $b'c'$. Each height of T is equidistant from the corresponding two vertices of T', so the fact that the three heights are concurrent is a consequence of the existence of a point equidistant from all three vertices of T'. □

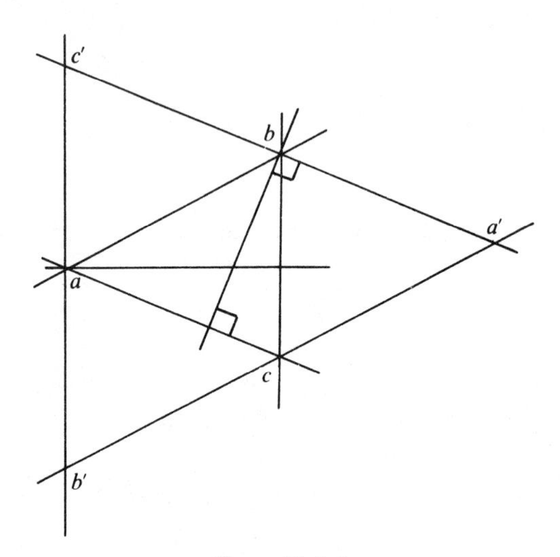

Figure 10.1.1.

(ii) We write $\overrightarrow{ab} = \overrightarrow{ac} + \overrightarrow{cb}$, and use the fact that the scalar product is bilinear to obtain $(\overrightarrow{ab}|\overrightarrow{cd}) = (\overrightarrow{ac}|\overrightarrow{cd}) + (\overrightarrow{cb}|\overrightarrow{cd}) = (\overrightarrow{ac}|\overrightarrow{cd}) + (\overrightarrow{dc}|\overrightarrow{bc})$, whence

$$(\overrightarrow{ab}|\overrightarrow{cd}) + (\overrightarrow{ac}|\overrightarrow{db}) + (\overrightarrow{ad}|\overrightarrow{bc}) = (\overrightarrow{ac}|\overrightarrow{cb}) + (\overrightarrow{ac}|\overrightarrow{bc}) = 0.$$

So now let a, b and c be the three vertices of a triangle T, and let d be the intersection of the heights dropped from a and b; we have $(\overrightarrow{ad}|\overrightarrow{bc}) = (\overrightarrow{bd}|\overrightarrow{ac}) = 0$, whence $(\overrightarrow{ab}|\overrightarrow{cd}) = 0$, which means that the line $\langle c, d \rangle$ is the height dropped from c, thus showing that the three heights are concurrent. □

(iii) Denote the vertices of T by a, b and c, and the feet of the heights dropped from a, b and c by a', b' and c', respectively. Then we get

$$\frac{\overrightarrow{a'b}}{\overrightarrow{a'c}} = \frac{\tan \widehat{aa'ab}}{\tan \widehat{aa'ac}},$$

whence

$$\frac{\overrightarrow{a'b}}{\overrightarrow{a'c}} \cdot \frac{\overrightarrow{b'c}}{\overrightarrow{b'a}} \cdot \frac{\overrightarrow{c'a}}{\overrightarrow{c'b}} = \frac{\tan \widehat{aa'ab} \cdot \tan \widehat{bb'bc} \tan \widehat{cc'ca}}{\tan \widehat{aa'ac} \cdot \tan \widehat{bb'ba} \tan \widehat{cc'cb}}.$$

The triangles $a'ac$ and $b'bc$ are similar because they share an angle at c and they both have one right angle; this implies $\widehat{aa'ac} = \widehat{bcbb'} = -\widehat{bb'bc}$. Since the tangent function is odd, we get $\overrightarrow{a'b}/\overrightarrow{a'c} \cdot \overrightarrow{b'c}/\overrightarrow{b'a} \cdot \overrightarrow{c'a}/\overrightarrow{c'b} = (-1)^3 = -1$. The theorem of Ceva (cf. 2.1) completes the proof. □

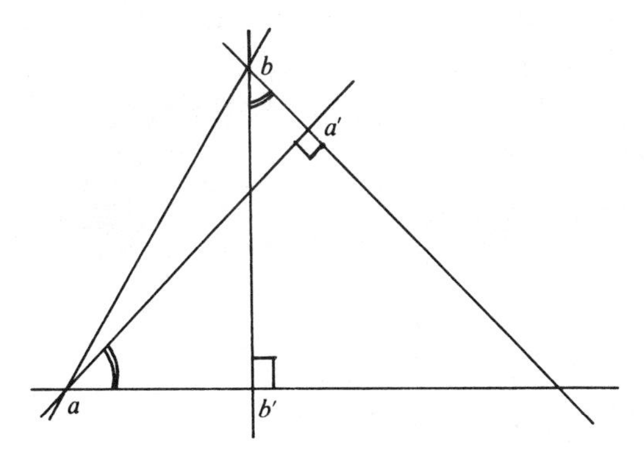

Figure 10.1.2.

(iv) Let a, b and c be the vertices of the triangle T, and let d be the intersection of the heights dropped from a and b. Then $\langle a, d\rangle \cup \langle b, c\rangle$ and $\langle b, d\rangle \cup \langle a, c\rangle$ are two conics which generate the pencil $\mathscr{F}$ of all conics passing through $\{a, b, c, d\}$. The line at infinity ∞ is proper for $\mathscr{F}$, so the theorem of Desargues (cf. 16.F) can be applied: there is an involution f of ∞ such that for any conic α in $\mathscr{F}$ the intersection of α with ∞ consists of a point m and its image $f(m)$. Since $\langle a, d\rangle \perp \langle b, c\rangle$, the two points at infinity of the conic $\langle a, d\rangle \cup \langle b, c\rangle$, which we'll denote by m and $f(m)$, satisfy $[m, f(m), I, J] = -1$ (where I and J are the cyclical points), and so do the points at infinity of the conic $\langle b, d\rangle \cup \langle a, c\rangle$. An involution is determined by its action on two points, so f can be defined by the formula $[m, f(m), I, J] = -1$ (cf. 6.D or [B, 6.7.2, 6.7.4]). The conic $\langle c, d\rangle \cap \langle a, b\rangle$ belongs to $\mathscr{F}$ since it passes through a, b, c, d; this shows that the third height also passes through d, completing the proof. □

Refer also to the solution of problem 10.3.

10.2 First part. The relation $R^2 - 2Rr = d^2$ is trickier to prove than it might seem, although it is nothing too deep. Suppose that the angles A, B, C of the triangle $\mathscr{T} = \{x, y, z\}$ are such that $B \geq C$. Recall first how to find the tangency points of the circle inscribed in $\mathscr{T}$: the lengths u, v, w in figure 10.2.1 satisfy $u + w = b$, $w + v = a$, $v + u = c$, where a, b, c are the sides of $\mathscr{T}$. With the notation $p = (a + b + c)/2$, we have $u = (p - a)/2$, $v = (p - b)/2$ and $w = (p - c)/2$.

Denoting by ω (resp. O) the center of the circle inscribed in $\mathscr{T}$ (resp. circumscribed around $\mathscr{T}$), we have

$$x\omega = \frac{p-a}{\cos A/2}, \qquad r = (p-a)\tan\frac{A}{2}.$$

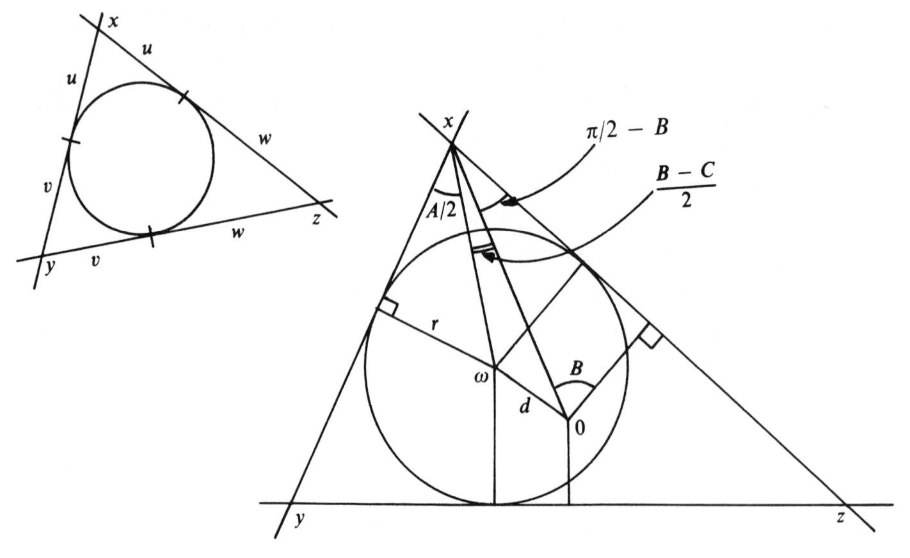

Figure 10.2.1.

The angle at x of the triangle $\{x, \omega, O\}$ has the value $(B-C)/2$, as can be seen from the figure (use $A+B+C=\pi$ and the fact that the central angle is twice the inscribed angle, 10.D or [B, 10.9.3]). We apply the fundamental relation in 10.A (or [B, 10.3.1]) to calculate $d=\omega O$:

$$d^2=(\omega O)^2=(x\omega)^2+(xO)^2-2(x\omega)(xO)\cos\frac{B-C}{2}$$

$$=R^2+\frac{(p-a)^2}{\cos^2 A/2}-2\frac{R(p-a)}{\cos A/2}\cos\frac{B-C}{2}.$$

This implies that the relation we want to show is equivalent to

$$\frac{2R(p-a)}{\cos A/2}\cos\frac{B-C}{2}-\frac{(p-a)^2}{\cos^2 A/2}=2Rr=2R(p-a)\tan\frac{A}{2},$$

or again to

$$2R\cos\frac{B-C}{2}\cos\frac{A}{2}-(p-a)-2R\sin\frac{A}{2}\cos\frac{A}{2}=0.$$

This is equivalent, given the formulas in 10.A (or [B, 10.3.2]) and the value of p, to the equality

$$2R\cos\frac{B-C}{2}\cos\frac{A}{2}-(\sin B+\sin C-\sin A)R-R\sin A=0,$$

$$2\cos\frac{B-C}{2}\cos\frac{A}{2}-\sin B-\sin C=0,$$

which follows directly from $A=\pi-(B+C)$. □

Second part. Now let C, C' be two circles whose radii R, r satisfy the relation $R^2 - 2Rr = d^2$, where d is the distance between the two centers. Take any point x in C. We have to show that we can determine two other points y, z such that the triangle $\{x, y, z\}$ is inscribed in C and circumscribed around C'. Draw the two tangents to C' from x; they exist because the relation $R^2 - 2Rr = d^2$ implies that $d < R - r$, so C' is inside C. Let y and z be the points where these two tangents intersect C. We have to show that $\langle y, z\rangle$ is tangent to C'.

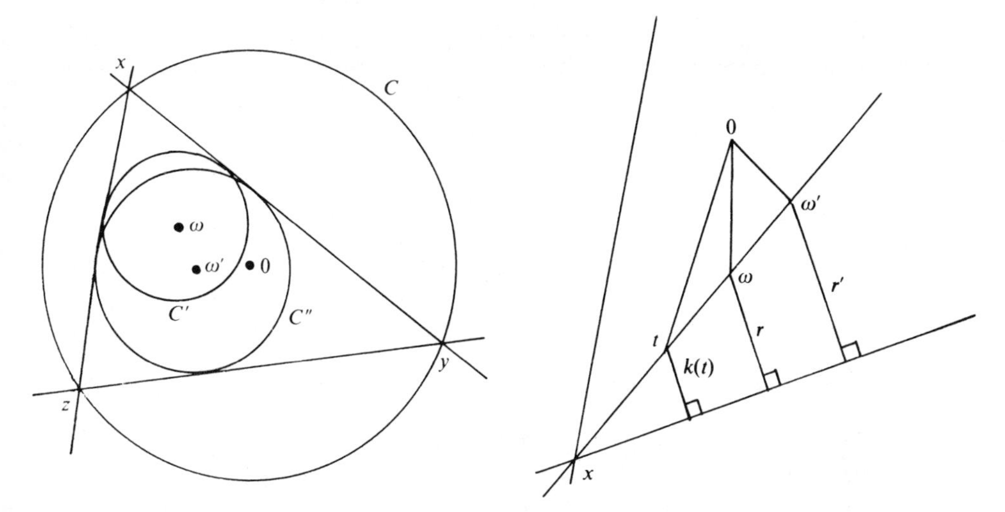

Figure 10.2.2.

The solution method we will use is typical; we utilize the necessary condition proved above and a uniqueness argument. Here, however, things are a bit more delicate. The circle C'' inscribed in $\{x, y, z\}$ has radius r', and the distance $d' = \omega' O$ from its center to the center O of C satisfies $R^2 - 2Rr' = d'^2$.

The centers ω, ω' are on the bisector D of the angle at x. Given a point t in D, denote by $k(t)$ its distance to $\langle x, z\rangle$ (equal to its distance to $\langle x, y\rangle$). This function is affine along D, so in particular the function $t \mapsto R^2 - 2Rd(t) - (tO)^2$ is of second degree; it must have at most two zeros. It vanishes at ω, evidently, but also at $t = x$; this shows that $\omega' = \omega$. □

10.3 Let $\mathscr{T} = \{a, b, c\}$ be the triangle in question; given a line D on the plane, denote by D' (resp. D'', D''') its reflection through the side $\langle b, c\rangle$ (resp. $\langle c, a\rangle, \langle a, b\rangle$) of $\mathscr{T}$. We shall say that the line D is *good* if these three reflections are concurrent. The name "line of the images" is justified in the following way: if $x = D' \cap D'' \cap D'''$, this implies that the reflections x', x'', x''' of x through the three sides of $\mathscr{T}$ are collinear points (lying on D, of course). If we consider the three sides of $\mathscr{T}$ as mirrors, the three images of x will be aligned on the "line of images".

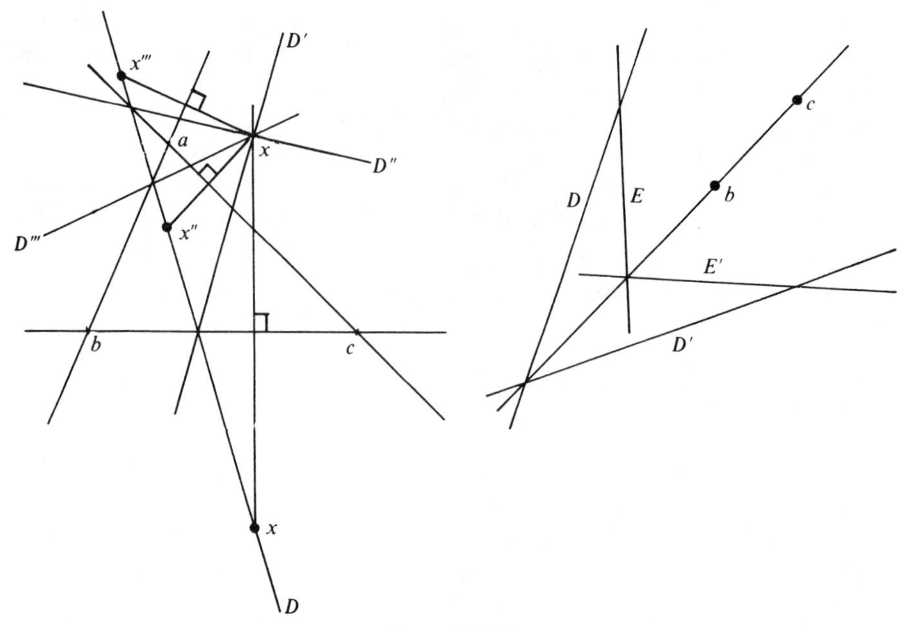

Figure 10.3.1.

The idea now is to use two arbitrary lines D, E of the plane and observe that the three oriented angles (cf. 8.F) $\widehat{D'E'}$, $\widehat{D''E''}$ and $\widehat{D'''E'''}$ are equal, since they all have the value $-\widehat{DE}$ by 8.F or [B, 8.7.7.3]. Then we apply the condition on the oriented angles formed by four cocyclic points (10.D or [B, 10.9.5]). Now suppose that D and E are both good, and their reflections intersect at points x and y, respectively. The reasoning above shows that the five points x, y, $D' \cap E'$, $D'' \cap E''$ and $D''' \cap E'''$ lie on the same circle. We obtain the following

Lemma. *If D and E are both good, any line passing through $D \cap E$ is also good.*

This is because the technique above shows that for any line F passing through $D \cap E$, the reflections F', F'', F''' pass through a point z on the circle defined by the five points we mentioned. This point z is well-determined, by the angle $\widehat{D'F'}$ for instance. □

Now it suffices to remark that not all lines in the plane can be good, and that we know three good lines already—the heights of the triangle. This means that the good lines are exactly those passing through the intersection point of the heights (in particular we recover the result that the three heights are concurrent).

The cocyclicity condition, mentioned above, implies that the points $D' \cap D'' \cap D'''$, for all good lines D, lie on the same circle. This circle must be circumscribed around the triangle, since if D is the height dropped from a, for

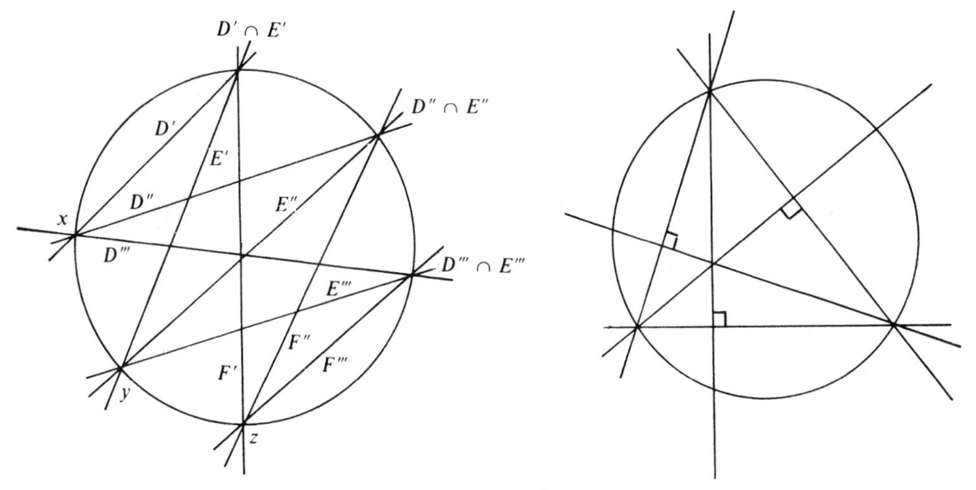

Figure 10.3.2.

instance, the intersection point of the three reflections is a itself. Conversely, every point of the circle is the intersection of the reflections of a good line. □

Bonus. We obtain three results for free. The first is the Simson line (cf. 10.D or [B, 10.9.7.1]). In fact, in order for the projections of a point x on the sides of a triangle to be collinear, it is necessary and sufficient that their images by a homothety of center x and ratio 2 be collinear. But this is exactly the line of images.

The second result is that the points obtained by reflection of the orthocenter through the sides of a triangle belong to the circle circumscribed around the triangle.

The third is that the Simson line, after a homothety of center x and of ratio 2, passes through the orthocenter.

10.4 The idea in both cases is that there is an obvious candidate for the absolute minimum, namely the intersection point of the diagonals. In the case of a convex quadrilateral, this point clearly exists; call it x. For any point z in the plane, we have

$$za + zb + zc + zd = (za + zc) + (zb + zd) \geq ac + bd.$$

Because of the strict triangle inequality, we can only have equality if z belongs to both the segment $[a, c]$ and the segment $[b, d]$, i.e. if $z = x$. Thus this point is the unique absolute minimum for the function

$$z \mapsto za + zb + zc + zd$$

giving the sum of the distances to the four vertices of a convex quadrilateral. □

To deal now with the hexagon circumscribed around an ellipse, it is enough to apply the theorem of Brianchon, which asserts that the three diagonals of

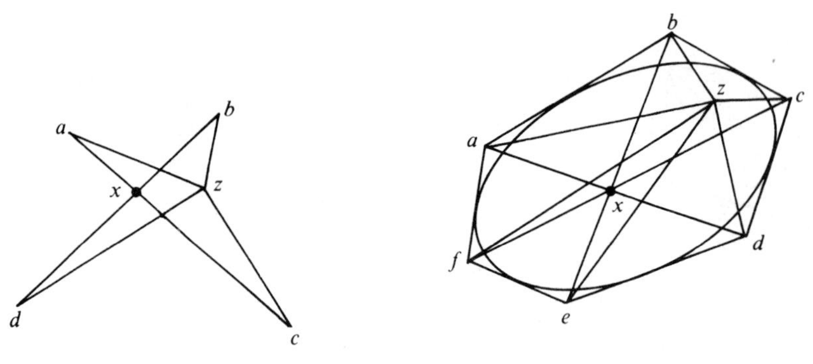

Figure 10.4.1.

such a hexagon are concurrent (cf. 16.C or [B, 16.2.13]). Letting x be the intersection point, and z an arbitrary point in the plane, we have

$$za + zb + zc + zd + ze + zf \geq ad + be + cf,$$

and equality holds only if $z = x$. □

REMARK. The condition that the "first variation" (cf. 9.G or [B, 9.10]) be zero at the point where the minimum is achieved is trivially verified in both cases, since at x the unit vectors $\overrightarrow{xa}/\|\overrightarrow{xa}\|$, $\overrightarrow{xb}/\|\overrightarrow{xb}\|$, $\overrightarrow{xc}/\|\overrightarrow{xc}\|$, $\overrightarrow{xd}/\|\overrightarrow{xd}\|$, for example, are arranged in opposite pairs.

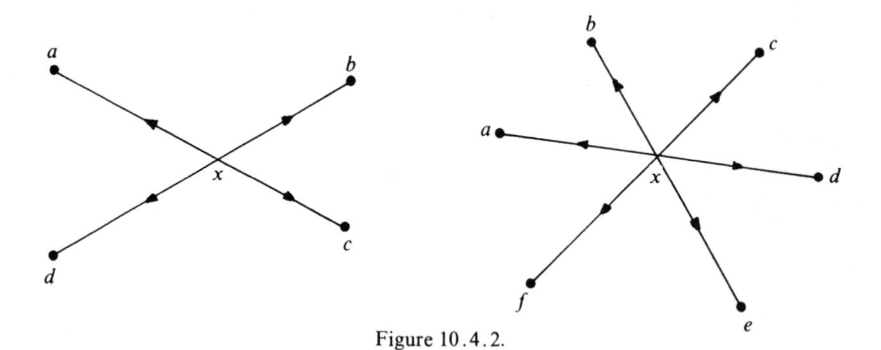

Figure 10.4.2.

10.5 The first idea is that, since the angle $\overline{Ad, Aa}$ has the value $\pi/2$, the unknown point A belongs to the circle Γ whose diameter is $[a, d]$. The same remark applies to the other vertices; unfortunately this is not enough. Next we investigate the diagonal $\Delta = \langle A, C \rangle$ of the square. In fact, the fact that the angle $\overline{\Delta, Aa}$ is equal to $\pi/4$ shows that Δ intersects Γ in a well-known point, the midpoint of the arc (a, d) of Γ which does not contain A (see the properties of inscribed angles, cf. 10.D or [B, 10.9]).

This gives the following construction, assuming the quadrilateral (a, b, c, d) is convex: draw the circles Γ and Γ' whose diameters are $[a, d]$ and $[b, c]$, and on the arcs (a, d) and (b, c) turned towards the interior of the quadrilateral locate the midpoints x and y. The diagonal Δ we are looking for is the line $\langle x, y\rangle$. The vertex A will be the second intersection point of Δ with the circle Γ, and so on. □

This construction doesn't always work: the angles of (a, b, c, d) shouldn't be too small, for instance. Sometimes the solution is not unique, e.g. when the (a, b, c, d) is itself a square.

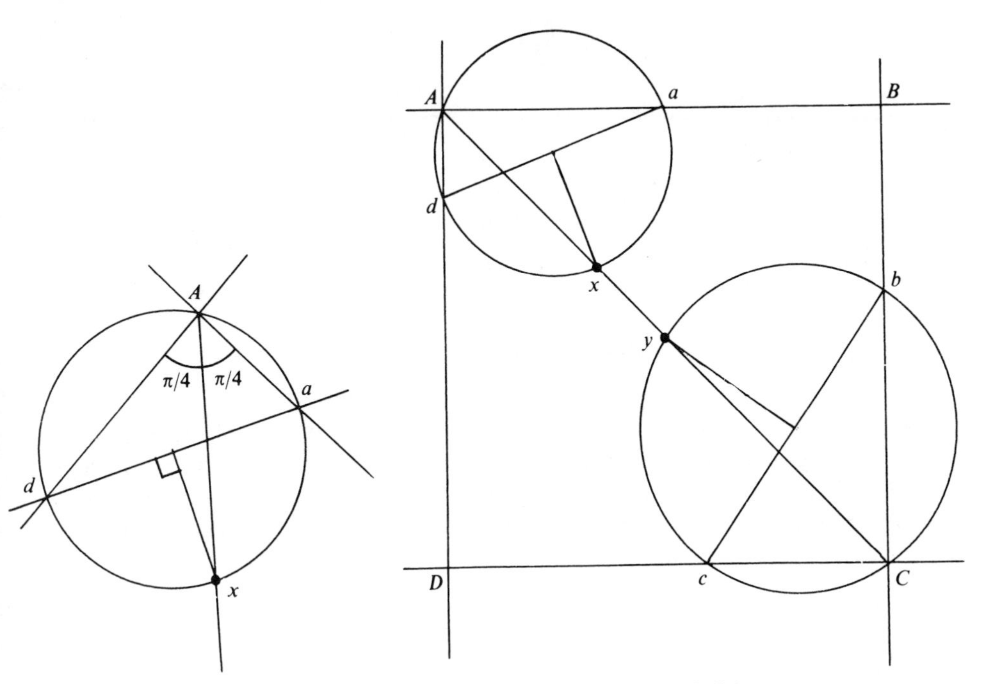

Figure 10.5.1.

Figure 10.5.2.

10.5 a) Construction of the image of a point by an inversion whose center i and fixed circle I are given

We'll give the construction of the image of a point m such that $im > r/2$, where r is the radius of I; this case is enough for what follows. Let c_1 and c_2 be the intersections of I with the circle of center m and passing through i, and let m' be the intersection of the two circles of centers c_1 and c_2 that pass through i (the points i, m and m' are collinear for reasons of symmetry). To see that m' is the image of m, one can compute the area of the triangle ic_1m':

$$A = \left(c_1 i \cdot c_1 m' \cdot \sin \overline{c_1 i, c_1 m'}\right)/2 = \left(r^2 \sin \overline{c_1 i, c_1 m'}\right)/2$$

and also

$$A = \left(im' \cdot mc_1 \cdot \sin \overline{mi, mc_1}\right)/2 = \left(im \cdot im' \cdot \sin \overline{mi, mc_1}\right)/2;$$

but the two angles in question are equal because imc_1 and ic_1m' are similar (being isosceles and having one angle in common at i), whence $im \cdot im' = r^2$. □

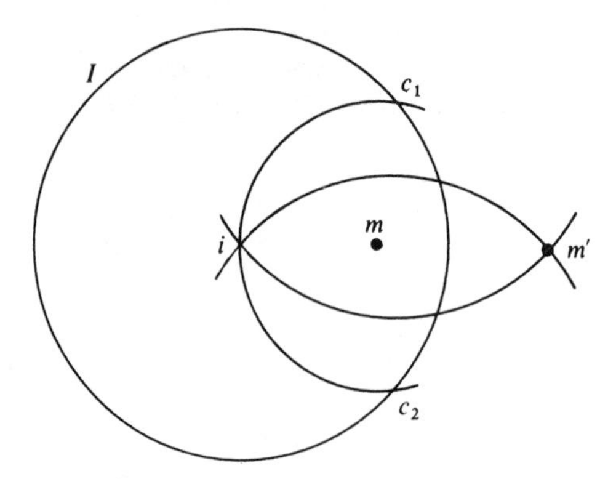

Figure 10.6.1.

b) Construction of the center of a given circle

Let C be the given circle, with center c (not given!). Choose a point i on C, and let I be a circle of center i intersecting C at c_1 and c_2. The construction in a) shows that the image c' of c by the inversion of center i and fixed circle I is the intersection of the circles of centers c_1 and c_2 and passing through i. We finish off by finding the image of c' (again using the same method) to recover c. □

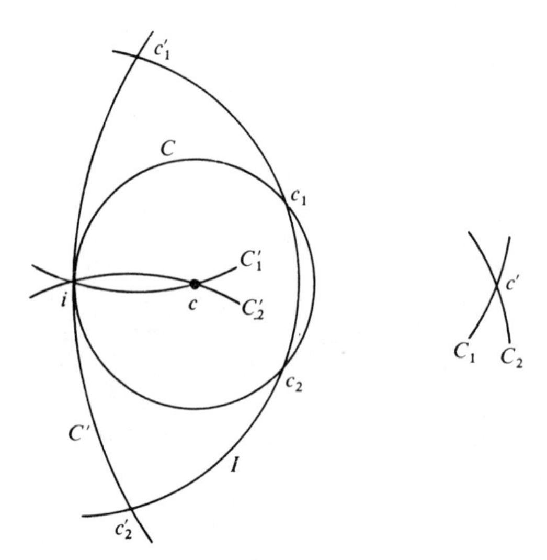

Figure 10.6.2.

c) Theorem of Mohr-Mascheroni

In geometric constructions using the ruler and the compass, we deal with two sorts of figures: circles (which can evidently be drawn with the compass alone) and straight lines, which are constructed from two of their points. We have to show that we can find the intersection points of such figures using only the compass. We start with a lemma.

Lemma. *Given three points, we can construct the circle that contains them using only the compass.*

Proof. Let i, j, k be the three points. Draw the circles I with center i and radius jk, J with center j and radius ik, and K with center k and radius ij. Then the points c_1 and c_2 shown in the figure lie on the circle that contains i, j and k. Consider now the inversion of center i and fixed circle I: the image c' of the desired center is in $C_1 \cap C_2$, where C_n is the circle of center c_n and passing through i. From this we can find the center c (cf. a)). □

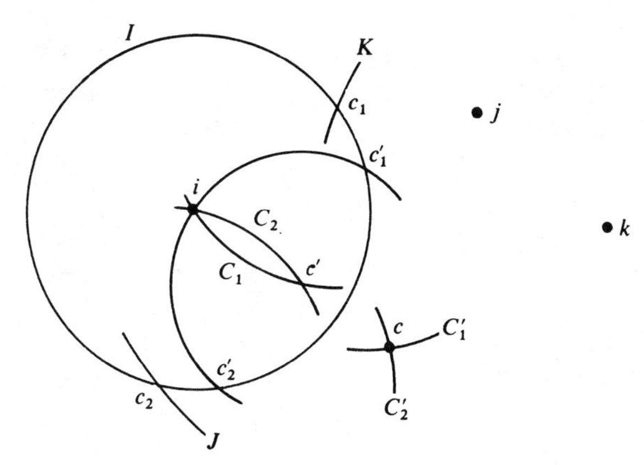

Figure 10.6.3.

Once we have this lemma, we prove the theorem as follows: let F_1 and F_2 be two figures (circles or lines), i_2 a point in the plane, I a circle of center i, and φ the inversion with center i and fixed circle I. Then we can construct $\varphi(F_1)$ and $\varphi(F_2)$ using the compass only: if F_n is a line $\langle a, b\rangle$, we find the images $\varphi(a)$ and $\varphi(b)$ of a and b, and $\varphi(F_n)$ will be the circle passing through $\varphi(a)$, $\varphi(b)$ and i; if F_n is a circle, we find the images of three of its points, and $\varphi(F_n)$ will be the circle passing through these three images. In this way we can obtain the points $\varphi(F_1) \cap \varphi(F_2)$ which are exactly the images under φ of the points of $F_1 \cap F_2$; we conclude by applying φ again to obtain $F_1 \cap F_2$. □

10.7 Let's first state the sequence of theorems to be shown:

Th_{2n}: Let $(D_i)_{i=1,\dots,2n}$ be $2n$ lines in a Euclidean plane, and let C_i be the circle associated with the family $(D_j)_{j\neq i}$ by theorem Th_{2n-1}; then the $2n$ circles C_i have a point p in common.

Th_{2n+1}: Let $(D_i)_{i=1,\dots,2n+1}$ be $2n+1$ lines in a Euclidean plane, and let p_i be the point associated with the family $(D_j)_{j\neq i}$ by theorem Th_{2n}; then the $2n+1$ points p_i lie on the same circle C.

We start with theorem Th_2, where we replace the assumption "C_i is the circle associated with D_j $(j\neq i)$ by theorem Th_1" by the assumption "C_i is the generalized circle $D_j(j\neq i)$" (cf. 20.A or [B, 20.1.5]). Then the theorems Th_2 and Th_3 are trivial: the first says that two lines "in general position", i.e. intersecting, have a point in common, and the second says that the three vertices of a triangle belong to the same circle (the circumscribed circle). Theorems Th_4 and Th_5 are those stated in the text. We shall show all theorems by induction.

a) Proof of theorem Th_{2n}

Denote by C_i the circle given by theorem Th_{2n-1} for the lines $(D_j)_{j\neq i}$, by $p_{i,j}$ the point given by theorem Th_{2n-2} for the lines $(D_k)_{k\notin\{i,j\}}$ by $C_{i,j,k}$ the circle given by theorem Th_{2n-3} for the lines $(D_l)_{l\notin\{i,j,k\}}$ etc. We know that C_1 and C_2 pass through the point $p_{1,2}$, so these two circles intersect at another point p; we'll show that all the circles C_i also contain this point.

Suppose we start with at least four lines D_1, D_2, D_i and D_j, and consider the four circles C_1, C_2, $C_{2,i,j}$ and $C_{1,i,j}$. We have

$$C_1\cap C_2=\{p_{1,2},p\},\quad C_2\cap C_{2,i,j}=\{p_{2,j},p_{2,i}\},$$
$$C_{2,i,j}\cap C_{1,i,j}=\{p_{1,2,i,j},p_{i,j}\}\quad\text{and}\quad C_{1,i,j}\cap C_1=\{p_{1,j},p_{1,i}\};$$

but $p_{1,2}$, $p_{2,j}$, $p_{1,2,i,j}$ and $p_{1,j}$ are cocyclic, since they all lie on $C_{1,2,j}$, so, by the theorem of the six circles (10.D or [B, 10.9.7.2]), the points p, $p_{2,i}$, $p_{i,j}$ and $p_{1,i}$ are cocyclic. This implies $p\in C_i$, since C_i is the only circle containing $p_{2,i}$, $p_{i,j}$ and $p_{1,i}$.

□

b) Proof of theorem Th_{2n+1}

We use the same notation as in *a*): p_i, $C_{i,j}$, etc. Points p_1, p_2 and p_3 determine a circle C; we show that all the points p_i lie on this circle. Suppose there are at least four lines D_1, D_2, D_3 and D_4 to start with, and consider the four circles $C_{1,2}$, $C_{1,3}$, $C_{3,i}$ and $C_{2,i}$. We have

$$C_{1,2}\cap C_{1,3}=\{p_{1,2,3},p_1\},\quad C_{1,3}\cap C_{3,i}=\{p_{1,3,i},p_3\},$$
$$C_{3,i}\cap C_{2,i}=\{p_{2,3,i},p_i\}\quad\text{and}\quad C_{2,i}\cap C_{1,2}=\{p_{1,2,i},p_2\};$$

but $p_{1,2,3}$, $p_{1,3,i}$, $p_{2,3,i}$ and $p_{1,2,i}$ are cocyclic, since they belong to $C_{1,2,3,i}$; so, by the theorem of the six circles, p_1, p_3, p_i and p_2 are cocyclic, implying that $p_i\in C$.

□

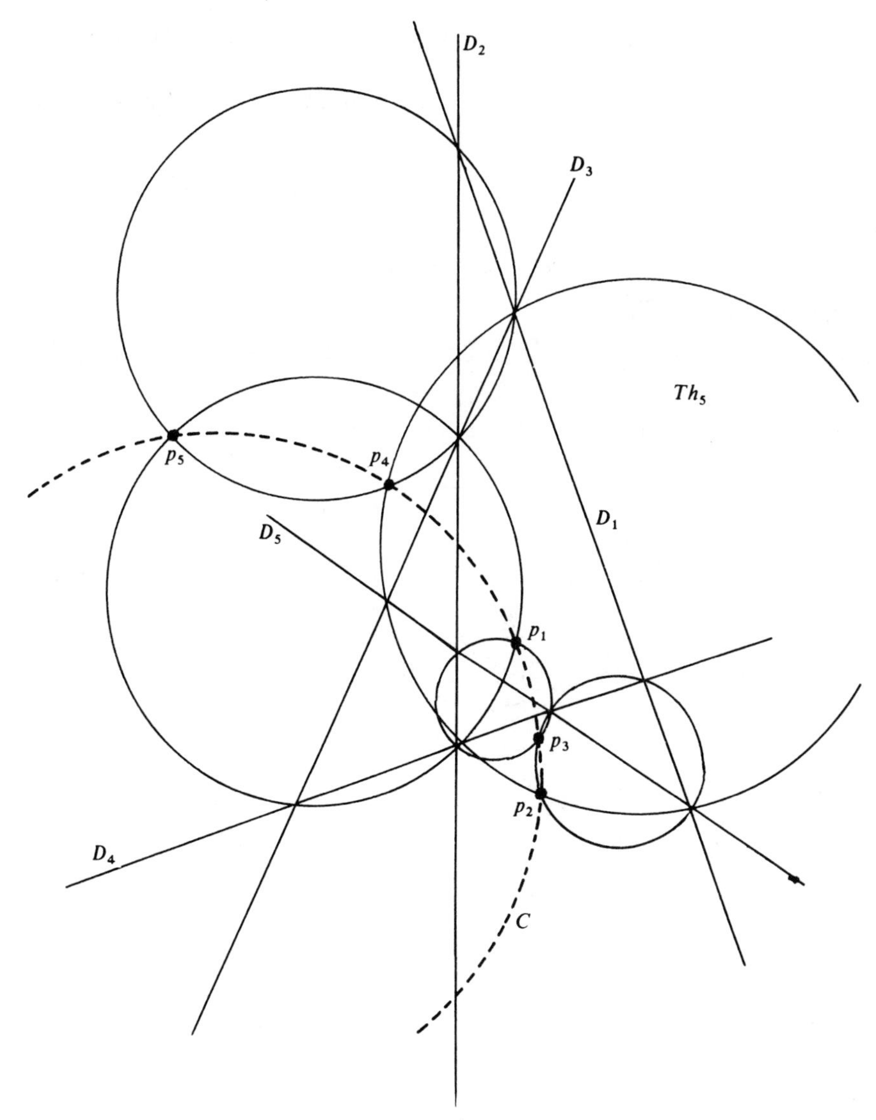

Figure 10.7.1.

c) Beginning of the induction

Observe first that in all of the above (including the theorem of the six circles) the word "cocyclic" stands for "cocyclic or collinear" (which justifies the proof in the case $2n+1=5$, where $C_{1,2,3,i}$ is a line). Because of this remark, we can see that to start the induction it is enough to prove theorems Th_2, Th_3 and Th_4. As mentioned at the beginning, theorems Th_2 and Th_3 are trivial. As for Th_4, it can be proved in the same way as the other theorems Th_{2n}, but replacing the theorem of the six circles by the following version, obtained by applying an inversion of pole c (see statement of the theorem in 10.4):

Let C_1 and C_2 be two circles and D_3 and D_4 two lines satisfying the conditions

$$C_1 \cap C_2 = \{a, a'\}, \quad C_2 \cap D_3 = \{b, b'\}, \quad D_3 \cap D_4 = \{c'\}$$

and

$$D_4 \cap C_1 = \{d, d'\};$$

then, if a, b, d are collinear, a', b', c', d' are cocyclic. □

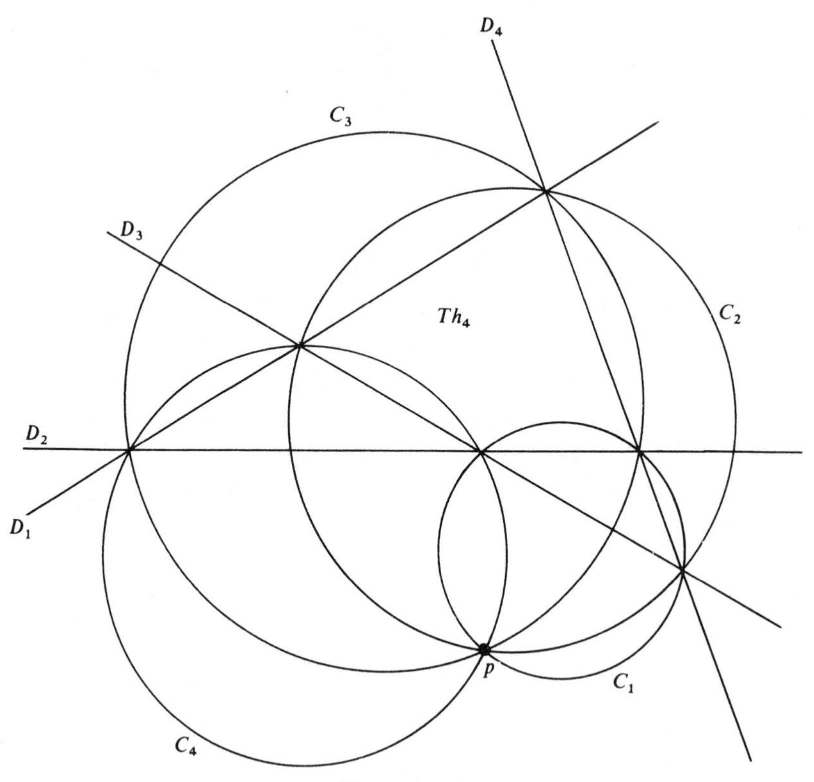

Figure 10.7.2.

REMARK. The careful reader will have remarked that our proof is not entirely correct, as can be seen from our remark in c) about the word "cocyclic". A detailed correction would require replacing all circles by generalized circles (cf. 20.A or [B, 20]) and using a generalized six-circle theorem.

10.8 a) Expression of ρ, ax and bx as functions of ab, r and s

Consider the shaded triangle below; the condition that the two circles tangent to D at a and x be tangent to one another can be expressed, by the

Pythagorean theorem, as $(r+\rho)^2 = ax^2 + (r-\rho)^2$, implying $ax^2 = 4\rho r$. Analogously, we get $xb^2 = 4\rho s$ and $ab^2 = 4rs$. Squaring the relation $ax + xb = ab$, on the other hand, gives $ax^2 + 2ax \cdot xb + bx^2 = ab^2$, whence $4\rho r + 2\sqrt{(4\rho r)(4\rho s)} + 4\rho s = 4rs$, or again $\rho(r + \sqrt{4rs} + s) = rs$; but $\sqrt{4rs} = ab$, so we finally get

$$\rho = rs/(r + ab + s).$$

Next we write $ax^2 = 4\rho r = 4r^2 s/(r + ab + s)$, multiply both numerator and denominator by $4r$ and replace $4rs$ by ab^2 everywhere, getting $ax^2 = 4r^2 \cdot ab^2/(4r^2 + 4ab \cdot r + 4rs)$, or

$$ax^2 = 4r^2 \cdot ab^2/(4r^2 + 4ab \cdot r + ab^2) = (2r \cdot ab/(2r + ab))^2,$$

or finally

$$ax = 2r \cdot ab/(2r + ab).$$

In the same way we obtain

$$bx = 2r \cdot ab/(2s + ab).$$

□

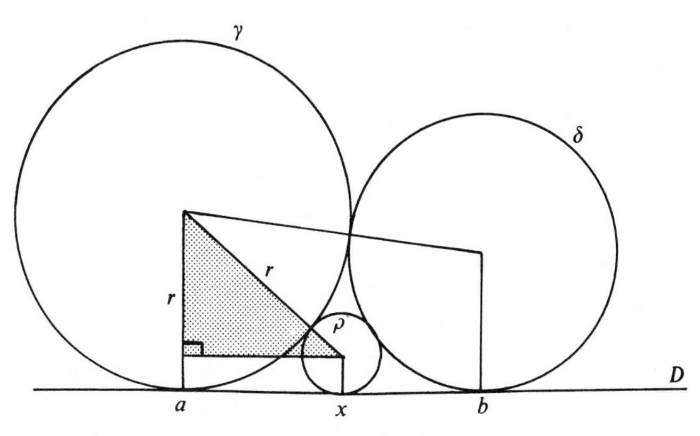

Figure 10.8.1.

b) The contact points of the Ford circles with the x-axis are rational

According to the formulas in a), the transformation from $\mathbf{R}^3$ into $\mathbf{R}^3$ defined by $(ab, r, s) \mapsto (\rho, ax, bx)$ maps $\mathbf{Q}^3$ into $\mathbf{Q}^3$. Since the two first circles have rational contact points with the line D (namely the points 0 and 1), this implies the same holds for all other circles. □

c) All rationals in [0, 1] are obtained in this way

As suggested by the hint, we show, by induction on l, that any point $i/l \in [0,1]$ (i and l relatively prime) is obtained as the tangency point of a Ford circle of radius $1/2l^2$. For $l = 1$, the points $i/l \in [0,1]$ are $0/1$ and $1/1$; they are obtained from the two original circles, which have radius $1/2$. For $l \geq 2$, the conditions $i/l \in [0,1]$ and i, l relatively prime imply $0 < i < l$, and consequently there are integers k and n such that $0 < k \leq i$, $0 < n < l$ and $kl - ni = 1$. Now put $j = i - k$ and $m = l - n$; we get

$$0 \leq (j/m) < (i/l) < (k/n) \leq 1$$

(the first inequality comes from the inequalities on k and n; the second and third come from $mi - jl = kl - ni = 1 > 0$; and the last follows from

$$k = (1 + ni)/l \leq (n + ni)/l = n((1 + i)/l) \leq n,$$

since $i < l$ implies $(1 + i)/l \leq 1$). Notice also that j and m are relatively prime, and so are k and n, since $mi - jl = kl - ni = 1$. Now since $0 < n < l$ and $0 < m < l$, we can apply the induction assumption to points a with abscissa j/m and b with abscissa k/n, obtaining a Ford circle γ of radius $r = 1/2m^2$ and tangent to D at a, and a Ford circle δ of radius $s = 1/2n^2$ and tangent to D at b. From a) we verify that γ and δ are tangent:

$$ab^2 = ((k/n) - (j/m))^2 = (km - nj)^2/n^2m^2 = 1/n^2m^2 = 4rs,$$

since

$$km - nj = kl - ni = 1.$$

Again using a), we calculate the radius and contact point of the small circle between γ and δ:

$$\rho = \left(4m^2n^2\left((1/2m^2) + (1/mn) - (1/2n^2)\right)\right)^{-1}$$
$$= 1/(2n^2 + 4mn + 2m^2) = 1/2(m+n)^2 = 1/2l^2,$$

and

$$ax = \left(m^3n\left((1/m^2) + (1/mn)\right)\right)^{-1} = 1/m(m+n) = 1/ml,$$

so the abscissa of x is equal to $(j/m) + (1/ml) = (jl + 1)/ml = mi/ml = i/l$. This shows the point x with abscissa i/l is obtained as a Ford circle of radius $1/2l^2$, completing the induction step. □

Chapter 11

11.1 We assume the two sets A, B of the partition are non-empty. Since there are only two of them, they have the same boundary

$$F = \mathrm{Fr}(A) = \bar{A} \setminus \mathring{A} = \mathrm{Fr}(B) = \bar{B} \setminus \mathring{B}$$

(the boundary of a set is the difference between its closure and its interior). Moreover $F = \bar{A} \cap \bar{B}$, so the boundary is convex, since the closure of a convex set is convex (cf. 11.A or [B, 11.2.1]), and so is the intersection of two convex sets.

The boundary of a set has empty interior, so F is a convex set with no interior. According to 11.A (or [B, 11.2]), this means the dimension of F (see 11.A) must be 0 or 1, i.e. F is either a point or a convex subset of a line D of the plane, hence an interval of D.

Let's first rule out the case of a point. According to the above, A cannot have empty interior; take $a \in \mathring{A}$ and an interval $[b, b']$ contained in B (which can be found, since B cannot consist of a single point). For any $x \in [b, b']$, the segment $[a, x]$ contains a point of the boundary of $[a, x] \cap A$ in $[a, x]$, and such a point is also in F; in particular, F must have more than one point.

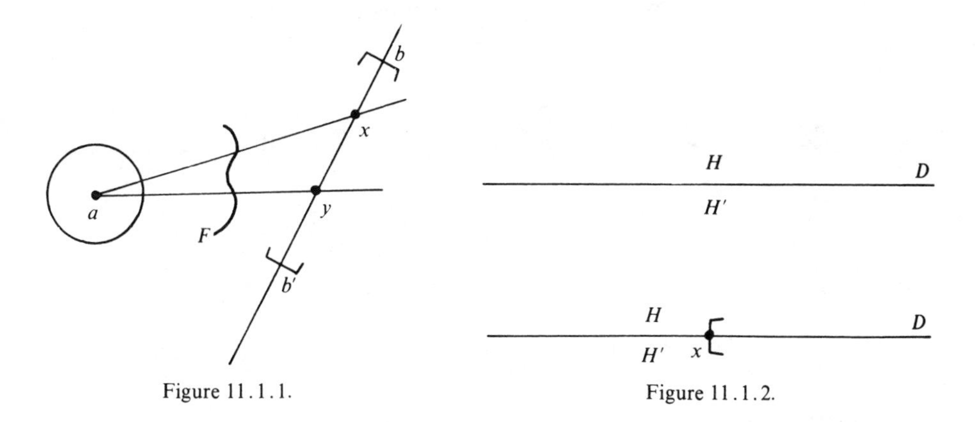

Figure 11.1.1.

Figure 11.1.2.

So far we know that the boundary F of A and B is a non-degenerate interval of a certain line D of the plane. But $A \cap D$ and $B \cap D$ are again convex, and they form a partition of D; this implies (relabeling A and B if necessary) that either $A \cap D = D$ or $A \cap D$ is a closed half-line whose origin is a point x of D. In either case $F = D$.

Finally, let H, H' be the two open half-planes determined by D. Again relabeling H and H' if necessary, we claim that $H \subset A$ and $H' \subset B$. This is again a consequence of the reasoning above: if $a \in A$, $b \in B$, the segment $[a, b]$ necessarily contains a point of $\mathrm{Fr}_{[a,x]}(A \cap [a, x]) \subset F = D$. Thus a, b belong to two distinct half-planes, and points of both A and B cannot lie in the same half-plane.

This leads to the following conclusion: there are only two possible types of partitions of a plane into two non-empty convex sets. The first is when A is a closed half-plane defined by a line D (and B is its complementary open half-plane); the second is when A is an open half-plane together with a closed half-line. □

11.2 To put this result into the proper light, recall that the set of extremal points of a convex set, even a compact convex one, is not closed in general; there are counterexamples already in dimension 3.

Let C be a closed convex set and (x_i) a sequence of extremal points of C, converging towards a point x in the plane; this point x belongs to the boundary of C, since the boundary is a closed set. Find a line of support D at x for the convex set C, and suppose that x is not extremal. Then by definition (11.C or [B, 11.6.4]) there are two points y and z in C with $x = (y + z)/2$ and

$y \neq z$. Notice first that y and z must necessarily belong to D, otherwise they would be on two different sides of D and D would not be a line of support (11.B or [B, 11.5]). Thus the segment $[y, z]$ is contained in D and in the boundary of C. Moreover the boundary C must coincide with $[y, z]$ in a neighborhood of x, since its interior is empty.

Since the sequence (x_i) converges towards x, for large values of i the points x_i will lie in the open segment $]y, z[$, contradicting their extremality. □

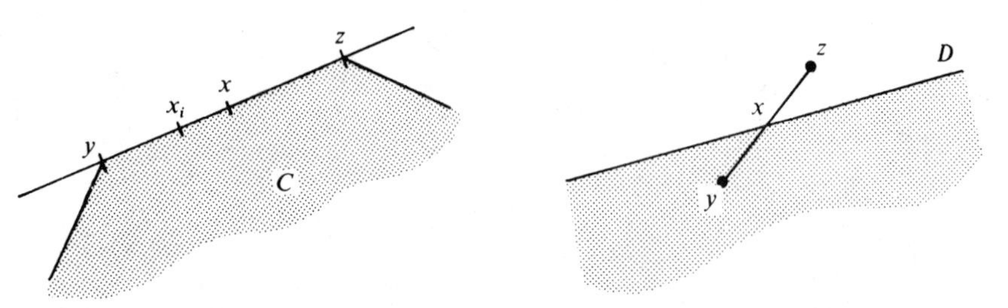

Figure 11.2.1. Figure 11.2.2.

11.3 Observe first that if v, x, y, u are arranged as in figure 11.3.1, the cross-ratio $[x, y, u, v]$ is greater than 1. In fact, sending v to infinity on the line in question (cf. 6.A or [B, 6.2.5]), we get $[x, y, u, v] = \overrightarrow{xu}/\overrightarrow{yu} > 1$. In particular $d(x, y) = \log[x, y, u, v]$, and we don't have to take absolute values.

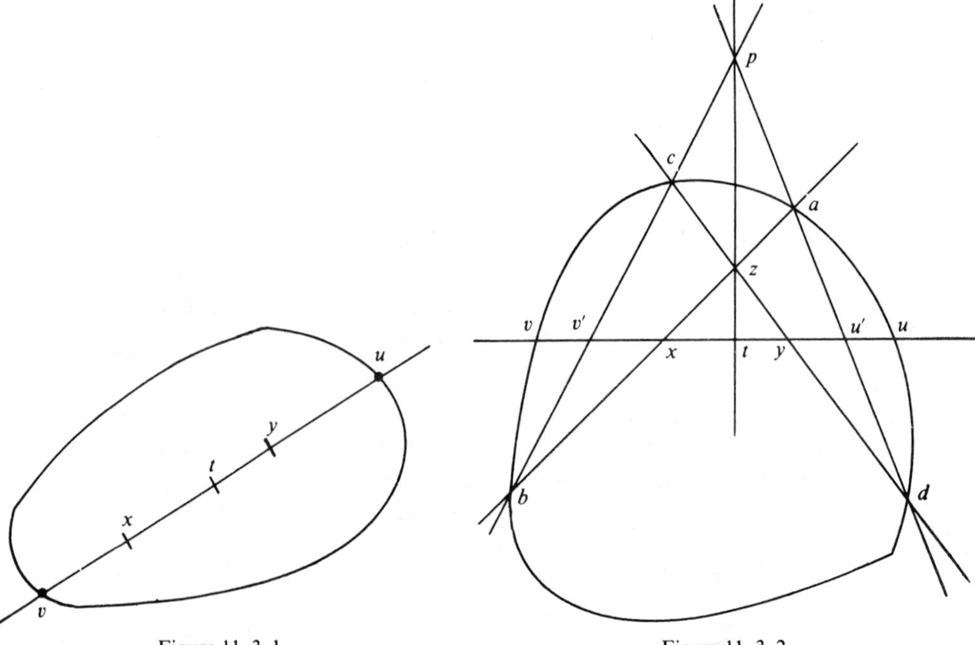

Figure 11.3.1. Figure 11.3.2.

Thus, if t belongs to the segment $[x, y]$, the remark above and problem 6.1 imply that

$$d(x,t)+d(t,y)=\log[x,t,u,v]+\log[t,y,u,v]$$
$$=\log([x,t,u,v][t,y,u,v])=\log[x,y,u,v]=d(x,y).$$

This essentially says that, on the open segment $]v,u[$, the metric we're considering coincides with the usual Euclidean metric of **R**. Now let z be an arbitrary point of $\mathring{A}$, not belonging to the line $\langle x, y\rangle$. We complete the picture as indicated in figure 11.3.2: the line $\langle x, z\rangle$ intersects the boundary of A at a, b, and $\langle y, z\rangle$ intersects it at c, d. Then we put $p=\langle a,d\rangle\cap\langle b,c\rangle$ and finally $v'=\langle x,y\rangle\cap\langle b,c\rangle$, $u'=\langle x,y\rangle\cap\langle a,d\rangle$ and $t=\langle p,z\rangle\cap\langle x,y\rangle$.

Consider the perspective of center p and mapping $\langle a,b\rangle$ to $\langle x,y\rangle$ (cf. 4.F or [B, 4.7]). Being a homography, it preserves cross-ratios, so that $[x,z,a,b]=[x,t,u',v']\geq[x,t,u,v]$, the latter inequality arising from the fact that the metric on $]u,v[$ is isometric to that of **R**, as we've seen above. Similarly, taking the perspective of center p and mapping $\langle c,d\rangle$ to $\langle x,y\rangle$ shows that

$$[z,y,d,c]=[t,y,u',v']\geq[t,y,u,v].$$

Multiplying the two inequalities obtained, applying problem 6.1 again and taking logarithms (all the quantities are greater than 1), we get

$$[x,z,a,b][z,y,d,c]\geq[x,t,u,v][t,y,u,v]=[x,y,u,v],$$
$$\mathrm{d}(x,z)+\mathrm{d}(z,y)\geq\mathrm{d}(x,y).$$

This metric is excellent because the segments $]v,u[$ are isometric to **R**. □

The study of the strict triangle inequality yields interesting results. We see that, if z does not belong to the segment $[x,y]$ and if $\mathrm{d}(x,z)+\mathrm{d}(z,y)=\mathrm{d}(x,y)$, we must perforce have $[t,y,u',v']=[t,y,u,v]$, implying $u=u'$ and $v=v'$. This can only take place when the whole segment $[b,c]$, as well as $[a,d]$, is

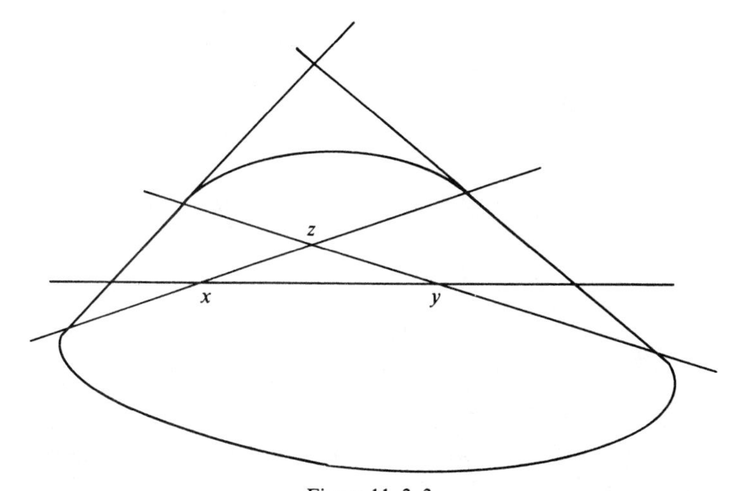

Figure 11.3.3.

contained in the boundary of A. Conversely, if there are two non-extremal points x, y each lying in one straight part of the boundary, we will have $d(x, z)+d(z, y)=d(x, y)$ for an infinite number of points z outside the segment $[x, y]$. On the other hand, we can be sure that our metric satisfies the strict triangle inequality if $\mathrm{Fr}(A)$ does not contain any non-extremal points. □

11.4 The idea is to use the logarithmic derivative of the polynomial P to find its derivative P'. We can factor P as a product of monomials $\prod_i (z-z_i)$, where z_i are its roots. Its logarithmic derivative is then

$$\frac{P'}{P}=\sum_i \frac{1}{z-z_i}.$$

Thus, if u is a root of P', we will have

$$\frac{P'(u)}{P(u)}=\sum_i \frac{1}{u-z_i}=\sum_i |u-z_i|^{-2}(\bar{u}-\bar{z}_i)=0,$$

whence

$$\sum_i |u-z_i|^{-2}(u-z_i)=0.$$

But this last relation means exactly that u is the barycenter of the points z_i with the masses $|u-z_i|^{-2}$, which are all positive or zero (cf. 3.A or [B, 3.4.6]). This evidently implies that u belongs to the convex hull of the points z_i (cf. 11.A or [B, 11.1.8.4]). □

Now suppose P has degree 3; we denote its roots by a, b, c and the roots of its derivative P' by u, v. To show the desired property, we use the second little theorem of Poncelet (cf. 17.B or [B, 17.6.3.6]). Indeed, it is enough to show that the oriented angles between lines $\widehat{ab\,au}$ and $\widehat{av\,ac}$ are equal, and the same holds for the other two vertices, for if this is the case, the ellipse having one focus at u and tangent to the three lines $\langle a, b\rangle$, $\langle b, c\rangle$, $\langle c, a\rangle$ must necessarily have v as the other focus.

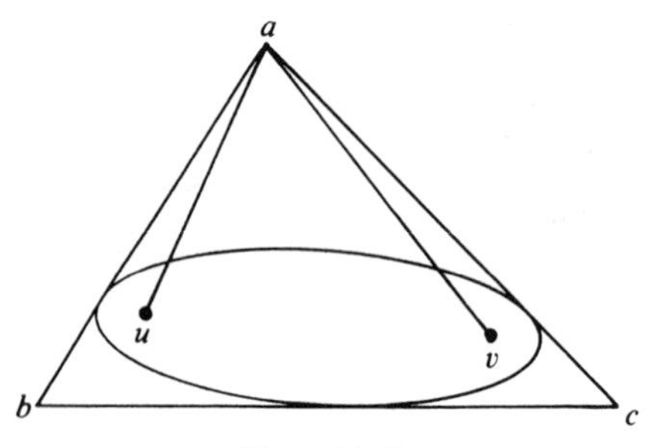

Figure 11.4.

In the complex plane, the two angles above will be equal if and only if the complex numbers $(b-a)/(u-a)$ and $(v-a)/(c-a)$ have the same argument, or, in other words, if their quotient $\Delta(a)=[(b-a)(c-a)]/[(u-a)(v-a)]$ is a real number (ditto for $\Delta(b)$ and $\Delta(c)$). We shall compute this ratio explicitly by using the relationship between the roots and the coefficients of P. The length of the calculation can be reduced by choosing P appropriately, since the property we want is invariant under plane isometries. Take $P=z^3+pz+q$, hence $P'=3z^2+p$; then $a+b+c=0$, $abc=-q$, $u+v=0$ and $uv=p/3$. Thus we find for the vertex a, using the fact that $a^3+pa+q=0$:

$$\Delta(a)=\frac{-q-2pa-2q}{\dfrac{pa}{3}-pa-q}=3 \qquad \square$$

11.5 Let x, y be two points in the set $N(A)$, and z an arbitrary point of A. The segment $[x, z]$ is entirely contained in A, since A is star-shaped at x. But A is also star-shaped at y, so the triangle $\{x, y, z\}$ is entirely contained in A. This implies that $[t, z]$ is contained in A for every point t of $[x, y]$; but since this is true for any $z \in A$, the set A is star-shaped at t. This shows that $N(A)$ is convex.

In the figure below, $N(A)$ is a single point for the butterfly-shaped region, and for the star it is the heavily shaded pentagon in the middle.

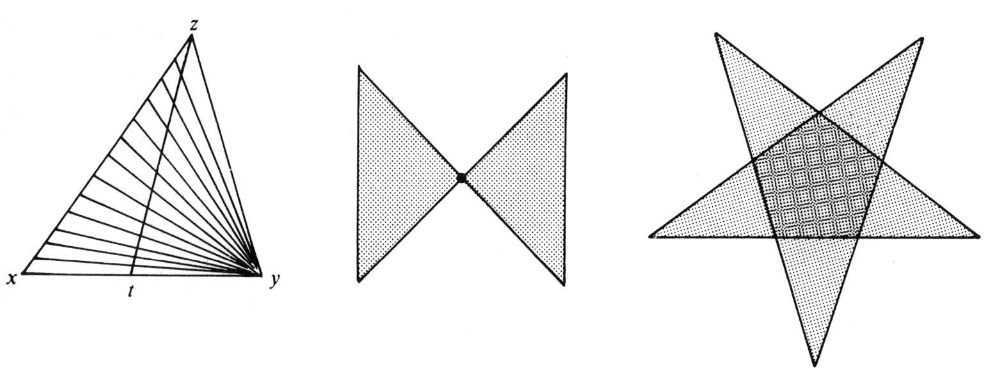

Figure 11.5.

Chapter 12

12.1 The method of figure 12.1.1 is a bit lengthy, but we hope it will please the reader because of its naturality and intuitiveness. We assume we have constructed a regular decagon, and we let R be the radius of the circle in which it is inscribed (cf. 12.C or [B, 12.4.3]). We also *put* $ab=L$ and $ac=M$, using the labels in figure 12.1.4.

Step 1. We claim that $LM=R^2$ and $M-L=R$. In fact, according to the properties of the inscribed and central angles (cf. 10.D or [B, 10.9]), and since the sum of the angles of a triangle is equal to π, we have:

$$\overline{Oa, Ob} = \overline{aO, ac} = \overline{cO, ca} = \frac{\pi}{5},$$

$$\overline{ba, bd} = \overline{Od, Oc} = \frac{2\pi}{5},$$

$$\overline{da, db} = \overline{dO, dc} = \pi - \overline{dO, da} = \pi - \left(\pi - \overline{Oa, Ob} - \overline{aO, ac}\right)$$

$$= \frac{\pi}{5} + \frac{\pi}{5} = \frac{2\pi}{5}.$$

This shows the two triangles $\{a, b, d\}$ and $\{c, d, O\}$ are isosceles, as they have two equal angles (cf. 10.A (watch out!) and [B, 10.2.2]). In particular $ab = ad = L$ and $M = ad + dc = ab + Oc = L + R$. On the other hand the two triangles $\{a, O, d\}$ and $\{a, c, O\}$ are similar; hence (cf. 10.A or [B, 10.2.7]), we get $ad/aO = aO/ac$, which implies

$$LM = ab \cdot ac = ad \cdot ac = R^2. \qquad \square$$

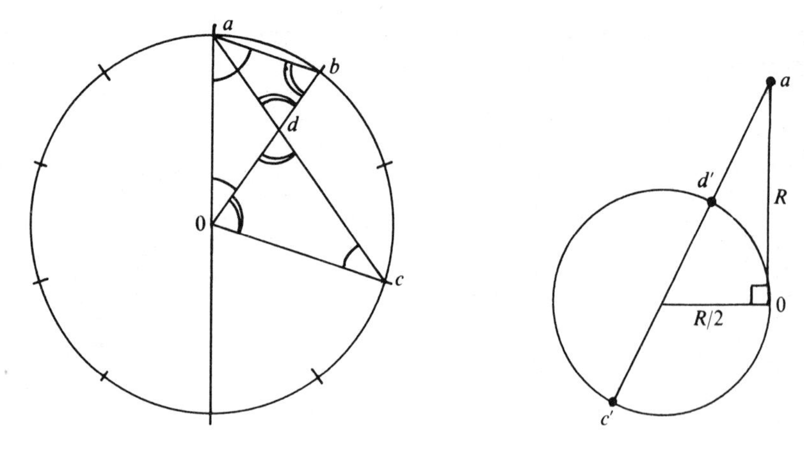

Figure 12.1.4.

Step 2. The two relations $LM = R^2$ and $M - L = R$ determine L, M uniquely as roots of the second-degree equation

$$x^2 - Rx - R^2 = 0;$$

their values are $L = [(\sqrt{5} - 1)/2]R$ and $M = [(\sqrt{5} + 1)/2]R$. The geometric construction of figure 12.1.2 can be justified in the following way: $ac' - ad' = R$ is clear, and

$$ad' \cdot ac' = aO^2 = R^2$$

holds because of the property of the power of a point relative to a circle (cf. 10.B or [B, 10.7.10]).

Conclusion of the first construction. It is enough to apply the uniqueness of regular polygons (12.C or [B, 12.4.2]). Since ab is necessarily equal to

$[(\sqrt{5}-1)/2]R$, the construction does indeed yield a regular decagon, from which we get a regular pentagon. □

Second method: the paper strip. This is not a real mathematical construction, and isn't even of practical use. We will however show that the figure we obtain includes a regular pentagon. Observe first that *folding* a strip of paper gives equal angles (fig. 12.1.3).

We now have to show that the five angles $\hat{a}, \hat{b}, \hat{c}, \hat{d}, \hat{e}$ are equal (and consequently have the value $3\pi/5$), and the five sides ab, bc, cd, de, ea are also equal. The three folds, along the sides cd, ae, bc, yield the following equalities:

$$\hat{a}=\hat{e}, \quad \hat{b}=\hat{c}, \quad \hat{c}=\hat{d}, \quad eb=ec.$$

The fact that ad is parallel to bc and that $eb=ec$ shows that the figure is symmetric along a line passing through e, and the symmetry maps b into c and a into d. This implies $ae=ed$ and $\hat{a}=\hat{d}$, showing that the five angles are indeed equal. As for the sides, the three folds give $bc=de$, $ab=cd$ and $bc=ea$; this, together with $ae=ed$ obtained above, shows that the sides are equal as well. □

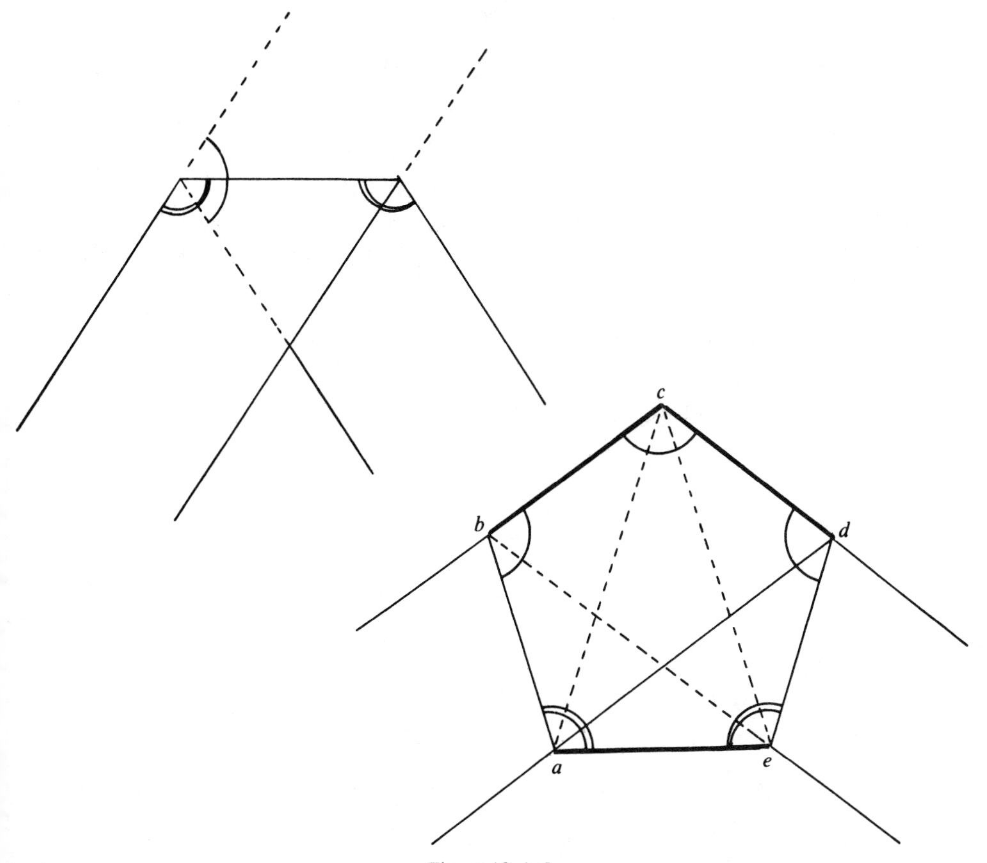

Figure 12.1.5.

12.2 We recall that the group $G(P)$ of a regular polyhedron acts simply transitively on the triples consisting of a vertex, an edge containing this vertex and a face containing this edge (cf. 1.D and 12.C or [B, 12.5.2.5]). Thus we only have to determine how many such triples contain a given edge a. Since an edge determines exactly two vertices s, s' and two faces f, f', we obtain the four triples (s, a, f), (s, a, f'), (s', a, f) and (s', a, f'). □

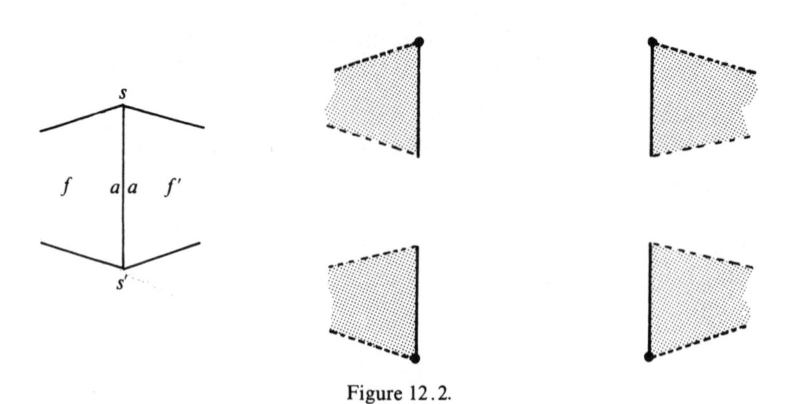

Figure 12.2.

12.3 Calculation of the volume. Denote by z the coordinate along the rotation axis D, and by x a coordinate in P such that the pair (x, z) is orthonormal and D corresponds to $x = 0$. Denote by $A(z)$ the area of the section of C by a horizontal plane whose height is z, and suppose that K is given by the two functions $f, h:[a, b] \to \mathbf{R}$ (i.e. K is the union of the graphs of f and h). Then we have

$$\mathscr{L}_E(C) = \int_a^b A(z)\,\mathrm{d}z \quad \text{(by Fubini's theorem)},$$

$$\mathscr{L}_P(K) = \int_K \mathrm{d}x\,\mathrm{d}y, \quad A(z) = \pi\left(h^2(z) - f^2(z)\right).$$

The x-coordinate of the centroid g of K is given by (cf. 2.G)

$$\mathrm{d}(g, D) = \frac{\int_K x\,\mathrm{d}x\,\mathrm{d}y}{\mathscr{L}_P(K)}.$$

But

$$\int_K x\,\mathrm{d}x\,\mathrm{d}y = \int_a^b \left(\int_{f(z)}^{h(z)} x\,\mathrm{d}x\right)\mathrm{d}z = \frac{1}{2}\int_a^b \left(h^2(z) - f^2(z)\right)\mathrm{d}z.$$

This shows that we indeed have $\mathscr{L}_E(C) = 2\pi\,\mathrm{d}(g, D)\mathscr{L}_P(K)$. □

EXAMPLES. 1. Take K to be a disc of radius r whose center lies at a distance R from D. The solid C is a torus; since the centroid of K coincides

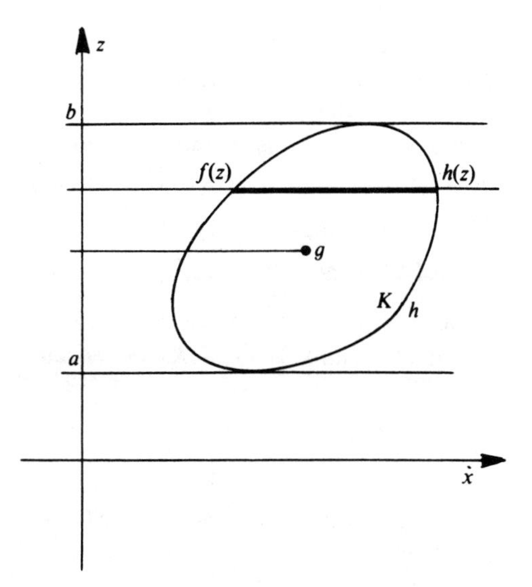

Figure 12.3.1.

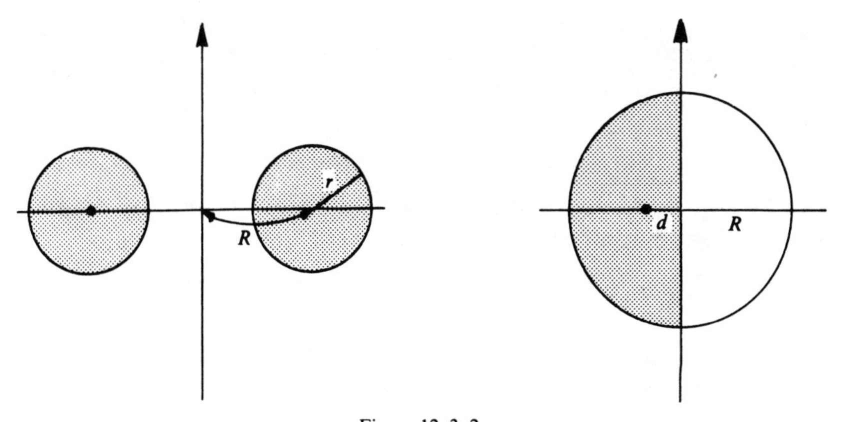

Figure 12.3.2.

with its center, we get for the volume of the solid torus C:

$$\mathscr{L}_E(C) = 2\pi R(\pi r^2) = 2\pi^2 R r^2.$$

2. Take K as a half-disc of radius R whose diameter lies in D. The solid obtained by rotating K around D is simply the sphere of radius R. We know its volume

$$\mathscr{L}_E(S) = \tfrac{4}{3}\pi R^2,$$

so the Guldin formula allows us to find the distance d from the centroid g of the half-disc K to the line D (this uniquely determines g by symmetry):

$$\frac{4}{3}\pi R^3 = 2\pi d \frac{\pi R^2}{2}, \quad \text{or} \quad d = \frac{4}{3\pi} R.$$

Calculation of the area. We could use the same method as above, but it is more instructive to use the "painting technique" (cf. 12.B or [B, 12.10.17]). In the proof, three dots $(\cdots)$ will denote terms in λ which can be neglected. The idea is to thicken K and C, using balls of radius λ, so as to transform them into a solid $C(\lambda)$ and a surface $K(\lambda)$, respectively. We then know that

$$\mathscr{L}_E(C(\lambda)) = \mathscr{L}_E(C) + \lambda \mathfrak{A}_E(C) + \cdots,$$
$$\mathscr{L}_P(K(\lambda)) = \mathscr{L}_P(K) + \lambda \mathfrak{A}_P(K) + \cdots.$$

By the associative property of barycenters, we will have the following relation between the centroids $g(\lambda)$, g and $h(\lambda)$ of $K(\lambda)$, K and $K(\lambda)\setminus K$, respectively:

$$g(\lambda) = \frac{\mathscr{L}_P(K)\cdot g + \mathscr{L}_P(K(\lambda)\setminus K)\cdot h(\lambda)}{\mathscr{L}_P(K(\lambda))}.$$

When λ approaches 0, the point $h(\lambda)$ tends toward the centroid h of the homogeneous wire along the boundary of K. Thus we have

$$\mathscr{L}_P(K(\lambda))\,\mathrm{d}(g(\lambda), D) = \mathscr{L}_P\,\mathrm{d}(g, D) + \lambda \mathfrak{A}_P(K)\,\mathrm{d}(h, D) + \cdots.$$

We now apply the Guldin formula for volumes: for every λ, we get $\mathscr{L}_E(C(\lambda)) = 2\pi\,\mathrm{d}(g(\lambda), D)\mathscr{L}_P(K(\lambda))$. Taking the term in λ in both sides, we obtain

$$\mathfrak{A}_E(C) = 2\pi\,\mathrm{d}(h, D)\mathfrak{A}_P(K). \qquad \square$$

EXAMPLES. 1. As above, the area of the torus C obtained by rotating a circle of radius r whose center lies at a distance R from D, the rotation axis, is given by

$$2\pi R(2\pi r) = r\pi^2 Rr.$$

2. For a homogeneous wire in the shape of a semicircle of radius R, we find that the distance d from the centroid to the diameter is given by

$$4\pi R^2 = 2\pi d(\pi R), \quad \text{or} \quad d = \frac{2}{\pi}R.$$

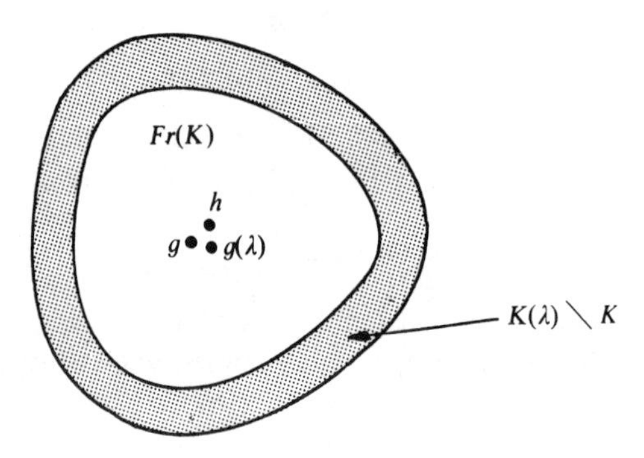

Figure 12.3.3.

12.4 Step 1. We start by using the Appolonius theorem (cf. 15.D or [B, 15.6.4]), that says that, in an ellipsoid F of center a, the volume of the parallelepiped built on a set $\{aa_i\}_{i=1,\dots,n}$ of conjugate half-diameters is a constant $A(F)$. In particular, this volume is the same as in the case when all the aa_i are the half-axes of F. Thus $\mathscr{L}(F)=\beta(n)A(F)$. This implies the following

Lemma. *Let $\{aa_i\}_{i=1,\dots,n}$ be a set of pairwise orthogonal radii of F. Then the following inequality holds:*

$$A(F)\geq \prod_i \overline{aa_i}.$$

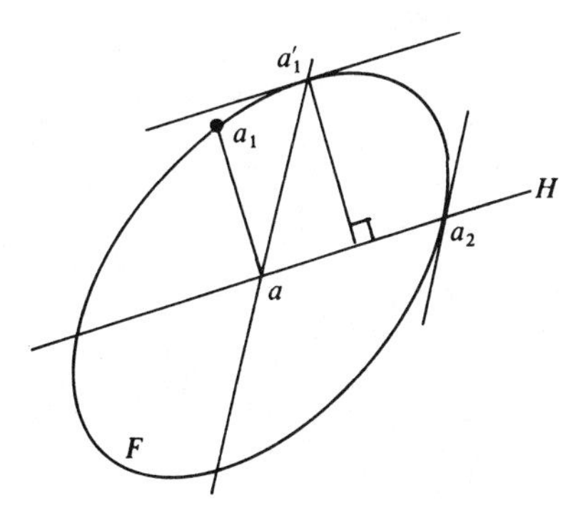

Figure 12.4.1.

The proof is by induction on n; the lemma is obvious from $n=1$. Let H be the hyperplane passing through the center a and containing $a_2,\dots,a_n$, and let aa_1' be a half-diameter conjugate to H. Observe that the distance d from a_1' to H satisfies $d\geq \overline{aa_1}$. Denote by F' the ellipsoid obtained by intersecting F with the hyperplane H, and notice that adding aa_1' to a system of conjugate half-diameters of F' gives a system of conjugate half-diameters of F. Finally, notice that the volume of the parallelepiped built on aa_1' and $n-1$ vectors of H is given by the product of d by the volume of the parallelepiped in H built on the same $n-1$ vectors. The induction hypothesis, applied to F', shows that

$$A(F)\geq dA(F')\geq d\prod_{i=2}^{n}\overline{aa_i}\geq \overline{aa_1}\prod_{i=2}^{n}\overline{aa_i}\geq \prod_i \overline{aa_i}. \qquad \square$$

Step 2. Let S be the unit sphere of X, centered at O, and consider one of the axes p_ip_i' of the given ellipsoid E; let $2\alpha_i=\overline{p_ip_i'}$ be the length of this axis. Denote by ω the center of E, and by ξ_i the distance from ω to the hyperplane passing through O and orthogonal to p_ip_i'.

The polar body E^* contains the poles b_i, b_i' of the tangent hyperplanes K_i, K_i' to E at p_i, p_i'. The points b_i, b_i' are on the line through O parallel to p_ip_i', and their distances to O are $\overline{Ob_i}=1/(\alpha_i-\xi)$ and $\overline{Ob_i'}=1/(\alpha_i+\xi)$.

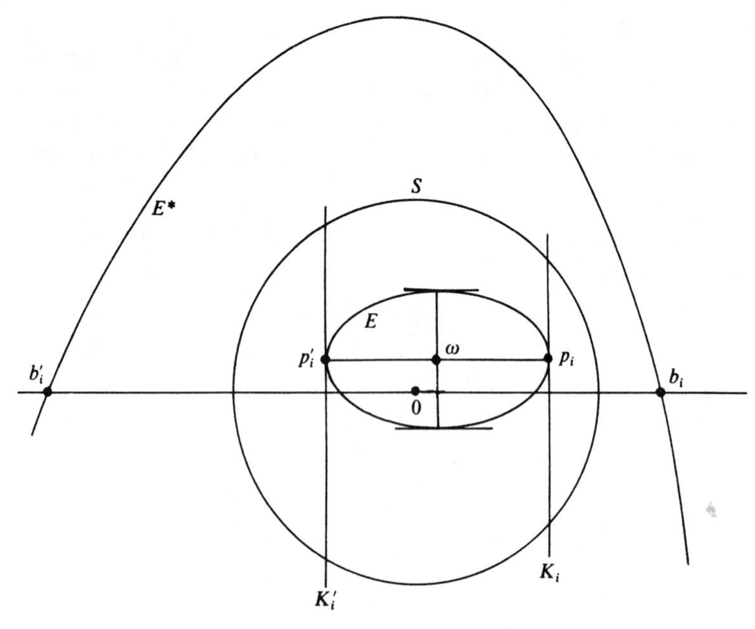

Figure 12.4.2.

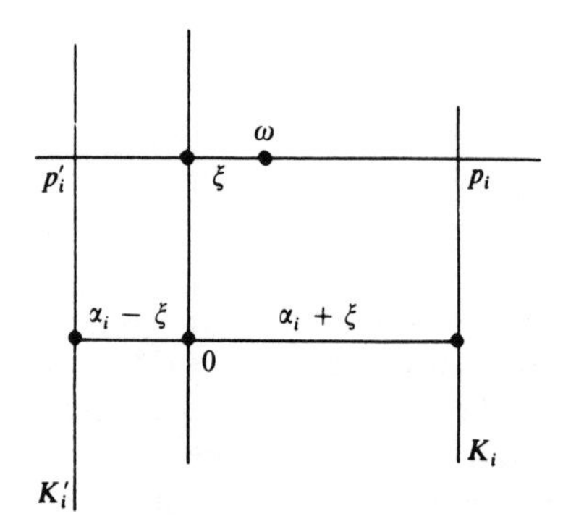

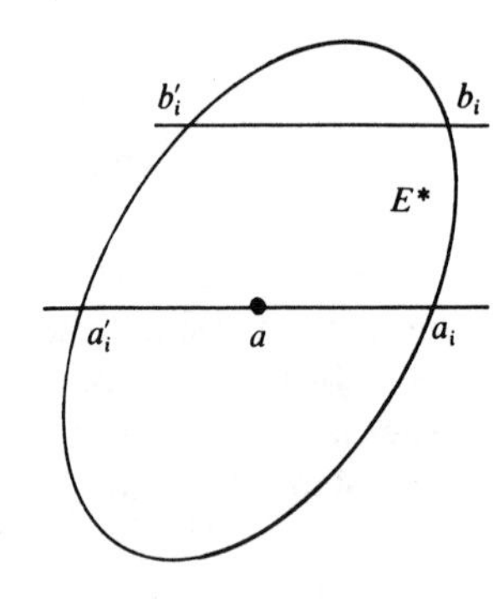

Figure 12.4.3.

In particular, the chord $b_i b_i'$ of E^* has length

$$\overline{b_i b_i'} = \frac{1}{\alpha_i - \xi_i} + \frac{1}{\alpha_i + \xi_i} = \frac{2\alpha_i}{\alpha_i^2 - \xi_i^2} \geq \frac{2}{\alpha_i}.$$

Take the chord $[a_i, a_i']$ passing through the center a and parallel to $\langle b_i, b_i' \rangle$. Using the inequality obtained above, we evidently get

$$\overline{aa_i} = \frac{\overline{a_i a_i'}}{2} \geq \frac{\overline{b_i b_i'}}{2} \geq \frac{1}{\alpha_i}.$$

Carrying our the operations above for all the axes $p_i p_i'$ of E, we finally get an orthogonal set $\{aa_i\}_{i=1,\ldots,n}$ of radii of E^*. Thus the lemma, together with the inequality above applied once for each i, shows that

$$\mathscr{L}(E^*) = \beta(n)A(E^*) \geq \beta(n)\prod_i \overline{aa_i} \geq \beta(n)\prod_i \frac{1}{\alpha_i}.$$

But, because the $(p_i p_i')$ are the orthogonal axes of E, we also have

$$\mathscr{L}(E) = \beta(n)\prod_i \alpha_i.$$

This implies the desired result, $\mathscr{L}(E)\mathscr{L}(E^*) \geq \beta(n))^2$.

If equality takes place, we must have for each i that $\overline{b_i b_i'} \geq 2/\alpha_i$, and consequently $\xi_i = 0$ means that $\omega = O$, the center of S. Conversely, if such is the case, the chords $a_i a_i'$ are in fact the orthogonal axes of E^*, whence $\mathscr{L}(E^*) = \beta(n)\Pi_i(1/\alpha_i)$, and $\mathscr{L}(E)\mathscr{L}(E^*) = (\beta(n))^2$. □

12.5 The idea here is to use the Euler equation, or, equivalently, the support function of the convex set (cf. [B, 11.8, 12.3]), and to find the curvature by means of this function. More precisely, let $h: \mathbf{R} \to \mathbf{R}_+$ be a function of class C^2 and periodic with period 2π; then the family of lines whose equations (in an orthonormal frame) are

$$\cos t \cdot x + \sin t \cdot y = h(t), \quad t \in \mathbf{R}$$

are the envelope of a closed plane curve, whose parametric representation is

$$t \mapsto (\cos t \cdot h - \sin t \cdot h', \sin t \cdot h + \cos t \cdot h').$$

Conversely, a strictly convex set whose boundary is of class C^2 can be obtained in this way as long as the origin O of the coordinate system belongs to C; in fact, $h(t)$ is given by the distance from O to the supporting line of C

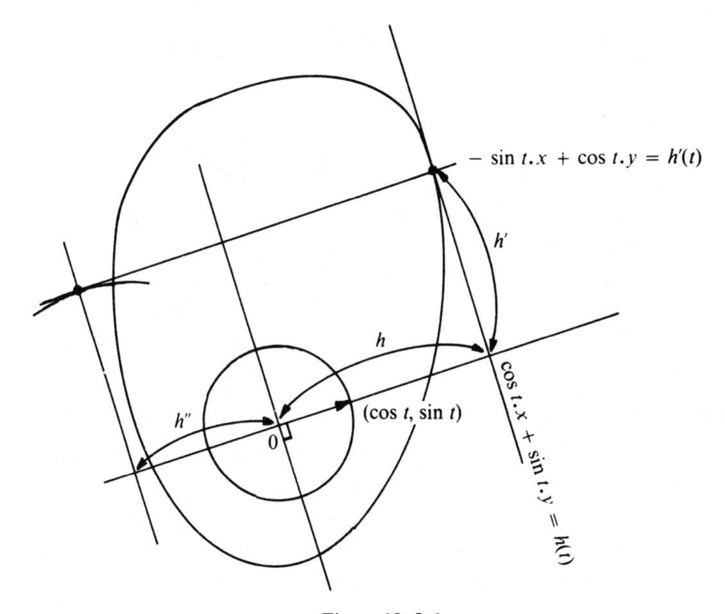

Figure 12.5.1.

whose direction is $(-\sin t, \cos t)$ and such that the vector $(\cos t, \sin t)$ points from the origin to the line.

In this context, the two Blaschke rolling theorems which we want to prove are consequences of the following more general result:

Lemma. *Let D and E be two convex sets defined by the functions $h(t)$ and $k(t)$ as above, so that both contain the origin O and their boundaries are tangent at O to the axis Oy. Then if for all t the curvature of E at the contact point with its supporting line of slope t is always less than or equal to the curvature of D at the contact point with its supporting line of same slope, the convex set E is contained in D.*

The radius of curvature D (resp. E) at the point parametrized by t is equal to $h(t)+h''(t)$ (resp. $k(t)+k''(t)$); to see this, it is enough to write the center of curvature as the contact point of the normal to the curve being considered with its envelope. Here the normal to D has the equation $-\sin t\cdot x+\cos t\cdot y = h'(t)$ (Euler equation), and the center of curvature is the intersection of this normal with

$$-\cos t\cdot x-\sin t\cdot y=h''(t)$$

(see above). This gives the desired value for the radius of curvature (see Figure 12.5.1).

We now know that

$$\rho(t)=k(t)+k''(t)\le h(t)+h''(t)=\sigma(t)$$

for every t. Since $h(0)=k(0)=0$ and $h'(0)=k'(0)=0$ by assumption, the solutions h, k of the differential equations $k+k''=\rho$, $h+h''=\sigma$ are given by

$$h(t)=\int_0^t\sigma(s)\sin(t-s)\,ds,\quad k(t)=\int_0^t\rho(s)\sin(t-s)\,ds.$$

Since $\rho\le\sigma$ for every t, we obtain

$$h(t)\ge k(t)\quad\text{for every}\quad t\in[0,\pi]$$

because $\sin(t-s)$ is necessarily positive. Moreover, replacing t by $-t$ shows that the inequality holds for all t.

Now we show by contradiction that E is contained in D. If there were a point of the boundary of D in the interior of E, we could take the supporting line Δ of D at this point, and the supporting line Θ of E parallel to Δ; for those two lines $h(t)<k(t)$. □

Proof of the rolling theorems. For the osculating circle of maximal curvature, take E as this circle and $D=C$ in the lemma. For the circle of minimal curvature, take $C=E$ and D as this circle. □

Figure 12.5.3 shows that this result is not true if we replace the circle of maximal (resp. minimal) radius of curvature by the circle of maximal (resp. minimal) radius contained in (resp. containing) C.

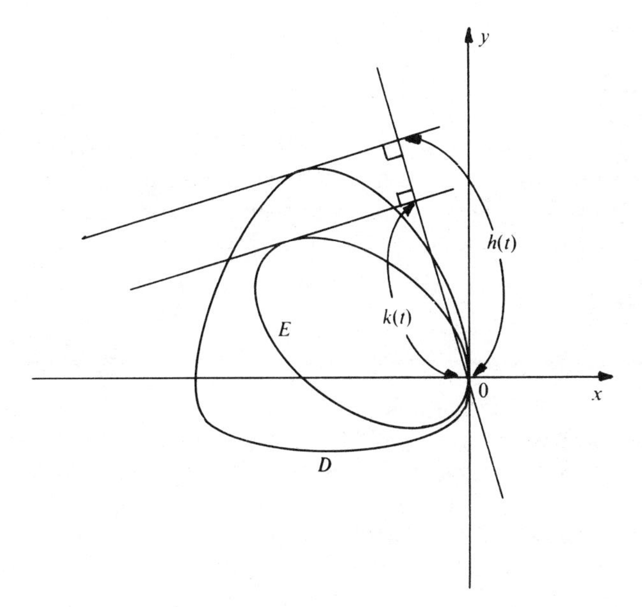

Figure 12.5.2.

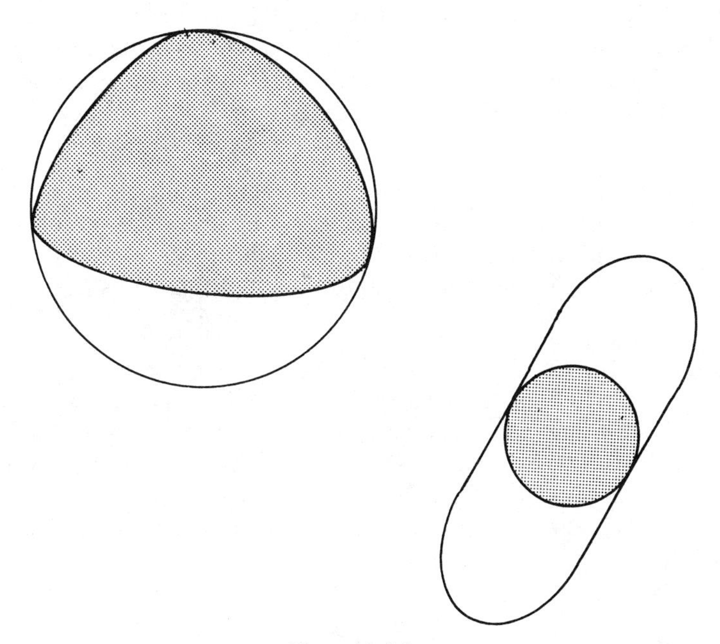

Figure 12.5.3.

Chapter 13

13.1 This is an immediate application of the existence of a non-singular completion (cf. 13.E or [B, 13.3.4]) and the Witt theorem (13.E or [B, 13.7.1]). In fact, let U be a null subspace of maximal dimension s contained in E. According to 13.E, there exists a non-singular completion F of U such that F is an Artin space Art_{2s}. Let $F^{\perp}$ be its orthogonal space; since F is non-singular, E can be written as an orthogonal direct sum

$$E = F \overset{\perp}{\oplus} F^{\perp}$$

(13.D or [B, 13.3.2]). We claim that $F^{\perp}$ contains no non-zero isotropic vector. For if x were such a vector, it would be orthogonal to U, so together with U it would span a null subspace of dimension greater than n, contradicting maximality.

This shows that if we can classify anisotropic spaces, such as $F^{\perp}$, we're done, since the Witt theorem says that two decompositions of the type $F \oplus^{\perp} F^{\perp}$ are always isomorphic, and $F = \mathrm{Art}_{2s}$ is well-known. □

13.2 We have seen (cf. 13.B or [B, 13.4.7]) that the number of classes of quadratic forms in dimension 1 over a field K is the number of elements of $K/(K^*)^2$. To find this number, consider the multiplicative group homomorphism $k \mapsto k^2$ from K^* into itself. Since the characteristic of K is assumed different from 2, the kernel of this homomorphism (which is isomorphic to $K^*/(K^*)^2$) is ± 1, so it has two elements and $K/(K^*)^2$ has three.

The three quadratic forms are the degenerate form 0, the form x^2 and a third one which, when -1 is not a square in K, is equal to $-x^2$. If -1 is a square the third form is different; for example for $K = \mathbf{Z}/5\mathbf{Z}$, it can be taken as $2x^2$.

13.3 We have to find an infinite number of elements in the quotient $\mathbf{Q}^*/(\mathbf{Q}^*)^2$. To do this, just notice that two distinct prime numbers p, q necessarily have different images in this quotient. The proof of this fact (i.e. that p/q is not a square in $\mathbf{Q}$) is classic, and is identical to the proof that $\sqrt{2}$ is irrational.

13.4 Let E be a two-dimensional space endowed with a non-degenerate quadratic form q. Recall first that if $f \in O^+(E)$ and $g \in O^-(E)$, we have $gfg^{-1} = f^{-1}$ (cf. 13.F or [B, 13.8.1]); we repeat the nice proof here. Since $g' = gf \in O^-(E)$, the map g', as well as g, is a reflection through a non-isotropic line. (This follows from the general result that every element of $O(E)$ is the product of at most $n = \dim E$ hyperplane reflections; if $n = 2$ and the map is orientation-reversing, the product has only one factor.) In particular, g, g' are involutions:

$$g^2 = g'^2 = \mathrm{Id}_E.$$

But now we have

$$gfg^{-1} = g(gg')g^{-1} = g'g^{-1} = g'^{-1}g^{-1} = (gg')^{-1} = f^{-1},$$

concluding the proof.

Now assume $O(E)$ is commutative; then $gfg^{-1} = f = f^{-1}$, so $f^2 = \mathrm{Id}_E$. But according to the classification of involutions (cf. 13.E or [B, 13.6.6]), f can only be the identity Id_E or reflection through the origin, $-\mathrm{Id}_E$. In particular $O^+(E)$ contains only the two elements $\pm\mathrm{Id}_E$ (which are distinct, since the characteristic is not 2). But $O^+(E)$ has index 2 in $O(E)$, so the order of $O^-(E)$ is also 2. Since $O^-(E)$ contains all reflections through non-isotropic lines, there can only be two such lines in E. But there are at most two isotropic lines as well (in fact either none or two, in which case $E = \mathrm{Art}_2$), so E contains exactly two or four lines. Thus K is finite, and the number of lines in E is equal to the order of K, plus 1. Ruling out $\#(K) = 1$, we end up with the field with three elements, and E is necessarily Art_2.

We still have to check that $O(\mathrm{Art}_2)$ over a three-element field is really commutative. This can be done directly, or just by remarking that there are only two groups of order 4, namely $\mathbf{Z}/4\mathbf{Z}$ and the Klein group (cf. [B, 0.2]). Only the Klein group contains only elements of square one, so $O(\mathrm{Art}_2)$ over the three-element field is isomorphic to it and in particular commutative. □

13.5 I'll first mention why this result is interesting. To show that the orthogonal group is generated by reflections through hyperplanes, the natural proof (which works very well if there are no isotropic vectors, for instance in the Euclidean case; see 8.C or [B, 8.2.12]) consists in taking a non-isotropic vector x and defining a hyperplane reflection switching x and $f(x)$, for a given $f \in O(E)$. This is done by considering the hyperplane orthogonal to $x - f(x)$, but for this to work the vector $x - f(x)$ has to be anisotropic. The result we will show says exactly that there are forms (E, q) for which this method does not work.

In order to guess what form f has, recall from [B, 13.6.7] that the condition we want for f (i.e. $x - f(x)$ is non-zero and isotropic for every non-isotropic vector x) implies that the kernel $U = \mathrm{Ker}(f - \mathrm{Id}_E)$ and the image $V = \mathrm{Im}(f - \mathrm{Id}_E)$ of the linear map $f - \mathrm{Id}_E$ are both null spaces, and that $E = U + V$. This shows that E must be an Artin space Art_{2s}, where $s = \dim U = \dim V$. It also shows that the matrix of f, in a basis compatible with the direct sum $U + V$, is of the form

$$\begin{pmatrix} I & S \\ O & I \end{pmatrix},$$

since U and V are the kernel and image of $f - \mathrm{Id}_E$, respectively. This guarantees that $x - f(x)$ is always isotropic (whether or not x is), for if $x = (a, b)$, then $x - f(x) = (a - S(a), 0) \in U$.

But we still have the condition $f \in O(E)$. The matrix of q, in a basis compatible with $U+V$, can be written as

$$\begin{pmatrix} O & I \\ I & O \end{pmatrix},$$

because $E = \mathrm{Art}_{2s}$ (cf. 13.A, 13.B or [B, 13.1.3.8]). The condition $f \in O(E)$ is equivalent to

$$\begin{pmatrix} I & S \\ O & I \end{pmatrix}\begin{pmatrix} O & I \\ I & O \end{pmatrix}\begin{pmatrix} I & S \\ O & I \end{pmatrix} = \begin{pmatrix} O & I \\ I & O \end{pmatrix}.$$

We find that the only condition is ${}^tS + S = 0$, i.e. S must be skew-symmetric.

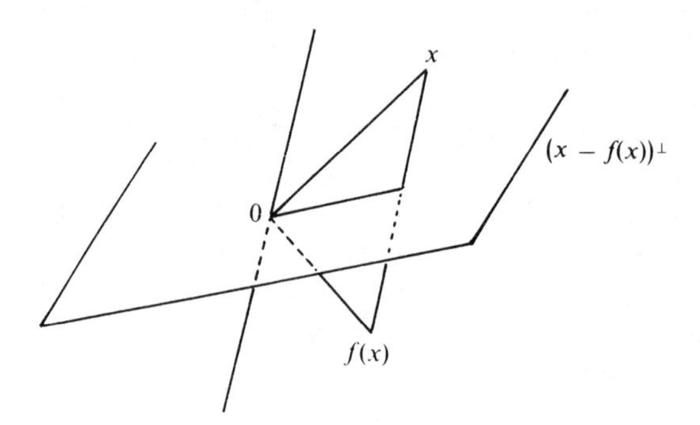

Figure 13.5.

Finally, we want $x - f(x)$ to be non-zero for every non-isotropic x; this implies $S(a) \neq 0$ for every non-zero a in U. Since S is skew-symmetric, this is only possible if the dimension of E is even (the determinant of a skew-symmetric $n \times n$ matrix is always zero if n is odd). We conclude that s is even: $s = 2s'$ and $E = \mathrm{Art}_{4s'}$. To find f, just take any skew-symmetric matrix S of rank $s = 2s'$. □

Chapter 14

14.1 Let G be the set of oriented lines of the n-dimensional real projective space $P^n(\mathbf{R})$. They are in one-to-one correspondence with the oriented two-dimensional vector subspaces P of $\mathbf{R}^n$ (cf. 4.B or [B, 4.6]). Considering in $\mathbf{R}^{n+1}$ its canonical Euclidean structure, each plane P can be defined by a pair of vectors (x, y) such that

$$\|x\|^2 = \|y\|^2 = 1 \quad \text{and} \quad (x|y) = 0. \tag{Q}$$

Two such pairs (x, y) and (x', y') correspond to the same element of G if and only if there is a real number t such that

$$x' = \cos t \cdot x + \sin t \cdot y, \quad y' = -\sin t \cdot x + \cos t \cdot y.$$

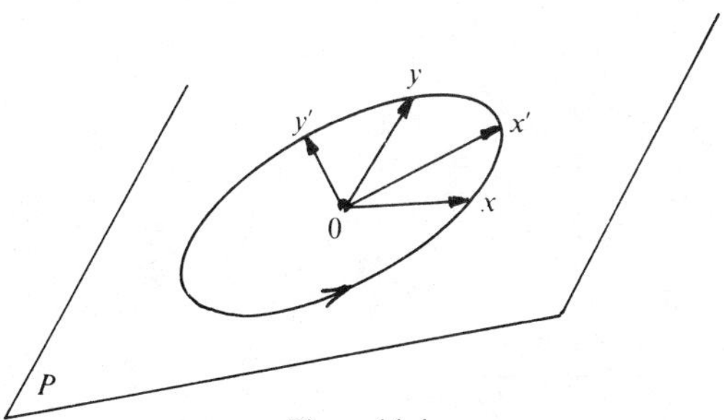

Figure 14.1.

On the other hand, $C(n)$ is the image in $P^n(\mathbf{C})$ of points whose homogeneous coordinates $(z_0, \ldots, z_n)$ satisfy $z_0^2 + z_1^2 + \cdots + z_n^2 = 0$ (where the z_i are complex numbers). The desired bijection $G \to C(n)$ is simply the quotient of the map from the pairs (x, y) above into $\mathbf{C}^{n+1}$ defined by

$$(x = (x_0, \ldots, x_n), y = (y_0, \ldots, y_n)) \mapsto (x_0 + iy_0, \ldots, x_n + iy_n).$$

The *image* of G is clearly contained in $C(n)$, for we have

$$\sum_k (x_k + iy_k)^2 = \sum_k x_k^2 - \sum_k y_k^2 + 2i \sum_k x_k y_k = \|x\|^2 - \|y\|^2 + 2i(x \mid y) = 0$$

for a pair (x, y) satisfying (Q).

The mapping is *injective*: if (x, y) and (x', y') have the same image, their projective coordinates are proportional in $\mathbf{C}$. Writing $z_k = x_k + iy_k$, $z_k' = x_k' + iy_k'$, this is equivalent to saying that $z_k' = \lambda z_k$ for every k, with λ fixed. Then $|z_k'| = |\lambda| \, |z_k|$; but

$$\sum_k |z_k|^2 = \sum_k x_k^2 + \sum_k y_k^2 = \|x\|^2 + \|y\|^2 = 2,$$

and the same holds for the z_k', so we get $|\lambda| = 1$, i.e. $\lambda = e^{it}$, and $x' = \cos t \cdot x + \sin t \cdot y$ and $y' = -\sin t \cdot x + \cos t \cdot y$, showing the element of G determined by the two pairs is the same.

The mapping is *surjective*: if $(z_k = x_k + iy_k)$ is an arbitrary $(n+1)$-tuple of complex numbers satisfying $\sum_k z_k^2 = 0$, we have $\|x\|^2 = \|y\|^2$ and $(x \mid y) = 0$. Dividing x and y by $\|x\|$ (which projectively changes nothing), we obtain a pair satisfying (Q).

Finally, the map is clearly continuous, and must be a homeomorphism since it is a bijection onto the compact set $P^n(\mathbf{C})$ (cf. 4.G or [B, 4.3.3]). □

14.2 We use the result about harmonically circumscribed quadrics: 14.E or [B, 14.5.4]. Let $\{a, b, c\}$ and $\{a', b', c'\}$ be two self-polar simplices with respect to conic α. Consider the conic α' passing through the five points a, b, c, a', b'. By construction, α' is harmonically circumscribed with respect to α, implying that the polar of a' with respect to α intersects α' in two points b', c''. Now this polar is by definition $\langle b', c' \rangle$, so b', c' and c'' lie on the same line, and both pairs (b', c') and (b', c'') are conjugate with respect to α. But this shows that $c' = c''$. □

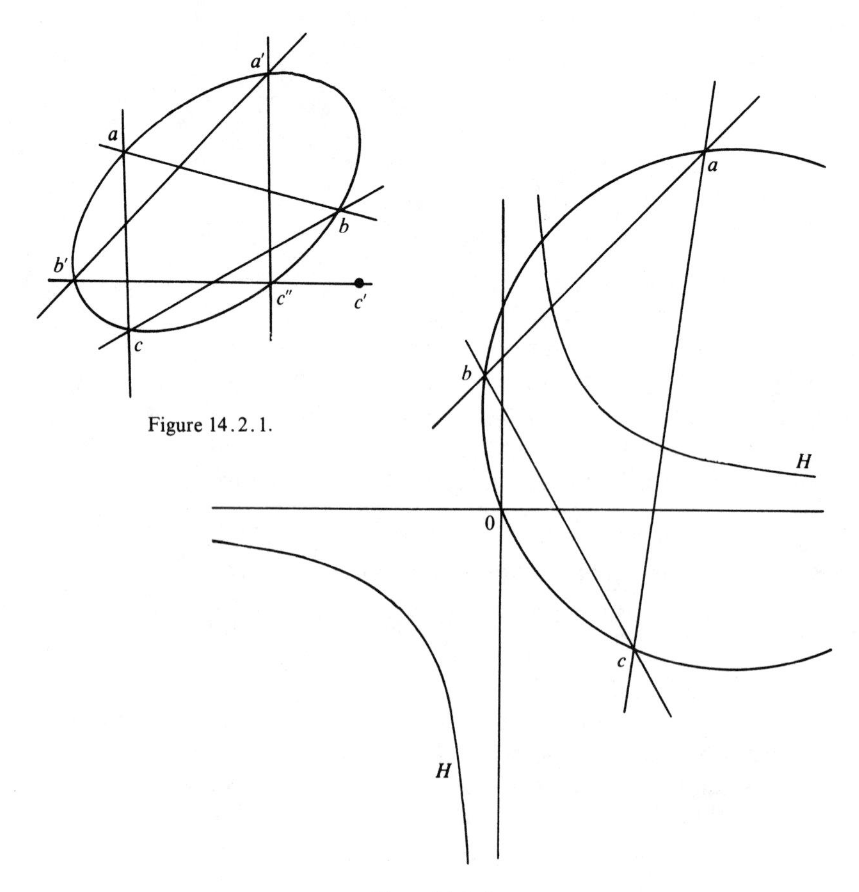

Figure 14.2.1.

Figure 14.2.2.

Application. Let X be a Euclidean plane, H an equilateral hyperbola (cf. 17.A) in X, and $\{a, b, c\}$ a self-polar triangle with respect to H. We claim that the circle circumscribed around $\{a, b, c\}$ passes through the center O of H.

We work in the projective completion of the complexification of X. Let $\overline{H}$ be the completion of H (cf. 15.A and 17.C or [B, 17.5]). The triangle formed by O and the two cyclical points (cf. 9.D or [B, 9.5.5]) is self-polar with respect to H, since H is equilateral, and this means by definition that its points at infinity correspond to orthogonal directions (cf. 17.A or [B, 17.1.3]). Thus the four points a, b, c, O, plus the two cyclical points, lie on the same conic, which must then be a circle (17.C or [B, 17.4.2]).

14.3 The geometric interpretation of this condition is that there exists a self-polar simplex $\{a_i\}$ with respect to the quadric α and all of whose faces of codimension 1 (i.e. $\langle a_2, \dots, a_n\rangle$, $\langle a_1, a_3, \dots, a_n\rangle$ etc.) are tangent to α' (we say such a simplex is *circumscribed* around α'). In fact, since envelope equations are given by the inverse of the matrices appearing in punctual equations (cf. 14.F or [B, 14.6.1]), the condition $\operatorname{trace}(\varphi'^{-1} \circ \varphi) = 0$ can be translated as $\operatorname{trace}(\varphi' \circ \varphi^{-1}) = 0$, which is the condition for α'^* to be harmonically circumscribed around α^*. But the notion of self-polar simplex is preserved by duality, while those of inscribed and circumscribed simplices are switched. □

The second property is the dual of the property in the preceding problem (14.2). Consider two triangles $\{a, b, c\}$ and $\{a', b', c'\}$ inscribed in the same conic. We first show that there exists a conic with respect to which the two triangles are self-polar. The condition for two fixed points to be conjugate with respect to a conic is linear on the coefficients of the conic in the projective space of conics $PQ(E)$, which has dimension 5 (cf. 14.D or [B, 14.2]). Thus we can find a conic for which the five pairs of points $\{a, b\}$, $\{b, c\}$, $\{c, a\}$, $\{a', b'\}$, $\{a', c'\}$ are conjugate. But this implies the points $\{b', c'\}$ are also conjugate, by problem 14.2.

By the observations above, two triangles inscribed in a conic and conjugate relative to another must be circumscribed around a third conic. From this we easily obtain many triangles inscribed in one conic and circumscribed around another, starting from one such triangle. We have proved the great Poncelet theorem for $n = 3$ (cf. 16.H, problem 10.2 or [B, 16.6]).

For a triangle $\{a, b, c\}$ circumscribed around a parabola P, we just apply the result above to the completion $\overline{P}$ of the complexification of P (cf. 15.A, 17.C or [B, 17.4.1]), and remark that we already know a triangle circumscribed around P, consisting of the focus of P and the two cyclical points. This implies that the circle passing through a, b, c also passes through the focus, by 17.C or [B, 17.4.3]. □

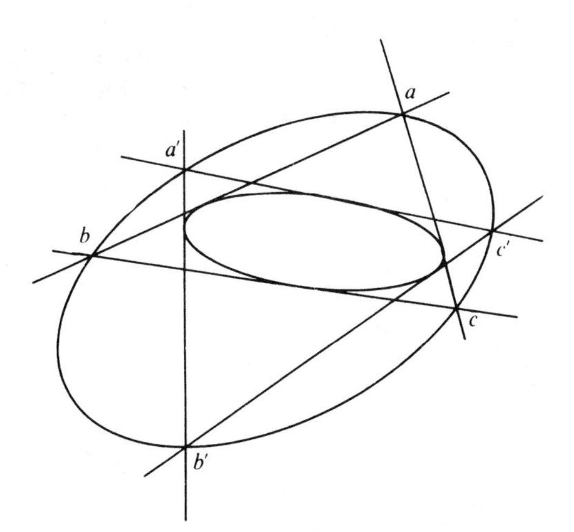

Figure 14.3.1.

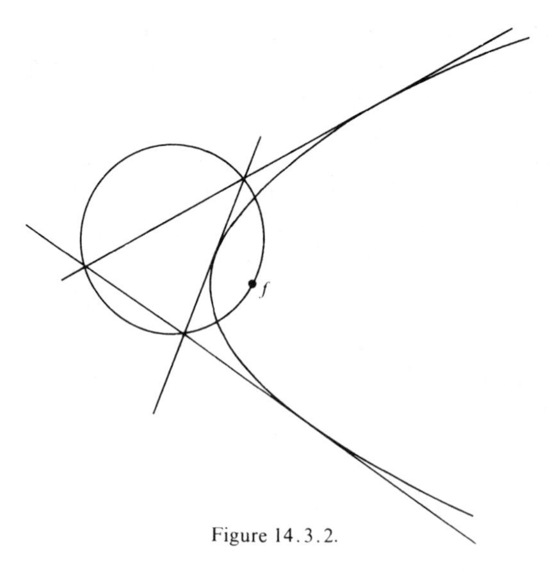

Figure 14.3.2.

Chapter 15

15.1 First part. Put $p=(a+b)/2$; we are going to be dealing with polarity with respect to C (cf. 15.C or [B, 15.5]). The polar line of m is $\langle a, b\rangle$, so the pole of the line $D=\langle m, p\rangle$ is the point x of $\langle a, b\rangle$ which is harmonically conjugate to p with respect to the pair of points (a, b). Since p is the midpoint of $[a, b]$, this means that $x=\infty_D$ is the point at infinity of D.

Since the pole of D is a point at infinity, D must contain the pole of the line at infinity ∞_X of the affine plane (cf. 14.E or [B, 14.5.1]). But this pole is exactly the center of C, by definition. By the construction of the polar line of a point by means of the tangency points of the tangents to the conic dropped from this point (cf. 14.E or [B, 14.5.2.6]), we see that the tangents to C at its intersection points with D (if they exist) must pass through the pole of D, i.e. through ∞_D. This says exactly that these tangents are parallel to $\langle a, b\rangle$.

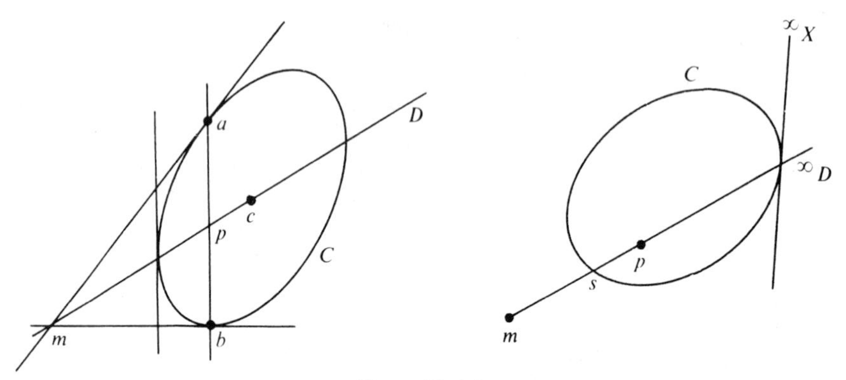

Figure 15.1.3.

Second part. Now if C is a parabola, the pole of ∞_X becomes the point ∞_C where C is tangent to the line at infinity, so that D is parallel to the axis of the parabola (cf. 15.C or [B, 15.5]). But D certainly intersects C since they have the point ∞_D in common. Since $[m, p, s, \infty_D] = -1$, we must have $s = (m + p)/2$ (see 6.B).

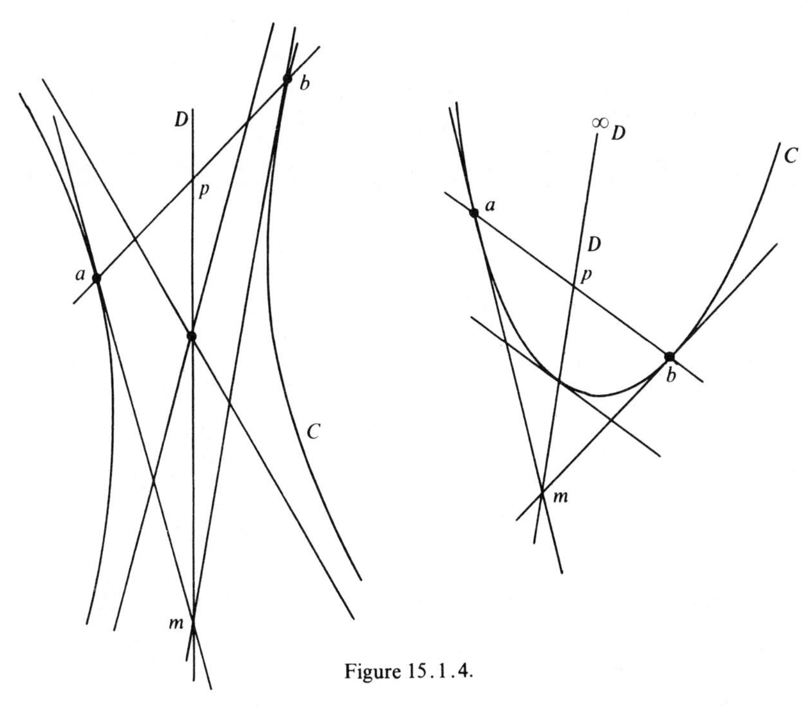

Figure 15.1.4.

Third part. If $\{m, a, b\}$ are given for a parabola, we successively construct $p = (a+b)/2$, $s = (m+p)/2$, $a' = (m+a)/2$, $b' = (m+b)/2$. The parabola will be tangent to $\langle a', b' \rangle$ at s. We can carry out the same construction for the triples $\{a', a, s\}$ and $\{b', b, s\}$ (see figure 15.1.5). We obtain $\text{area}(\{m, a', b'\}) = \frac{1}{4}\text{area}(\{m, a, b\})$. Further,

$$\begin{aligned}\text{area}(\{a', s, a\}) &= \text{area}(\{a', s, m\}) = \tfrac{1}{2}\text{area}(\{m, a', b'\}) \\ &= \tfrac{1}{8}\text{area}(\{m, a, b\}), \text{ and so on.}\end{aligned}$$

Thus the desired shaded area will be equal to

$$\begin{aligned}&\text{area}(\{m, a, b\})\left(1 - \frac{1}{4} - 2\cdot\frac{1}{4}\frac{1}{8} - 2\cdot\frac{1}{4}\frac{1}{32} - \cdots\right) \\ &= \text{area}(\{m, a, b\})\left(1 - \frac{1}{4} - \frac{1}{16} - \frac{1}{64} - \cdots\right) \\ &= \text{area}(\{m, a, b\})\left(2 - \frac{1}{1 - \frac{1}{4}}\right) = \frac{2}{3}\text{area}(\{m, a, b\}). \qquad \square\end{aligned}$$

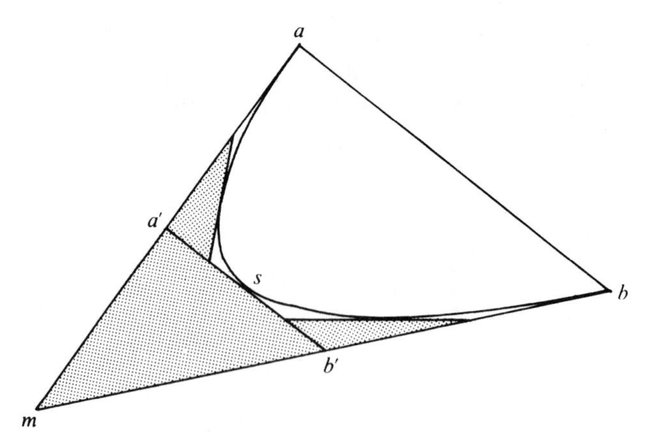

Figure 15.1.5.

15.2 The idea is to transform the given parallelogram into a square, for which the result is obvious. We use an appropriate affine transformation, and the ratio $\overrightarrow{c\alpha}/\overrightarrow{c\beta}$ is invariant for such transformations. Its value is $\sqrt{2}$ for a circle inscribed in a square; so it must be the same for the ellipse obtained by transforming back the circle. □

Note. This value $\sqrt{2}$ gives a construction for eight points of an ellipse of which two conjugate semi-diameters are known (15.C or [B, 15.5.10]). Using the eight points as a guide to fill in the rest of the curve, one avoids having to construct the axes (cf. [B, 17.9.22]).

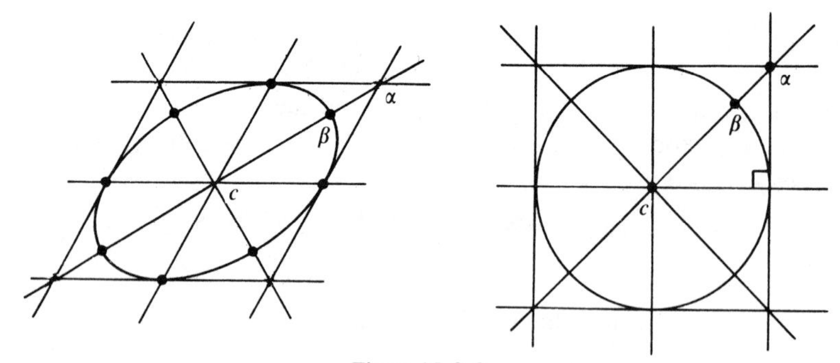

Figure 15.2.2.

15.3 Let $q=1$ be the equation of Q, where q is a quadratic form; we show the second property. Let u, v, w be three unit vectors whose directions are conjugate with respect to Q, and let a, b be the points where the parallel to u passing through x intersects Q. Then, up to a sign, the lengths $\overline{xa}, \overline{xb}$ are the roots of the equation in λ given by $q(x+\lambda u)=1$. If P is the polar form of q

(cf. 13.A or [B, 13.1.1]), we have

$$q(u)\lambda^2+2P(x,u)\lambda+q(x)-1=0.$$

The product $\overline{xa}\cdot\overline{xb}$ is thus the absolute value of the product of the roots, namely

$$\overline{xa}\cdot\overline{xb}=\frac{1-q(x)}{q(u)}.$$

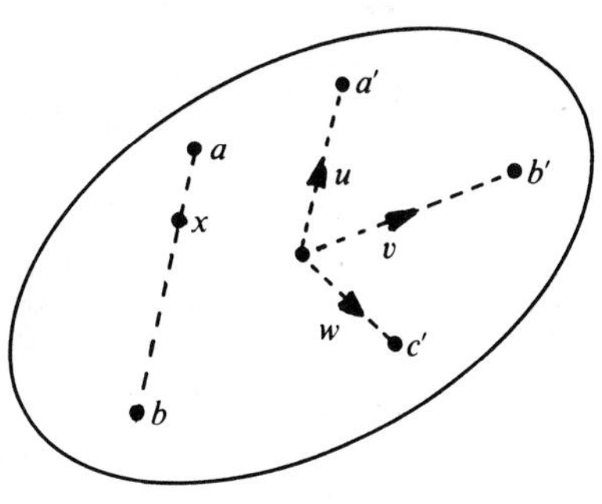

Figure 15.3.

We apply this to the case $x=O$, the center of Q. We obtain $q(u)=(Oa')^2$, where $[O,a']$ is a half-diameter in the direction of u. Doing the same for v and w, we finally get

$$\overline{xa}\cdot\overline{xb}+\overline{xc}\cdot\overline{xd}+\overline{xe}\cdot\overline{xf}=(1-q(x))\big((Oa')^2+(Ob')^2+(Oc')^2\big),$$

where Oa', Ob', Oc' are half-diameters parallel to u, v, w, respectively. By the theorem of Appolonius, the sum $(Oa')^2+(Ob')^2+(Oc')^2$ is a constant (cf. 15.D or [B, 15.6.4]). Its value is $\alpha^2+\beta^2+\gamma^2$ if the equation of Q is $x^2/\alpha^2+y^2/\beta^2+z^2/\gamma^2=1$ in an orthonormal frame. □

To show the first property, one proceeds exactly as above, except that now the three vectors u, v, w form an orthonormal frame, and one uses problem 15.4 below instead of the theorem of Appolonius. □

15.4 Let q be the quadratic form defining the ellipsoid $\mathscr{E}$ in the n-dimensional affine Euclidean space X, so that $\mathscr{E}=\{x\in X: q(x)=1\}$. The essential observation here is that the matrices representing q in different orthonormal bases are all similar (cf. [B, 15.6.7]); in fact, if A is the matrix representing q in one orthonormal basis, its matrix B in another basis will be ${}^tSAS$. But the matrix S is orthogonal since we're passing from one orthonormal basis to another, so ${}^tS=S^{-1}$ (cf. 8.A or [B, 8.2.1]), and we conclude $B=S^{-1}AS$ is indeed similar to A. So we can denote the trace of A by trace(q), and call it the *trace of* q, since the notion is independent of the basis in which A is expressed.

Now let $\{u_i\}$ be an arbitrary orthonormal basis, and $\{a_i\}$ the points where the half-lines $\langle O,u_i\rangle$ intersect $\mathscr{E}$. If $a_i=x_iu_i$, we must have $q(a_i)=x_i^2q(u_i)=1$, so $(Oa_i)^2 = x_i^2 = 1/[q(u_i)]$, and consequently

$$\sum_{i=1}^{n} \frac{1}{(Oa_i)^2} = \sum_{i=1}^{n} q(u_i) = \operatorname{trace}(q). \qquad \square$$

If $q = \sum_{i=1}^{n}(x_i^2/\alpha_i^2)$ in a basis that diagonalizes q, we will have, in particular:

$$\sum_{i=1}^{n} \frac{1}{(Oa_i)^2} = \sum_{i=1}^{n} \frac{1}{\alpha_i^2}.$$

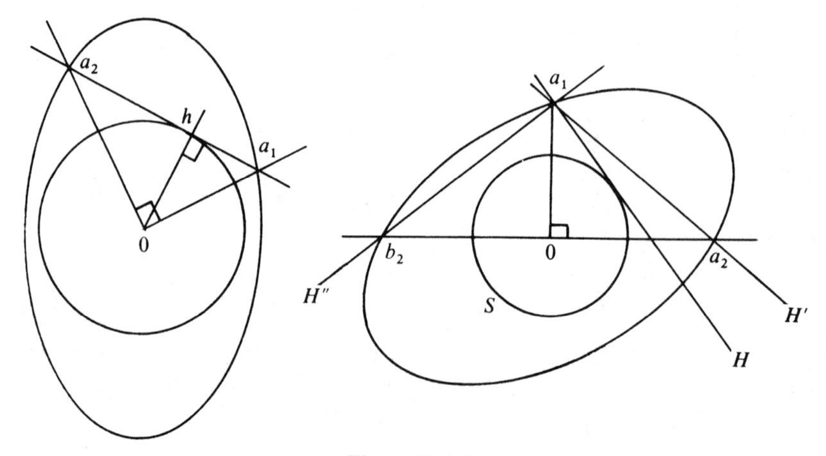

Figure 15.4.1.

From 9.B or [B, 9.2.6.4], the distance h from the center O to the affine hyperplane determined by the n points a_i must be constant and equal to

$$\frac{1}{h^2} = \operatorname{trace}(q) = \sum_{i=1}^{n} \frac{1}{(Oa_i)^2} = \sum_{i=1}^{n} \frac{1}{\alpha_i^2}.$$

The conclusion is that this hyperplane is always tangent to the sphere S of radius h centered at O.

Finally, we want to show that this family of hyperplanes envelopes S, i.e. every tangent hyperplane to S is obtained from an appropriate set (a_i). Let H be an arbitrary hyperplane tangent to S. Since h is smaller than all the α_i, the hyperplane H necessarily intersects $\mathscr{E}$; we can thus find points $(a_i)_{i=1,\ldots,n-1}$ in $\mathscr{E}$, by induction, so that each a_i is in H and the vectors $\overrightarrow{Oa_i}$ are pairwise orthogonal. Finally, let a_n, b_n be the two points where the line passing through the origin, and orthogonal to the $n-1$ vectors $\overrightarrow{Oa_i}$, intersects $\mathscr{E}$. According to the preceding result, the distance from O to the hyperplane defined by the a_i $(i=1,\ldots,n-1)$ and by a_n is equal to h, and thus equal to the distance from O to H. This shows that H passes through a_n or b_n. $\square$

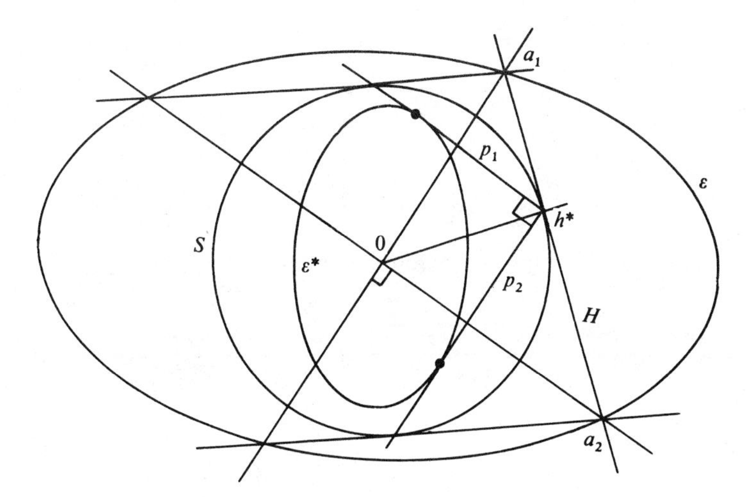

Figure 15.4.2.

Let's apply a polarity transformation with respect to a sphere centered at O to the above property. The simplest thing to do is to use the sphere S itself. (See 10.B, 14.E, 15.C or [B, 10.7.11, 14.5, 15.5] for polarity transformations with respect to quadrics.) Starting from the n points a_i and the hyperplane H spanned by them, the polarity gives us n hyperplanes H_i and a point h^*. The properties above say that each H_i is tangent to the ellipsoid $\mathscr{E}^*$, the polar body of $\mathscr{E}$ with respect to S, and that the H_i are pairwise orthogonal and their intersection point h^* belongs to S.

If we now start from the ellipsoid $\mathscr{E}^*$, which can be assumed arbitrary, we obtain the following result: the locus of points through which pass n pairwise orthogonal hyperplanes tangent to an ellipsoid is a sphere, concentric with the ellipsoid, and called its *orthoptic sphere*. □

15.5 The problem of the normals to a plane conic was studied geometrically, so we'll solve this problem analytically (cf. [B, 17.5.5.6]). Let's consider for instance the case of a quadric of equation $ax^2 + by^2 + cz^2 + d = 0$. Its tangent plane at the point (x_0, y_0, z_0) has the following equation:

$$axx_0 + byy_0 + czz_0 + d = 0 \quad (\text{cf. 15.C or [B, 15.5.5]}).$$

The normal direction is given by the vector (ax_0, by_0, cz_0), so the line joining (x_0, y_0, z_0) to the point $m = (u, v, w)$ will be normal to Q if and only if

$$\frac{x_0 - u}{ax_0} = \frac{y_0 - v}{by_0} = \frac{z_0 - w}{cz_0}.$$

Let this ratio be λ. A cone with vertex $m(u, v, w)$ and containing parallels to the coordinate axes passing through m will have an equation of the form

$$\xi(y - v)(z - w) + \eta(z - w)(x - u) + \zeta(x - u)(y - v) = 0.$$

Since

$$x_0=\frac{u}{1-\lambda a},\quad y_0=\frac{v}{1-\lambda b},\quad z_0=\frac{w}{1-\lambda c},$$

we get

$$x_0-u=\frac{-\lambda au}{1-\lambda a},\quad y_0-v=\frac{-\lambda bv}{1-\lambda b},\quad z_0-w=\frac{-\lambda cw}{1-\lambda c},$$

and substituting this into the desired equation for the cone, we get the condition

$$\frac{\xi bcvw}{(1-\lambda b)(1-\lambda c)}+\frac{\eta cawu}{(1-\lambda c)(1-\lambda a)}+\frac{\zeta abuv}{(1-\lambda a)(1-\lambda b)}=0,$$

which becomes

$$\lambda abc(\xi vw+\eta wu+\zeta uv)=\xi bcvw+\eta wuca+\zeta uvab.$$

This is satisfied for any value of λ if

$$\xi vw+\eta wu+\zeta uv=0\quad\text{and}\quad \xi vwbc+\eta wuca+\zeta uvab=0.$$

We thus find

$$\xi=a(b-c)u,\quad \eta=b(c-a)v,\quad \zeta=c(a-b)w.$$

The equation of the cone will be

$$\begin{aligned}a(b-c)u(y-v)(z-w)&+b(c-a)v(z-w)(x-u)\\&+c(a-b)w(x-u)(y-v)=0.\end{aligned}$$

It is easy to see that this is indeed the desired cone, and that it contains the center $(0,0,0)$ of Q. □

15.6 For given values of x, y, z, the equation

$$Q(\lambda)=\frac{x^2}{a^2+\lambda}+\frac{y^2}{b^2+\lambda}+\frac{z^2}{c^2+\lambda}-1=0$$

is of third degree in λ, so it has at most three roots. We want to show that it has exactly three roots whenever x, y, z are all three non-zero and the tangent planes to the corresponding quadrics are pairwise orthogonal. However, a direct attempt to solve the problem leads to very involved calculations.

The idea is to work backwards. Let u,v,w be arbitrary real numbers, subject only to the conditions $u\in\,]-c^2,+\infty[$, $v\in\,]-b^2,-c^2[$, $w\in\,]-a^2,b^2[$. Then (see 15.B or [B, 15.3.3]) the quadric $Q(u)$ (resp. $Q(v),Q(w)$) is an ellipsoid (resp. a one- or two-sheeted hyperboloid), and they intersect in a single point $(x(u,v,w),y(u,v,w),z(u,v,w))$ with $x>0$, $y>0$, $z>0$. To show this we only have to solve the linear system in x^2, y^2, z^2 which is given by the three equations $Q(u)=0$, $Q(v)=0$ and $Q(w)=0$. We find, for u and v

for instance:

$$x^2(u,v,w) = (a^2+u)(a^2+v)(a^2+w)/(a^2-b^2)(a^2-c^2)$$
$$y^2(u,v,w) = (b^2+u)(b^2+v)(b^2+w)/(b^2-c^2)(b^2-a^2)$$
$$z^2(u,v,w) = (c^2+u)(c^2+v)(c^2+w)/(c^2-b^2)(c^2-a^2).$$

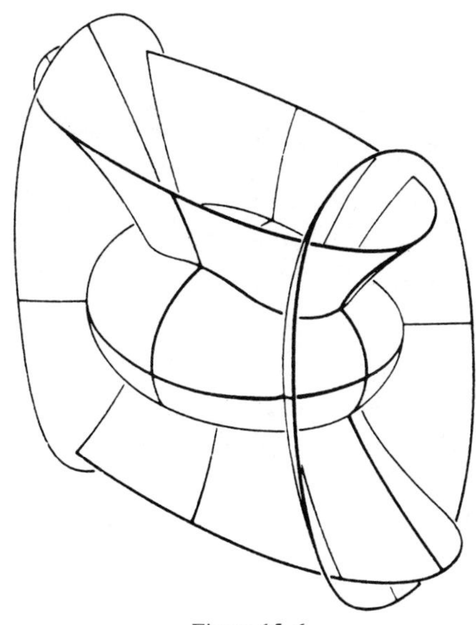

Figure 15.6.

The map $(u,v,w) \mapsto (x(u,v,w), y(u,v,w), z(u,v,w))$ is thus surjective onto the first octant $x>0$, $y>0$, $z>0$. But since, for fixed (x,y,z), there are at most three possible values of λ, they must be exactly u, v, w, and the mapping is bijective. We conclude that through each point of the first octant there pass exactly three quadrics $Q(\lambda)$: an ellipsoid, a one-sheeted hyperboloid and a two-sheeted hyperboloid.

Orthogonality of the tangent planes at this point is easily proved using the formulas above. At (x,y,z) the tangent plane to $Q(\lambda)$ has a normal vector $(x/(a^2+\lambda), y/(b^2+\lambda), z/(c^2+\lambda))$. Using the formulas given above for x^2, y^2 and z^2, we get

$$\left(\left(\frac{x}{a^2+u}, \frac{y}{b^2+u}, \frac{z}{c^2+u}\right)\middle|\left(\frac{x}{a^2+v}, \frac{y}{b^2+v}, \frac{z}{c^2+v}\right)\right)$$
$$= \frac{x^2}{(a^2+u)(a^2+v)} + \frac{y^2}{(b^2+u)(b^2+v)} + \frac{z^2}{(c^2+u)(c^2+v)}$$
$$= \frac{a^2+w}{(a^2-b^2)(a^2-c^2)} + \frac{b^2+w}{(b^2-c^2)(b^2-a^2)} + \frac{c^2+w}{(c^2-b^2)(c^2-a^2)} = 0. \square$$

REMARK. The coordinate change $(x, y, z) \mapsto (u, v, w)$, given by the (non-explicit) inversion of the formulas above, is sometimes essential for the study of several questions involving the ellipsoid

$$\frac{x^2}{a^2} + \frac{y^2}{b^2} + \frac{z^2}{c^2} - 1 = 0.$$

The case of conics is much easier; it is studied geometrically in 17.B, 17.D.1 or [B, 17.6.3].

Chapter 16

16.1 Recall that we gave a geometric proof in 16.C (and [B, 16.2.12]). Here we will work analytically, with the parametric representation $t \mapsto (t, t^2, 1)$ of the conic C given by $x^2 - yz = 0$, in projective coordinates such that $a = (1,0,0)$, $\beta = (0,1,0)$ and $\gamma = (0,0,1)$. The tangent to C at point $\alpha = (u, v, w)$ has the equation

$$ux - \tfrac{1}{2}(vz + wy) = 0$$

(cf. 14.E or [B, 14.5.3]), which here becomes $tx - \frac{1}{2}(t^2 z + y) = 0$. Thus this tangent intersects $\langle a, \beta \rangle$ in $c = (\frac{1}{2}, t, 0)$, and it intersects $\langle a, \gamma \rangle$ in $b = (t/2, 0, 1)$. The lines $\langle a, \alpha \rangle$, $\langle b, \beta \rangle$ and $\langle c, \gamma \rangle$ then have the following equations: $y - t^2 z = 0$, $x - \frac{1}{2}tz = 0$ and $tx - \frac{1}{2}y = 0$, respectively. Thus the point $(t/2, t^2, 1)$ belongs to all three. □

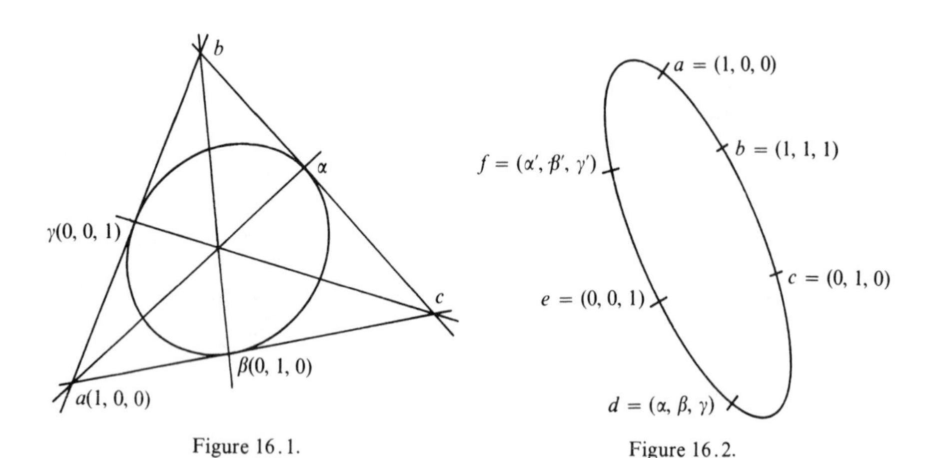

Figure 16.1. Figure 16.2.

16.2 Consider the six points a, b, c, d, e, f with projective coordinates $a = (1,0,0)$, $b = (1,1,1)$, $c = (0,1,0)$, $d = (\alpha, \beta, \gamma)$, $e = (0,0,1)$, $f = (\alpha', \beta', \gamma')$. The six lines $\langle a, b \rangle$, $\langle b, c \rangle$, $\langle c, d \rangle$, $\langle d, e \rangle$, $\langle e, f \rangle$, $\langle f, a \rangle$ have the following equations: $y = z$, $x = z$, $\gamma x = \alpha z$, $\beta x = \alpha y$, $\beta' x = \alpha' y$, $\gamma' y = \alpha' z$, respectively; their intersection points $\langle a, b \rangle \cap \langle d, e \rangle$, $\langle b, c \rangle \cap \langle e, f \rangle$, $\langle c, d \rangle \cap \langle f, a \rangle$ have coordinates (α, β, β), $(\alpha', \beta', \alpha')$ and $(\alpha\gamma', \alpha'\gamma, \gamma\gamma')$, respectively. Thus the three points will be collinear if and only if the determinant

$$\begin{vmatrix} \alpha & \alpha' & \alpha\gamma' \\ \beta & \beta' & \alpha'\gamma \\ \beta & \alpha' & \gamma\gamma' \end{vmatrix}$$

is zero.

To determine whether the initial six points are on the same conic, start by observing that this conic belongs to the pencil determined by the four points $(1,0,0)$, $(0,1,0)$, $(0,0,1)$ and $(1,1,1)$ which form the projective base. There are two degenerate conics with simple equations in this pencil, $y(x-z)$ and $x(y-z)$, so every conic of the pencil must be of the form $y(x-z)=kx(y-z)$, where k is any scalar in the base field (cf. 16.F or [B, 16.4.10]). The points (α,β,γ) and (α',β',γ') will both belong to such a conic if and only if

$$\frac{\beta(\alpha-\gamma)}{\alpha(\beta-\gamma)}=\frac{\beta'(\alpha'-\gamma')}{\alpha'(\beta'-\gamma')}.$$

Finally, we verify that this condition coincides with the condition on the determinant given above. □

16.3 In a projective coordinate system for which $p=(1,0,0)$, $q=(0,1,0)$ and $r=(0,0,1)$, we can assume that the conic has equation $x^2-yz=0$. Let $m=(s,s^2,1)$ and $n=(t,t^2,1)$; the four lines $\langle p,q\rangle$, $\langle p,r\rangle$, $\langle p,m\rangle$ and $\langle p,n\rangle$ have equations $z=0$, $y=0$, $s^2y-z=0$ and $t^2y-z=0$, respectively. The cross-ratio of the four lines is thus given by that of the pairs $(0,1)$, $(1,0)$, $(s^2,-1)$ and (t^2-1), or of the points $0,\infty,-s^2,-t^2$ (cf. 19.C or [B, 16.2]); its value is t^2/s^2. The cross-ratio of the four points q,r,m,n on C, on the other hand, is equal to that of their parameters (for any good parametrization; see 16.C or [B, 16.2.5]), which here have the values $0,\infty,s,t$; thus the cross-ratio is t/s. □

Recall (cf. 17.C or [B, 17.4.2]) that a very pretty application of this result is obtained by taking C as a circle in the Euclidean plane, complexifying and completing it into $\overline{C}$, and taking q,r as the two cyclical points (cf. 9.D or [B,

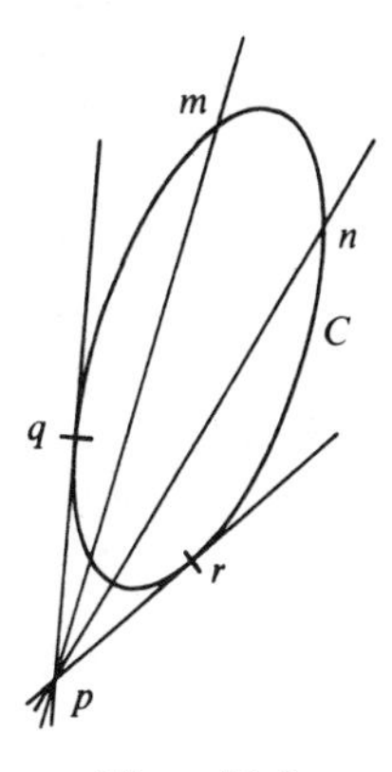

Figure 16.3.

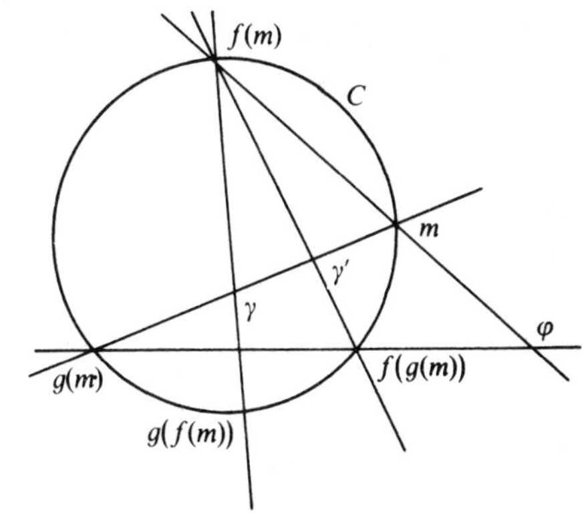

Figure 16.4.

9.5.5]). The Laguerre formula, followed by a return to C (cf. 8.H or [B, 8.8]), transforms the result we've just proved into the fact that the central angle is twice the inscribed angle (cf. 10.D or [B, 10.9.3]).

16.4 Let f, g be the two involutions of a conic C, and let φ, γ be their respective Frégier points. According to the defining property of Frégier points, we have that, for an arbitrary point $m \in C$, the image $f(m)$ is the other intersection point of C with the line $\langle\varphi, m\rangle$. Analogously, $g(m)$ is in $C \cap \langle\gamma, m\rangle$ and $f(g(m))$ is in $C \cap \langle\varphi, g(m)\rangle$, and finally $g(f(m))$ is in $C \cap \langle\gamma, f(m)\rangle$. If f and g commute, we get $f(g(m)) = g(f(m))$ and φ and γ are indeed conjugate with respect to C, by the construction mentioned in 14.E or [B, 14.5.2.6].

Conversely, if φ and γ are conjugate with respect to C, the same geometric construction implies that the point $\gamma' = \langle g(g(m)), f(m)\rangle \cap \langle m, g(m)\rangle$ is conjugate to φ with respect to C. If $\gamma = \gamma'$, we're done: $f(g(m)) = g(f(m))$. If not, the line $\langle\gamma, \gamma'\rangle$ is the polar of φ; but then $m = f(m)$, and $g(m) = f(g(m))$, whence $f(g(m)) = g(m) = g(f(m))$. □

16.5 In the case when the field of scalars is $\mathbf{R}$, we start by working with the complexification $\overline{C}$ of the conic C (cf. 17.C or [B, 17.4.1]) and the complexification $\bar{f}$ of the homography f. The extension $\bar{f}$ of f to $\overline{C}$ has thus exactly two fixed points, by assumption. We can assume these points are $(0,1,0)$ and $(0,0,1)$, and that $\overline{C}$ has equation $x^2 - yz = 0$ (cf. 16.A or [B, 16.1]). A good parametrization is given by $(s,t) \mapsto (st, s^2, t^2)$ (cf. 16.B or [B, 16.2]).

An arbitrary homography of C is expressed, in this parametrization, by $(s,t) \mapsto (as + bt, cs + dt)$; but since its two fixed points are $(0,1,0)$ and $(0,0,1)$, it must in fact have the form $(s,t) \mapsto (as, dt)$.

The line joining $m = (st, s^2, t^2)$ to $f(m) = (adst, a^2s^2, d^2t^2)$ will have the following equation:

$$(a^2 - d^2)s^2t^2x + (ad - d^2)st^3y + (a^2 - ad)s^3tz = 0,$$

or, more simply, $(a+d)stx + dt^2y + as^2z = 0$. Since the coefficients of x, y, z have degree two in (s,t), we conclude (cf. 14.F or [B, 14.6]) that this line, for a and d variable, forms the envelope of a conic $\overline{\Gamma}$. More precisely, in the dual coordinates (u, v, w) of (x, y, z), the conic $\overline{\Gamma}$ is parametrized by $(u = (a+d)st, v = dt^2, w = as^2)$. The envelope equation of $\overline{\Gamma}$ is obtained by eliminating s and t, so we get $adu^2 - (a+d)vw = 0$. The equation of the points is obtained by taking the inverse of the matrix that gives the tangents (cf. 14.F or [B, 14.6.1]), namely:

$$\begin{pmatrix} ad & 0 & 0 \\ 0 & 0 & -\dfrac{(a+d)^2}{2} \\ 0 & -\dfrac{(a+d)^2}{2} & 0 \end{pmatrix}^{-1} = \begin{pmatrix} a^{-1}d^{-1} & 0 & 0 \\ 0 & 0 & \dfrac{-2}{(a+d)^2} \\ 0 & \dfrac{-2}{(a+d)^2} & 0 \end{pmatrix},$$

which gives the equation $a^{-1}d^{-1}x^2 - [4/(a+d)^2]yz = 0$. Since this equation is of the form $x^2 - yz + kx^2$, the conic is indeed bitangent to $\bar{C}$, and the contact points are the fixed points of $\bar{f}$.

If we started with **R** as the field of scalars, we now have to come back to C. But since $\bar{C}$ and $\bar{f}$ come from a real conic and a real homography, the line of fixed points of $\bar{f}$ (whose equation is $x = 0$) is the complexification of a real line, and $\bar{\Gamma}$ is the complexification of a real conic Γ, which is indeed bitangent to C. □

We note en passant that bitangence in the complex case could have been heuristically deduced. In fact, if m is a fixed point of f, the line $\langle m, f(m)\rangle$ more or less represents the tangent to C at m (see box at the beginning of chapter 16).

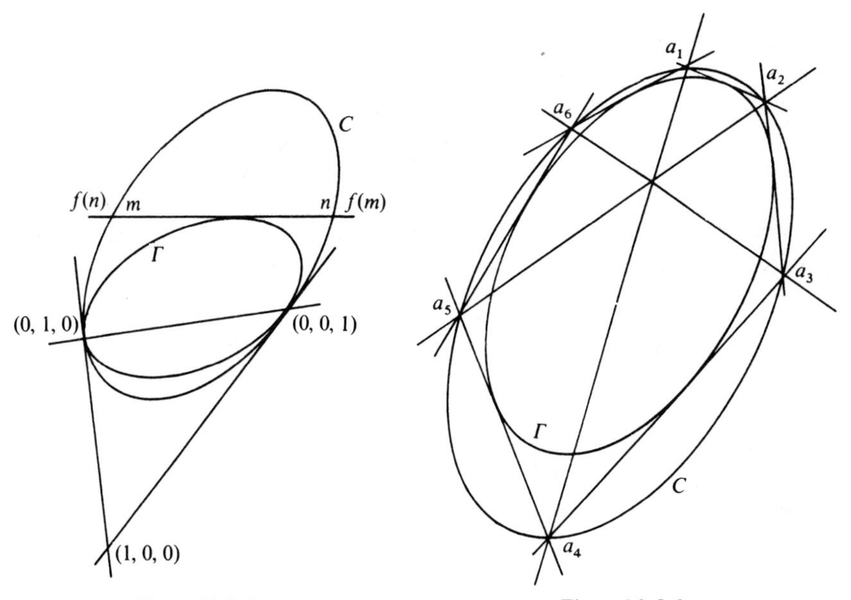

Figure 16.5.1. Figure 16.5.2.

Conversely, if C and Γ are bitangent, they can always be written in the form above (after complexifying, if necessary). We find the coefficients a and d which determine the homography f. Then we use the fact that if a tangent to Γ intersects C in two distinct points m and n, we either have $n = f(m)$ or $m = f(n)$. Moreover, if m and n are distinct, the two cases $n = f(m)$ and $m = f(n)$ are mutually exclusive; this is because a non-involutive homography has no point of period two (cf. 6.D or [B, 6.7.3]), or it can be proved here by a direct argument. □

Now suppose that $(a_i)_{i=1,\ldots,n}$ is a polygon inscribed in C and circumscribed around Γ, where Γ is bitangent to C. Up to a permutation of indices, we can assume that the homography f defining Γ satisfies $f(a_1) =$

$a_2, f(a_2) = a_3, \ldots, f(a_{n-1}) = a_n, f(a_n) = a_1$. The idea is to introduce the n-th iterate of f:

$$f^n = f \circ f \circ \cdots \circ f.$$

The homography f^n fixes each a_i. Since any homography of C with three fixed points is the identity, we have $f^n = \mathrm{Id}_C$.

But then, for any point m on C, the polygon $(m, f(m), f^2(m), \ldots, f^{n-1}(m))$ is inscribed in C and circumscribed around Γ. □

Note. If the number of sides is even, say $n = 2k$, we see that $f^k \circ f^k = \mathrm{Id}_C$, so f^k is an involution of C. According to 16.D (or [B, 16.3.6]), this implies that the lines $\langle a, a_{k+i} \rangle$ are concurrent, and their intersection point does not depend on the polygon. More generally, for every h between 2 and $n-2$, the lines $\langle a_i, a_{h+i} \rangle$ joining every h-th a_i are tangent to a conic bitangent to C.

16.6 For a given tangential pencil of real conics, we want to study the existence of conics in this pencil that pass through a fixed point in the plane. We will only treat the most complicated case, namely a pencil formed by the conics which are tangent to four lines in general position. Using a projective transformation, we can always reduce the problem to the case when the four lines have equations $x = \pm 1$, $y = \pm 1$.

It is instructive to use two different methods to determine the equations of the conics in the pencil. First we proceed directly: a conic

$$ax^2 + 2bxy + cy^2 + 2dx + 2ey + f = 0$$

has the lines $x = \pm 1$ as tangents if and only if the equations

$$a \pm 2by + cy^2 \pm 2d + 2ey + f = 0$$

have each a double root, which implies $(e \pm b)^2 = c(f \pm 2d + a)$. Analogously, having $y = \pm 1$ as tangents is equivalent to $(d \pm b)^2 = a(f \pm 2e + c)$. These four conditions are equivalent to the following:

$$eb = cd, \quad db = ae, \quad e^2 + b^2 = c(f + a), \quad d^2 + b^2 = a(f + c).$$

We conclude that $(ac - b^2)d = (ac - b^2)e = 0$. In the case $ac - b^2 \neq 0$, we find $d = e = 0$, then $a = c$ and $b^2 = a^2 + af$, whence the following general equation:

$$a^2(x^2 + y^2) + 2abxy + b^2 - a^2 = 0. \qquad \text{(C)}$$

Observe that a geometrical investigation of the diameters (cf. 15.C or [B, 15.5.8]) shows that the center of the conic, if it exists, must be at the origin.

In the case $ac - b^2 = 0$, we find equations of the form $(ux + vy + w)^2$, which correspond to a pencil of degenerate tangential conics, i.e. two pencils of lines (cf. [B, 14.6.3]).

A conic of form (C) passes through the point (x, y) if and only if its coefficients (a, b) satisfy the following equation:

$$(x^2 + y^2 - 1)a^2 + 2xyab + b^2 = 0,$$

and this equation has roots if and only if $(x^2-1)(y^2-1)\geq 0$. The solution set of this inequality is the complement of the shaded region in the figure below; in the shaded region itself there are no conics tangent to the four lines. □

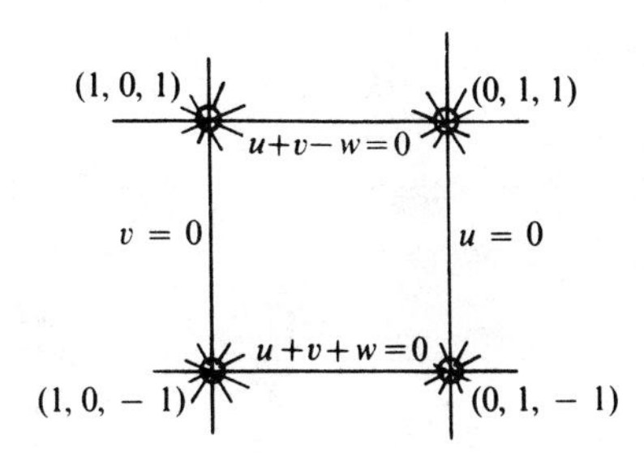

Figure 16.6.1.

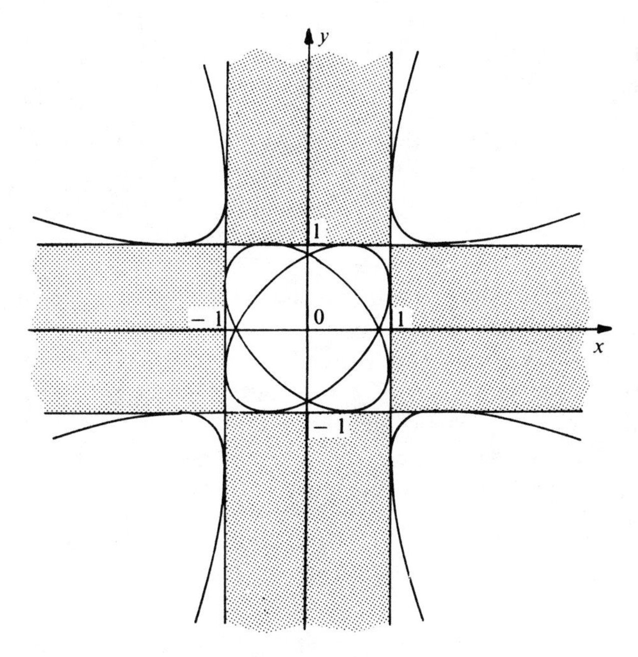

Figure 16.6.2.

The second method uses envelope equations. In homogeneous coordinates (u,v,w) dual to the cartesian coordinates $(x,y,1=z)$, the lines $x=\pm 1$, $y=\pm 1$ become points with coordinates $(1,0,\pm 1)$ and $(0,1,\pm 1)$, respectively. This gives rise to two pairs of opposite lines joining these points: $v=0$, $u=0$, $u+v+w=0$ and $u+v-w=0$.

The envelope equation of the corresponding pencil is (cf. 16.F or [B, 16.4.10]):

$$(u+v+w)(u+v-w)+\lambda uv=0,$$

or again

$$u^2+(2+\lambda)uv+v^2-w^2=0,$$

which corresponds to the matrix

$$A=\begin{pmatrix}1 & \lambda & 0\\ \lambda & 1 & 0\\ 0 & 0 & -1\end{pmatrix}$$

(substituting λ for $\lambda+2$). The punctual equation in the variables $(x, y, z=1)$ is then given by the inverse inverse matrix A^{-1}, which is equal to

$$A^{-1}=\begin{pmatrix}\dfrac{-1}{\lambda^2-1} & \dfrac{\lambda}{\lambda^2-1} & 0\\ \dfrac{\lambda}{\lambda^2-1} & \dfrac{1}{\lambda^2-1} & 0\\ 0 & 0 & -1\end{pmatrix}.$$

This gives us the equation $-x^2+2\lambda xy-y^2+1-\lambda^2=0$, which coincides with (C). □

16.7 For $y=0$, we must have $z=0$ and we get the point $(1,0,0)$. For $z=0$ we get $y=0$ and the same point. Otherwise we must have

$$x=\frac{y^2}{z}=\frac{z^2}{y},\quad \text{whence}\quad \left(\frac{y}{z}\right)^3=1.$$

But 1 is the only cube root of unity in the field with three elements, so we get just the point $(1,1,1)$.

We investigate the tangents to the two conics at their intersection points. At $(1,0,0)$ both tangents are given by $z=0$ (see 14.E or [B, 14.5.3]); and at $(1,1,1)$ the equations are $(\frac{1}{2}x+z)=y$ and $(\frac{1}{2}x+y)=z$, which are both the same as $x+y+z=0$ in a field of characteristic three. We conclude that the two conics are tangent at both their intersection points. □

Chapter 17

17.1 Let C be a conic and a an arbitrary point on it. Using the Laguerre formula (8.H or [B, 8.8.7]) and duality of pencils of lines (4.B or [B, 6.5.1]), we observe that saying that the oriented angle between the lines D, D' is constant is equivalent to saying that D, seen as a point in the projective dual a^*, is taken to D' by a fixed homography f (apply the formula in 6.D or [B, 6.6.1]). But, by 16.D (or [B, 16.3]), this means that there is a homography $\bar{f}$ of C such that $m'=\bar{f}(m)$, where m (resp. m') is the second point in the intersection of C and D (resp. C and D'). Now we apply problem 16.5: the family of lines $\langle m, f(m)\rangle$ envelopes a conic Γ, whose complexification is bitangent to the complexification of C. In real space Γ can never be bitangent to C because D and D' are always distinct. □

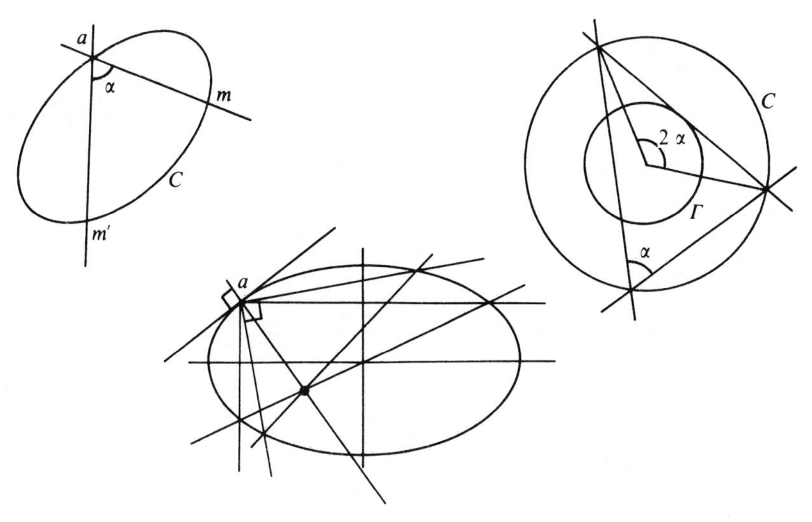

Figure 17.1.

When the two lines form a right angle, f is an involution, and $\langle m, f(m)\rangle$ passes through a fixed point, the Frégier point of f (cf. 16.D or [B, 16.3.6]). This point can be found by using the normal to C at a and parallel lines to the axes. Observe that if C is a circle, the conic Γ is a circle concentric with C. This can be proved either by noting that Γ is bitangent to C at the two cyclical points (17.C or [B, 17.4.2]), or directly, by considering the fact that the central angle is twice the inscribed angle (10.D or [B, 10.9.3]). □

17.2 Cocyclic points on an ellipse. We first give a geometric proof. We know from 17.C (or [B, 17.5.5]) that m_1, m_2, m_3 and m_4 are cocyclic if and only if the chords $\langle m_1, m_2\rangle$ and $\langle m_3, m_4\rangle$ have the same slope with respect to the axes. Considering the ellipse as the image of the circle under an affine map (cf. 17.A or [B, 17.7.1]), and denoting by $\langle m_1', m_2'\rangle$ and $\langle m_3', m_4'\rangle$ the chords of this circle whose images are $\langle m_1, m_2\rangle$ and $\langle m_3, m_4\rangle$, we get that the two chords of the ellipse have the same slope if and only if the same happens for the chords of the circle. But for a circle the angle formed by the chord $\langle m_1', m_2'\rangle$ and the axis is given by $(t_1+t_2)/2$ modulo π, since the parametrization on the circle (which corresponds to that on the ellipse) is just given by the angle with the axis. So we get the desired condition:

$$\frac{t_1+t_2}{3} \equiv -\frac{t_3+t_4}{2} \pmod{\pi},$$

or again

$$t_1+t_2+t_3+t_4 \equiv 0 \pmod{2\pi}. \qquad \square$$

The analytical proof is interesting. Observe first that it is sufficient to prove that the relation is necessary. In fact, if it is necessary it uniquely determines t_4 as a function of t_1, t_2, t_3; on the other hand, these three points define a circle,

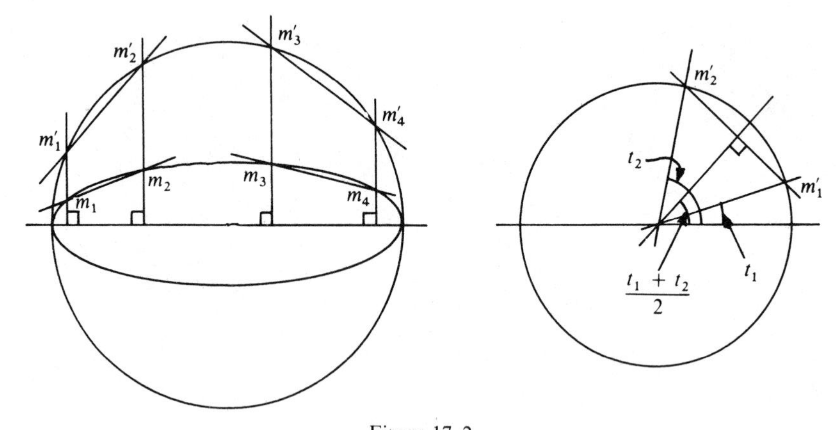

Figure 17.2.

which must intersect the ellipse in a well-defined fourth point (cf. 16.F or [B, 16.4.8]). Thus, let $x^2+y^2+2\alpha x+2\beta y+\gamma=0$ be the general equation of a circle in the plane; if we put $\vartheta=\tan(t/2)$, we get

$$x=a\frac{1-\vartheta^2}{1+\vartheta^2}\quad\text{and}\quad y=2b\frac{\vartheta}{1+\vartheta^2}.$$

The desired values of ϑ are the roots of the equation

$$a^2(1-\vartheta^2)^2+4b^2\vartheta^2+2\alpha a(1-\vartheta^2)(1+\vartheta^2)+4\beta b\vartheta(1+\vartheta^2)$$
$$+\gamma(1+\vartheta^2)^2=0,$$

which is of fourth degree in ϑ. Its important property is that the coefficients of ϑ and ϑ^3 are opposite to one another. Thus, denoting by $\sigma_k(k=1,2,3,4)$ the symmetric polynomials in $\vartheta_i(i=1,2,3,4)$, i.e.

$$\sigma_1=\vartheta_1+\vartheta_2+\vartheta_3+\vartheta_4,\quad \sigma_2=\vartheta_1\vartheta_2+\cdots+\vartheta_3\vartheta_4,$$
$$\sigma_3=\vartheta_1\vartheta_2\vartheta_3+\vartheta_1\vartheta_2\vartheta_4+\vartheta_1\vartheta_3\vartheta_4+\vartheta_2\vartheta_3\vartheta_4,\quad \sigma_4=\vartheta_1\vartheta_2\vartheta_3\vartheta_4,$$

we find that for any circle the parameters ϑ_i satisfy the relation $\sigma_1+\sigma_3=0$.

It is not difficult to see that

$$\tan\left(\frac{t_1}{2}+\frac{t_2}{2}+\frac{t_3}{2}+\frac{t_4}{2}\right)=\frac{\sigma_1+\sigma_3}{1-\sigma_2+\sigma_4}.$$

This implies that

$$\tan\left(\frac{t_1}{2}+\frac{t_2}{2}+\frac{t_3}{2}+\frac{t_4}{2}\right)=0,$$

or again

$$\frac{t_1}{2}+\frac{t_2}{2}+\frac{t_3}{2}+\frac{t_4}{2}\equiv 0\,(\operatorname{mod}\pi).$$

□

Application. If a circle osculates an ellipse at a point whose parameter is t, this implies (16.E or [B, 16.4.3]) that the value t is a triple root. The remaining intersection point will have a parameter t' such that $3t + t' \equiv 0 (\text{mod}\, 2\pi)$, e.g. $t' = -3t$.

Thus when $t_1 + t_2 + t_3 + t_4 \equiv 0 (\text{mod}\, 2\pi)$, we also have

$$t_1' + t_2' + t_3' + t_4' \equiv -3(t_1 + t_2 + t_3 + t_4) \equiv 0 (\text{mod}\, 2\pi). \qquad \square$$

Normals to an ellipse. Let (u, v) be the point from which we want to drop normals to the ellipse. Such a normal will intersect the curve in the point

$$(a \cos t, b \sin t) = \left(a \frac{1 - \vartheta^2}{1 + \vartheta^2}, b \frac{2\vartheta}{1 + \vartheta^2} \right)$$

if and only if the vector $(u - a \cos t, v - b \sin t)$ is perpendicular to the tangent vector $(-a \sin t, b \cos t)$ at this point. This gives the condition

$$-a \sin t (u - a \cos t) + b \cos t (v - b \sin t) = 0,$$

from which we get the following equation in ϑ:

$$bv\vartheta^4 + 2(au + a^2 - b^2)\vartheta^3 + 2b(au + b^2 - a^2)\vartheta - bv = 0.$$

The symmetric polynomials σ_i must then satisfy the following conditions:

$$\sigma_2 = 0 \quad \text{and} \quad \sigma_4 = -1.$$

In particular, we deduce that the expression

$$\tan\left(\frac{t_1}{2} + \frac{t_2}{2} + \frac{t_3}{2} + \frac{t_4}{2} \right) = \frac{\sigma_1 + \sigma_3}{1 - \sigma_2 + \sigma_4}.$$

is infinite. This implies

$$\frac{t_1 + t_2 + t_3 + t_4}{2} \equiv \frac{\pi}{2} (\text{mod}\, \pi), \quad \text{or} \quad t_1 + t_2 + t_3 + t_4 \equiv \pi (\text{mod}\, 2\pi).$$

Conversely, if $\sigma_2 = 0$ and $\sigma_4 = -1$, we can clearly find a point (u, v) satisfying the equation above, starting from the values of σ_1 and σ_3. $\square$

Application. We obviously obtain the theorem of Joachimstal as a consequence, for diametrically opposite points have parameters differing by π, and if

$$t_1 + t_2 + t_3 + t_4 \equiv \pi (\text{mod}\, 2\pi),$$

the point with parameter $t_4 + \pi$ satisfies $t_1 + t_2 + t_3 + (t_4 + \pi) \equiv 0 (\text{mod}\, 2\pi)$. $\square$

17.3 The idea here is to use problem 15.4. Let $\langle a, 0, a' \rangle$ and $\langle b, 0, b' \rangle$ be two conjugate diameters of the ellipse E. The centers of the desired circles lie on the bisectors D, D' of the two diameters. Pick one intersection of each bisector with the ellipse, and call them m and n, as in figure 17.3.1. One gnawing doubt immediately manifests itself: are the distances from m and n to $\langle a, a' \rangle$ the same?

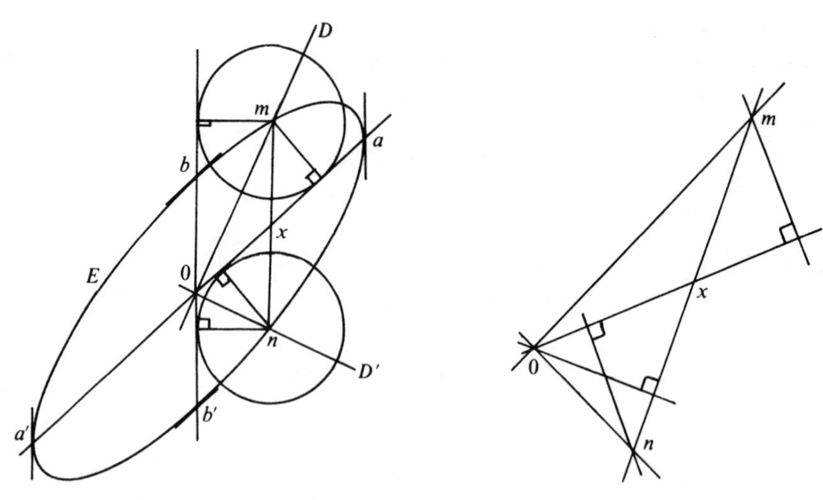

Figure 17.3.1. Figure 17.3.2.

We settle this question by showing that $\langle m, n\rangle$ is parallel to $\langle b, b'\rangle$. In fact, the lines $\langle 0, a\rangle$ and $\langle 0, b\rangle$ are harmonic conjugates with respect to D, D'. But then, denoting by x the intersection point of $\langle m, n\rangle$ with $\langle a, a'\rangle$, we get that the harmonic conjugate of x with respect to m, n on the line $\langle m, n\rangle$ must lie both on $\langle b, b'\rangle$ and on the line at infinity; this shows that x is the midpoint of $[m, n]$.

Now the distance from either point m, n to $\langle a, a'\rangle$ is equal to the distance from O to $\langle m, n\rangle$, from the elementary properties of the right triangle $\{0, m, n\}$. Thus the radius of the circles we are studying is equal to the height of the triangle $\{0, m, n\}$, which by problem 15.4 is equal to $\alpha^2\beta^2/(\alpha^2+\beta^2)$, where α, β denote the half-axes of E. □

17.4 We proceed analytically. Take coordinate axes for which the circle C has equation $x^2+y^2-2ax-b^2=0$, and where the two points we're considering are $(0, \pm b)$. Let $ux+vy+w=0$ be the equation of the tangent to C at the point where a given ellipse is also tangent. Considering the degenerate conic formed by the two lines $ux+vy+w=0$ and $x=0$, and taking the pencil passing through this conic and the circle C, we see that the desired equation for the ellipse is of the form

$$x^2+y^2-2ax-b^2+kx(ux+vy+w)=0$$

(cf. 16.F or [B, 16.4.10]).

Now let $r=\sqrt{a^2+b^2}$ be the radius of C, and $(a+r\cos t, r\sin t)$ its contact point with the tangent $ux+vy+w=0$. We can write

$$ux+vy+w=\cos t(x-a)+\sin t\cdot y-r=0$$

as the equation of this tangent. Then the ellipse can be written as

$$x^2+y^2-2ax-b^2+kx((x-a)\cos t+\sin t\cdot y-r)=0,$$

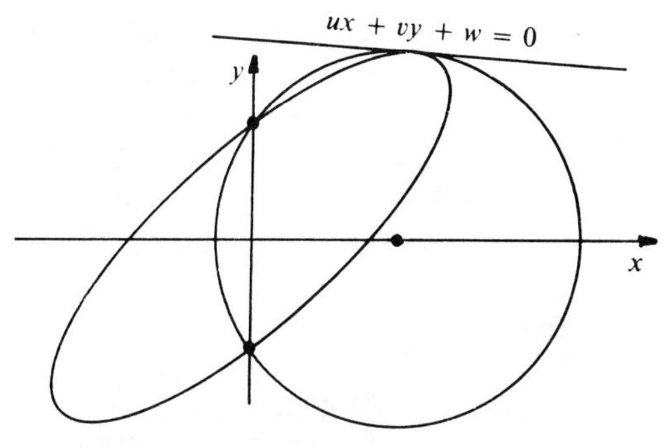

Figure 17.4.

and its center can only be at the origin if the term in x vanishes, i.e. if $k=[-2a/(r+a\cos t)]$. Substituting in the equation, we get

$$Ax^2-2Bxy+Cy^2=1,$$

where

$$A=\frac{r-a\cos t}{b^2(r+a\cos t)},\quad B=\frac{a\sin t}{b^2(r+a\cos t)},\quad C=\frac{1}{b^2}.$$

From 17.B or [B, 17.2.1], the eccentricity of this ellipse will be known if we have the ratio between $(A+C)^2$ and $AC-B^2$, since that's the ratio between the square of the trace and the volume (i.e. the determinant) of the matrix of the defining quadratic form q such that the ellipse is given by $q=1$. Performing the calculations, we obtain a constant:

$$\frac{AC-B^2}{(A+C)^2}=\frac{\dfrac{r-a\cos t}{r+a\cos t}-\dfrac{a^2\sin^2 t}{(r+a\cos t)^2}}{\left(\dfrac{r-a\cos t}{r+a\cos t}+1\right)^2}=\frac{r^2-a^2}{4r^2}=\frac{b^2}{4(a^2+b^2)}.\quad\square$$

17.5 We give an analytical proof. Let $y^2=2px$ be the equation of the parabola P, and (a,b) the point from which we're drawing normals to P. The equation of the tangent to P at point (x_0,y_0) is given by $yy_0=p(x+x_0)$ (cf. 15.C or [B, 15.5.5]), so the normal at this point will go through (a,b) if and only if $p/(x_0-a)=-y_0/(y_0-b)$. Plugging this into $y^2=2px$, we find the following third-degree equation for the y-coordinate y_0 of the desired points:

$$y_0^3+2p(p-a)y_0-2p^2b=0.$$

The sum of the three roots y, y', y'' of this equation is clearly equal to 0, which is equivalent to saying that the centroid of the three points $(y^2/2p,y),(y'^2/2p,y'),(y''^2/2p,y'')$ lies on the axis Ox of the parabola. $\square$

The converse follows from what we've just proved, since the relation

$$y+y'+y''=0$$

uniquely determines the third point as a function of the first two.

The equivalent property with the circle passing through m, m', m'' derives from 17.C (or [B, 17.5.5.3]) and the associative property of barycenters (3.B or [B, 3.4.8]). We take the opportunity to give an analytical proof of 17.C ([B, 17.5.5.3]). If $x^2 + y^2 + 2\alpha x + 2\beta y + \gamma = 0$ is the equation of an arbitrary circle, its intersections with the parabola $y^2 = 2px$ are calculated by solving for the y-coordinate. We find the fourth-degree equation

$$\frac{y^4}{4p^2} + y^2\left(\frac{\alpha}{p} + 1\right) + 2\beta y + \gamma = 0,$$

whose term in y^3 is zero, implying that the sum of the four roots y, y', y'', y''' is zero. Thus, if $y + y' + y'' = 0$, we certainly have $y''' = 0$! □

Chapter 18

18.1 In the notation of figure 18.1.1 (see page 110), we will have $r = a/\sqrt{3}$, then $R^2 = (R - e)^2 + r^2$, whence $R = (a^2 + 3e^2)/6e$.

The system of levers is used to control the pressure applied by the needle at point B on the surface of the sphere under examination. Accurate measurements require taking into account the deformation caused by this pressure, as well as the pressure at points A. In taking successive measurements, for instance, one should make sure that the pressure changes very little, and this is accomplished if the needle g always points to the same mark on the scale. □

18.2 Rhumb lines intersect meridians at a constant angle (cf. [B, 18.1.8.2]). Stereographic projection preserves angles (cf. 18.A or [B, 18.18.6]), and maps meridians onto lines passing through the origin in the plane. Thus the images of rhumb lines under stereographic projection intersect lines through the origin at a constant angle, showing that they must be logarithmic spirals (cf. 9.E or [B, 9.6.9]). A rigorous proof of this last implication uses the uniqueness of trajectories of a differentiable vector field, and the fact that logarithmic spirals do intersect lines through the origin at constant angles (since they are invariant under a group of similarities centered at the origin). One can also solve directly the differential equation that corresponds to the condition of intersection at a constant angle. □

18.3 We have to show that, for every spherical "triangle" with sides a, b, c, we have

$$|b - c| \leq a \leq b + c,$$

and equality holds if and only if the three vertices x, y, z of the triangle are in the same plane through the origin. The two inequalities above are equivalent to

$$\cos(b + c) \leq \cos a \leq \cos(b - c),$$

or again to

$$\left|\frac{\cos a - \cos b \cos c}{\sin b \sin c}\right| \leq 1.$$

The value of the Gram determinant of the three vertices x, y, z is, by definition,

$$\Delta = \begin{vmatrix} (x|x) & (x|y) & (x|z) \\ (y|x) & (y|y) & (y|z) \\ (z|x) & (z|y) & (z|z) \end{vmatrix}.$$

We know from 8.J (or [B, 8.11.6]) that Δ is equal to the square of the volume of the parallelepiped constructed on the vectors $\overrightarrow{Ox}, \overrightarrow{Oy}, \overrightarrow{Oz}$ (where O is the center of the sphere). Since $(x|x) = (y|y) = (z|z) = 1$, and $\cos a = (y|z)$, $\cos b = (z|x)$, $\cos c = (x|y)$, we get

$$\Delta = \begin{vmatrix} 1 & \cos c & \cos b \\ \cos c & 1 & \cos a \\ \cos b & \cos a & 1 \end{vmatrix} = 1 + 2\cos a \cos b \cos c - \cos^2 a - \cos^2 b - \cos^2 c,$$

which can be written as $\Delta = \sin^2 b \sin^2 c - (\cos a - \cos b \cos c)^2$. But $\Delta \geq 0$, and equality only holds if x, y, z are linearly dependent. □

18.4 As in the preceding problem, it is enough to notice that the Gram determinant of the points (x_i) is equal to the determinant

$$\det\left(\cos\left(\overline{x_i x_j}\right)\right) = \det\left((x_i|x_j)\right).$$

But since the $d+2$ points lie in $\mathbf{R}^{d+1}$, the volume of the parallelepiped they span when considered in $\mathbf{R}^{d+2}$ is zero. This shows the Gram determinant vanishes. □

18.5 On a sphere S' of radius R, the length of the arc yz determined by the central angle a is equal to Ra:

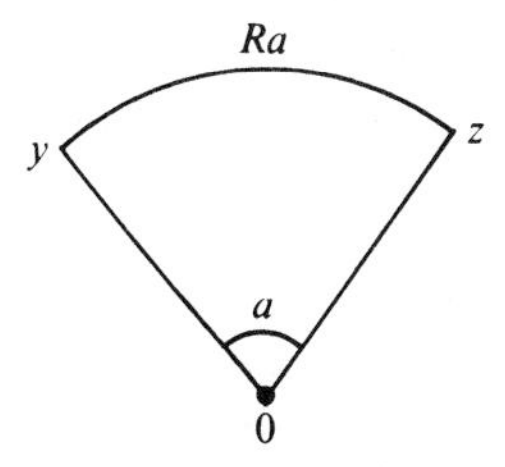

Figure 18.5.

Thus, for a triangle $\{x, y, z\}$ in S', the vertex angles α, β, γ will be given by the analogue of the formulas of 18.C (or [B, 18.6.13]) obtained by replacing a, b, c by a/R, b/R, c/R, respectively. Put $\varepsilon = 1/R$, thus ε approaches zero when R tends to infinity. We can thus consider the first terms of a series development of the formulas

$$\cos \varepsilon a = \cos \varepsilon b \cos \varepsilon c + \sin \varepsilon b \sin \varepsilon a \cos \alpha,$$

$$\frac{\sin \alpha}{\sin \varepsilon a} = \frac{\sin \beta}{\sin \varepsilon b} = \frac{\sin \gamma}{\sin \varepsilon c}.$$

Passing to the limit, we obtain the two formulas

$$a^2 = b^2 + c^2 - 2bc\cos\alpha, \quad \frac{\sin\alpha}{a} = \frac{\sin\beta}{b} = \frac{\sin\gamma}{c}$$

(cf. 10.A or [B, 10.3]). □

18.6 The idea is to apply the fundamental formula of spherical trigonometry (18.C). Define the angular position of the first (resp. second) shaft by the function $b: t \mapsto b(t)$ (resp. $c: t \mapsto c(t)$); then $b(t)$ and $c(t)$ are the sides of a spherical triangle $\{a, m(t), n(t)\}$ where $\overline{am(t)} = b(t)$, $\overline{an(t)} = c(t)$ and the angle at a is equal to the angle ϑ between the axes of the two shafts. Thus we have

$$\cos b \cos c + \sin b \sin c \cos\vartheta = \cos\frac{\pi}{2} = 0,$$

which implies

$$\tan b \tan c = -\frac{1}{\cos\vartheta}.$$

To obtain the angular velocities of the shafts, which are given by the derivatives of the functions b and c with respect to t, we differentiate the preceding equation:

$$b'(1+\tan^2 b)\tan c + c'(1+\tan^2 c)\tan b = 0.$$

Thus the desired ratio of angular velocities is

$$\left|\frac{b'}{c'}\right| = \left|\frac{\tan b}{\tan c}\,\frac{1+\tan^2 c}{1+\tan^2 b}\right| = \frac{1}{\cos\vartheta}\,\frac{\cos^2\vartheta\tan^2 b + 1}{\tan^2 b + 1}$$

The function $x \mapsto (1/\cos\vartheta)[(\cos^2\vartheta \cdot x^2 + 1)/(x^2+1)]$ has its maximum $1/\cos\vartheta$ for $x = 0$, and its minimum $\cos\vartheta$ for $x = \infty$. Thus the worst possible ratio of the two velocities is $\cos\vartheta$, and it is achieved when $m(t)$ or $n(t)$ coincide with a or its antipodal point. For $\vartheta = \pi/6, \pi/4, \pi/3$ we obtain the values 0.866, 0.707 and 0.5, respectively. It is clear that the ratio deteriorates rapidly as the angle increases.

When two Hooke joints are coupled in a homokinetic joint, however, symmetry requires that the two end shafts have the same angular velocities. The velocity of the middle shaft is variable, but this does not bother the passengers of the vehicle... □

18.7 The idea is to transform the three given spheres into spheres that are inside and tangent to a torus of revolution. We shall use the essential properties of inversions (10.C or [B, 10.8]) without mentioning them explicitly.

Let P be the plane containing the centers of $\Sigma, \Sigma', \Sigma''$, and let $\sigma, \sigma', \sigma''$ be the great circles $\sigma = P \cap \Sigma$, $\sigma' = P \cap \Sigma'$ and $\sigma'' = P \cap \Sigma''$. Assume that $\sigma, \sigma', \sigma''$ are so disposed that there are two nonintersecting circles γ, γ' tangent to all three circles $\sigma, \sigma', \sigma''$. Then, according to 10.D (or [B, 10.10.2]), there is an inversion with center in P that transforms γ, γ' into concentric circles. This inversion leaves P fixed; let it take γ, γ' into δ, δ' and $\sigma, \sigma', \sigma''$ into $\vartheta, \vartheta', \vartheta''$.

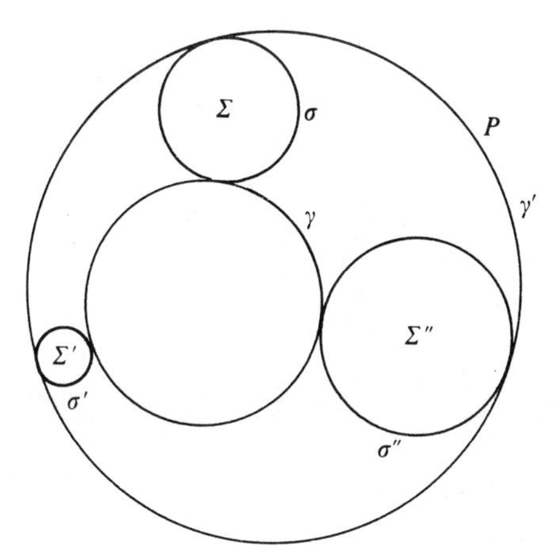

Figure 18.7.1.

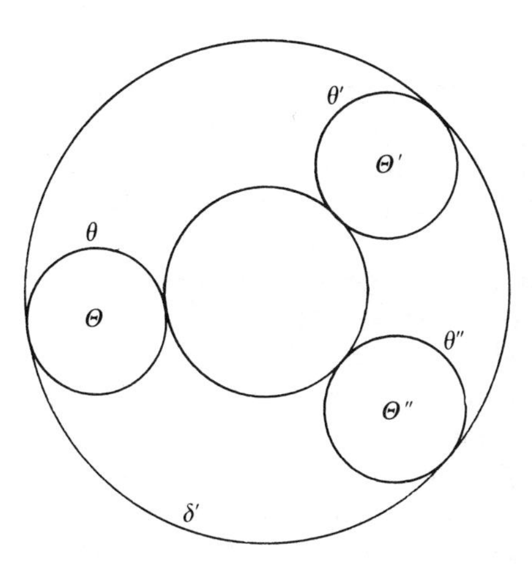

Figure 18.7.2.

We consider the torus of revolution T in $\mathbf{R}^3$ which has P as its equatorial plane and which intersects P in δ and δ'; T is well-defined by these conditions. Similarly, let $\Theta, \Theta', \Theta''$ be the spheres of $\mathbf{R}^3$ having P as their equatorial plane and such that $\Theta \cap P = \vartheta$, $\Theta' \cap P = \vartheta'$, $\Theta'' \cap P = \vartheta''$. Let S be a sphere tangent to $\Theta, \Theta', \Theta''$ and lying outside them. Since Θ, Θ' and Θ'' have the same radius, the center s of S lies at the same distance from the three centers p, p', p'' of the small spheres; thus s belongs to the axis Δ of the torus T.

Thus any sphere S, tangent to Θ and centered at $s \in \Delta$, is tangent to T along the whole parallel of latitude passing through the point where it touches Θ. We can then certainly say that this one-parameter family of spheres

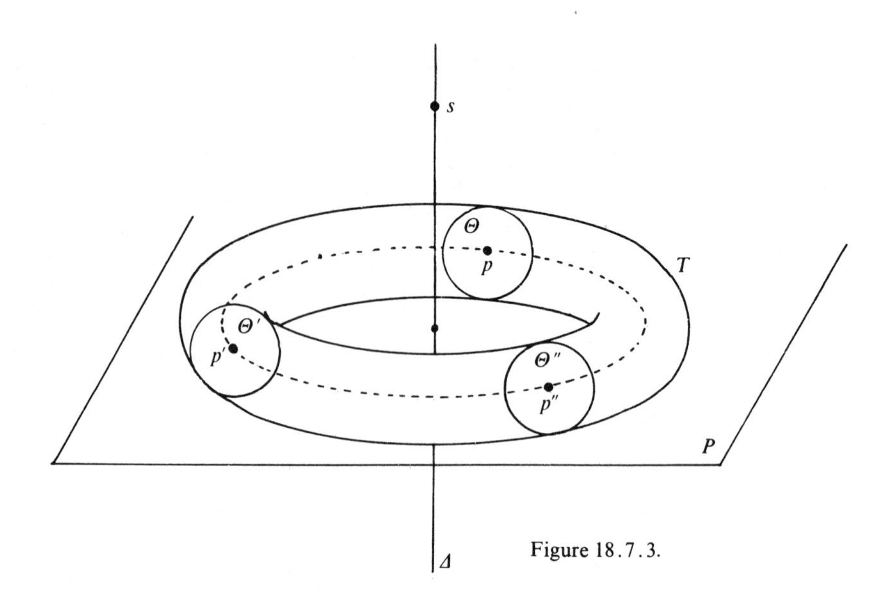

Figure 18.7.3.

envelopes the torus T. Observe that there is a second one-parameter family of spheres with the same property, namely the images of Θ by rotation around the axis Δ; each of these spheres lies inside T and is tangent to it along a meridian. Moreover, each sphere of the first family is tangent to each sphere of the second.

Now we go back to our three original spheres $\Sigma, \Sigma', \Sigma''$ by applying again the inversion considered above. The image of T under this inversion is the desired Dupin cyclid. This surface is the one-parameter envelope of each of two one-parameter families of spheres, and it is tangent to each sphere in these families along a whole circle.

So far we know two families of circles on the cyclid, which are formed by the circles where each enveloping sphere (of either family) touches the cyclid (corresponding to the meridians and longitudes of T). These two families intersect at right angles. But there are on T another two families of circles, the circles of Villarceau (cf. 18.D or [B, 18.9]), which have the property of intersecting the meridians (and hence the parallels of latitude) at a constant angle α, the same for both families. By inversion, one sees that the Dupin cyclid will also contain four families of circles making the angles shown below at each point, and so that through each point there is exactly one circle of each family:

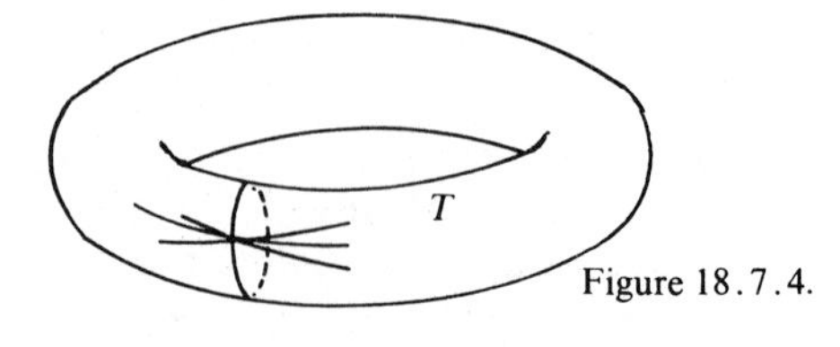

Figure 18.7.4.

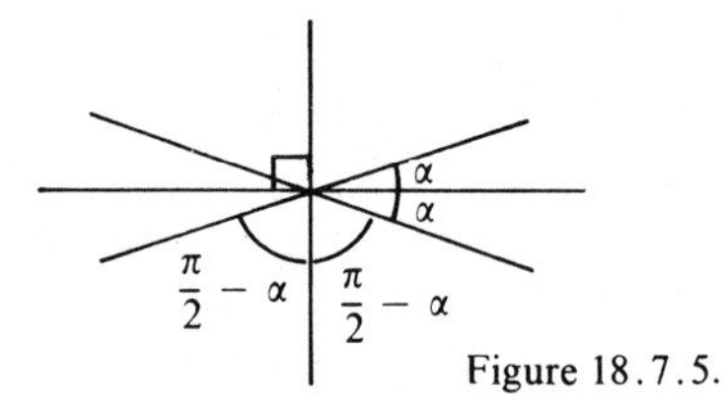

Figure 18.7.5.

Finally, by [B, 10.12.3, 18.9.3], we know that two circles c, c' of a family of circles of Villarceau in a cyclid bound a paratactic annulus. This means that every sphere containing c intersects c' at a constant angle β, and every sphere containing c' intersects c at the same angle β. □

Note. If $\Sigma, \Sigma', \Sigma''$ are not arranged in such a way as to satisfy the condition we set at the beginning of the solution, one obtains analogous results but the cyclids are of different types. For instance, if γ, γ' are secant, they can be transformed by inversion into two intersecting lines (cf. 10.D). The torus T is then replaced by a cone of revolution. If γ, γ' are tangent, T is replaced by a cylinder of revolution. Applying inversion again one gets Dupin cyclids of different shapes (but containing only two families of circles).

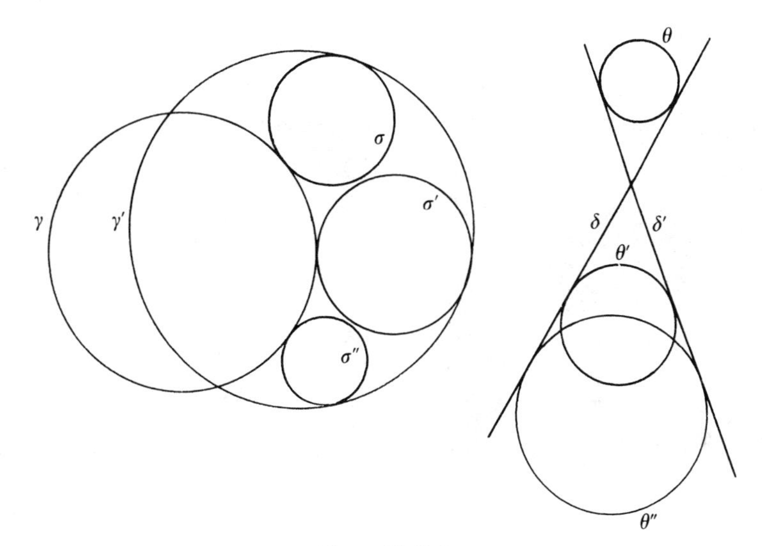

Figure 18.7.6.

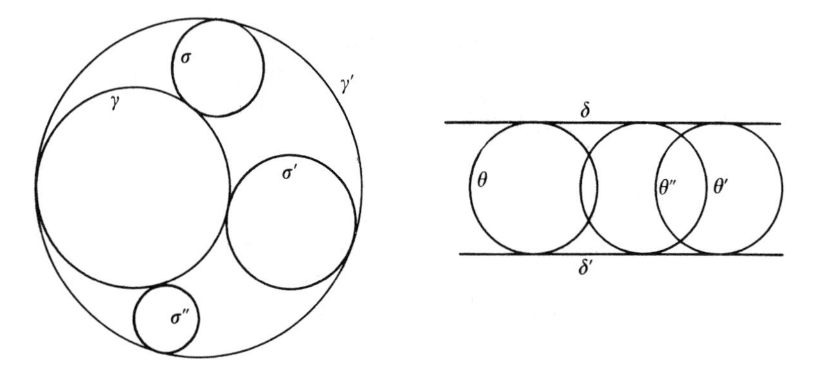

Figure 18.7.7.

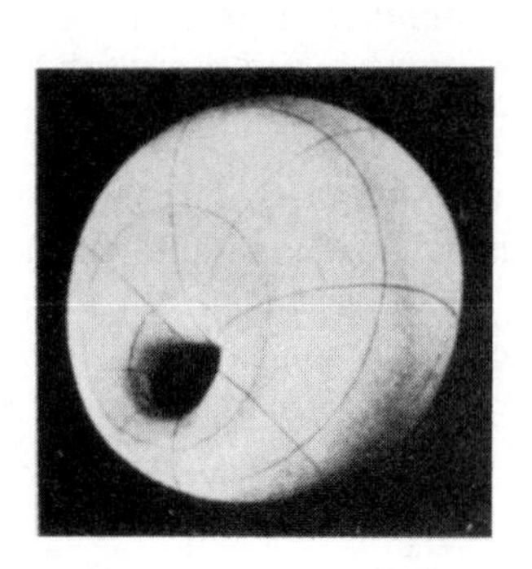

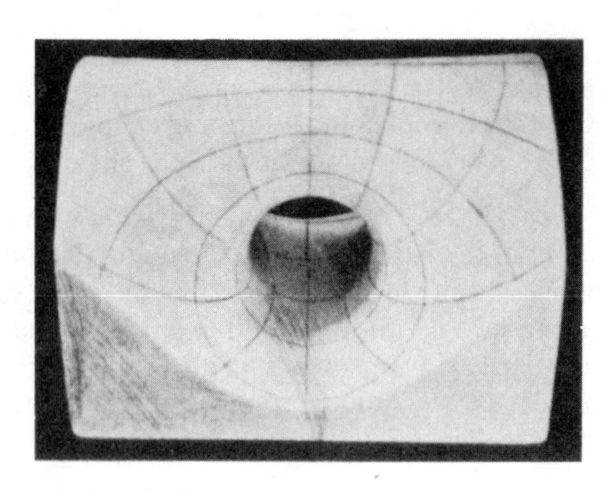

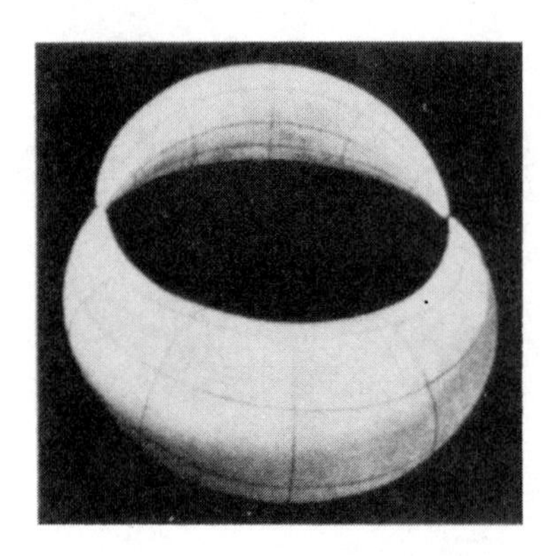

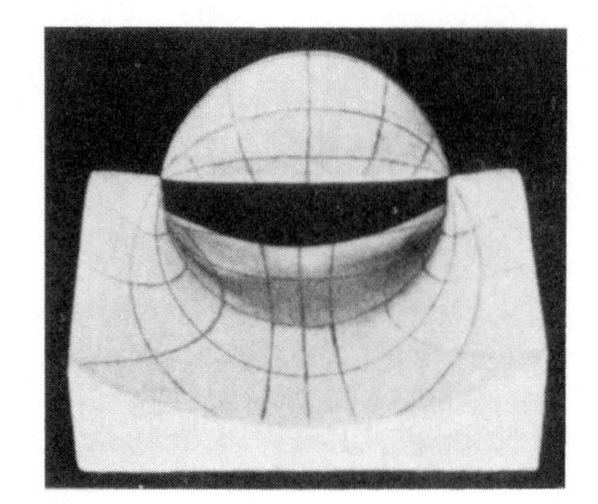

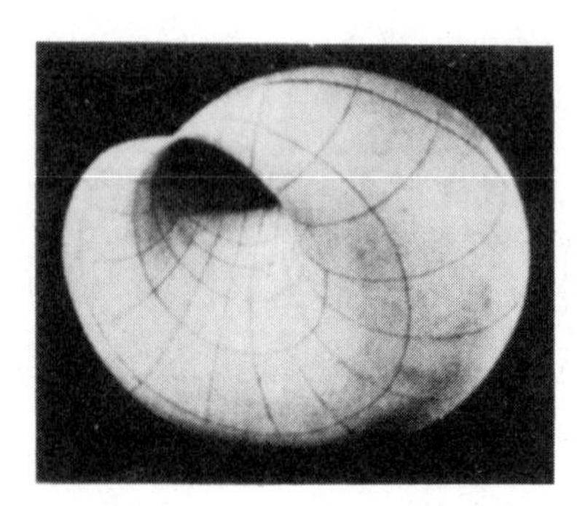

Figure 18.7.8.

Chapter 19

19.1 Preliminary note. In n-dimensional Euclidean or hyperbolic space, every equilateral set contains at most $n+1$ points, and for every length $a>0$ there is exactly one such set, up to isometries. This follows from 9.F or [B, 9.7.1]. We shall see that the situation is different in the case of spherical or elliptic geometry. The distance between two arbitrary points of an equilateral set will be always *denoted* by a.

Three-point sets (triangles). Let $\{m_1, m_2, m_3\}$ be an equilateral triangle in P, and let $\mathscr{T}=\{x, y, z\}$ be a triangle in S which projects onto the *points* m_1, m_2, m_3, but not necessarily onto the triangle in P (it is not always possible to find a triangle in S that projects onto a triangle in P; see 19.A or [B, 19.1.3]). The sides of $\mathscr{T}$ can only have lengths a or $\pi - a$. We can always replace a point of $\mathscr{T}$ by its antipodal point, which amounts to changing the lengths of two sides into their complements relative to π, and leaving the third side alone; thus we can suppose we're in one of the following situations:

(i) the three sides of $\mathscr{T}$ have length a;
(ii) the three sides of $\mathscr{T}$ have length $\pi - a$.

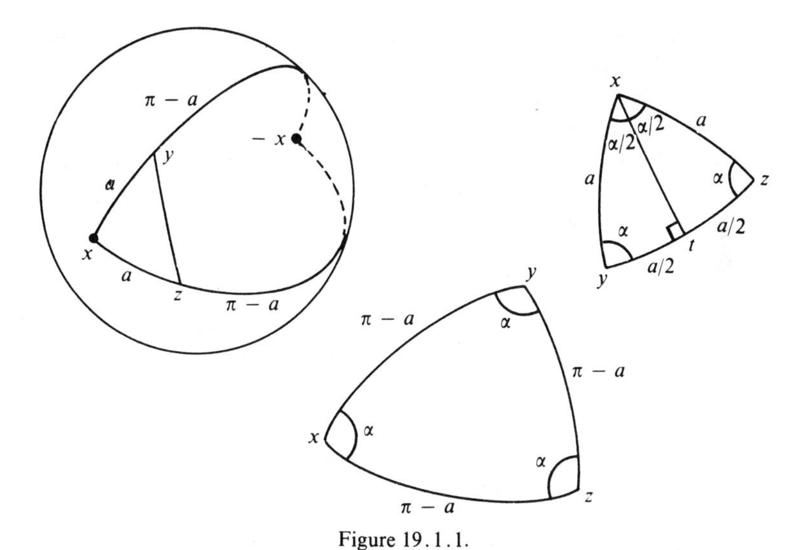

Figure 19.1.1.

Remark first that in either case the value of a completely determines $\mathscr{T}$ up to an isometry (congruence of spherical triangles, cf. 19.A or [B, 18.6.13.10]). In particular, the three angles of $\mathscr{T}$ are equal and we shall denote them by α in either case. Now consider the midpoint t of the segment $[y, z]$; the congruence criteria show that the angles at t of the triangles $\{x, y, t\}$ and $\{x, z, t\}$ are the same, and thus equal to $\pi/2$. By the same token the angle at x of $\{x, y, t\}$ is equal to $\alpha/2$. Applying the formulas of spherical trigonometry (18.C or [B,

18.6.13]), we conclude that in case (i)

$$\frac{\sin\frac{\pi}{2}}{\sin a} = \frac{\sin\frac{\alpha}{2}}{\sin\frac{a}{2}}.$$

This implies

$$2\sin\frac{\alpha}{2}\cos\frac{a}{2} = 1, \tag{i}$$

which uniquely determines a as a function of α and vice versa.

Case (ii) is treated by substituting $\pi - a$ for a, whence

$$2\sin\frac{\alpha}{2}\sin\frac{a}{2} = 1. \tag{ii}$$

In case (i), the side a varies between 0 and $\pi/2$ for values of α between $\pi/3$ and $\pi/2$. In case (ii), a varies from $\pi/2$ to $\pi/3$ while α varies from $\pi/2$ to π.

To summarize: up to isometries, for each $a \in]0, \pi/3[$ there is one equilateral triangle and for each $a \in [\pi/3, \pi/2]$ there are two (non-isometric) triangles. Remark that case (i) corresponds to triangles of type I and case (ii) to triangles of type II (see 19.A). □

Four-point sets: case (ii). If $\{m_1, m_2, m_3, m_4\}$ is equilateral, so is every three-element subset of this set. Suppose that all those subsets fall into case (ii) above. In particular, let $\mathscr{T} = \{x, y, z\}$ be a triangle in S which projects onto $\{m_1, m_2, m_3\}$, and all of whose sides have length $\pi - a$. Let t be a point in S that projects onto m_4; the distances from t to x, y, z all measure a or $\pi - a$. In fact the first possibility is excluded, since then, taking one of the antipodal points of x, y, z, we would obtain a triangle of type (i), contradicting our assumption.

By the congruence criteria for spherical triangles (cf. 19.A or [B, 18.6.3.10]), and since $t \neq x$, we know that t is the reflection of x through the line $\langle y, z\rangle$. Moreover, if p is the midpoint of $[y, z]$, we know that the angles of $\{x, y, p\}$ and $\{x, p, z\}$ are equal to $\pi/2$ (at p) and $\alpha/2$ (at x). We conclude also that t

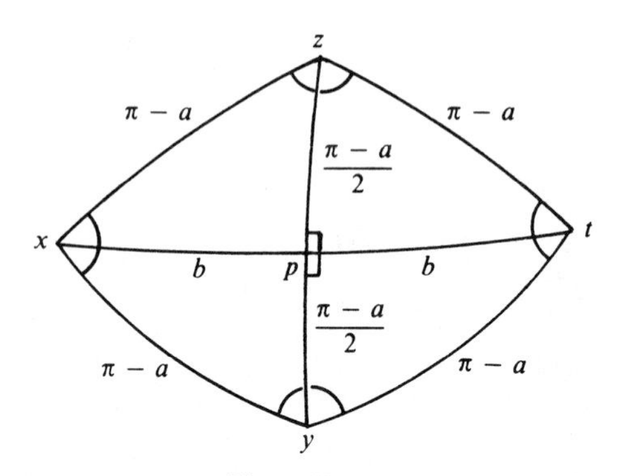

Figure 19.1.2.

lies on the great circle passing through x and p. However, the distance from x to t is not twice the distance $\overline{xp}$.

In fact, writing $b = \overline{xp}$, we get (cf. 18.C or [B, 19.1.1]):

$$\cos b = \cos(\pi - a)\cos\frac{\pi - a}{2} + \sin(\pi - a)\sin\frac{\pi - a}{2}\cos\alpha.$$

Using the formula $2\sin(\alpha/2)\sin(a/2) = 1$ obtained above (case (ii)), we find

$$\cos b = -\frac{\cos a}{\sin\dfrac{a}{2}},$$

whence $b > \pi/2$ and $\overline{xt} = 2\pi - 2\overline{xp} = 2\pi - 2b$. But we must have $\overline{xt} = \pi - a$; this implies $b = (\pi + a)/2$, and consequently $\sin(a/2) = [\cos a/\sin(a/2)]$, which gives the values

$$\sin\frac{a}{2} = \frac{1}{\sqrt{3}}, \quad \sin\frac{\alpha}{2} = \frac{\sqrt{3}}{2}, \quad \text{and} \quad \alpha = \frac{2\pi}{3}$$

since $\alpha > \pi/2$. The final conclusion is that the four points x, y, z, t are the vertices of a regular tetrahedron in S. Indeed, a regular tetrahedron satisfies the required conditions, including $\sin(a/2) = 1/\sqrt{3}$, and it is unique up to an isometry. In P we can say for example that the equilateral four-point set is formed by the four diagonals of a cube.

Moreover, there is no equilateral n-point set, for $n \geq 5$, containing a subset of the above type; this is because the three points x, y, z uniquely determine the fourth point t. In fact, t is the reflection of x through $[y, z]$, the reflection of y through $[z, x]$, or the reflection of z through $[x, y]$. □

Four-point sets: case (i). Now let $\{m_1, m_2, m_3, m_4\}$ be an equilateral set and $\mathscr{T} = \{x, y, z\}$ which projects onto $\{m_1, m_2, m_3\}$ and whose sides have length a. If t projects onto m_4, the distances from t to the points x, y, z are either a or $\pi - a$. Taking antipodal points if necessary, we can assume that the three distances are (a, a, a) or $(a, a, \pi - a)$. Assume that $\overline{ty} = \overline{tz} = a$. The triangles $\{x, y, z\}$ and $\{y, z, t\}$ are isometric, so t is the reflection of x through $[y, z]$, since we cannot have $t = x$! Also, $[x, t]$ intersects $[y, z]$ at its midpoint p, and forming a right angle.

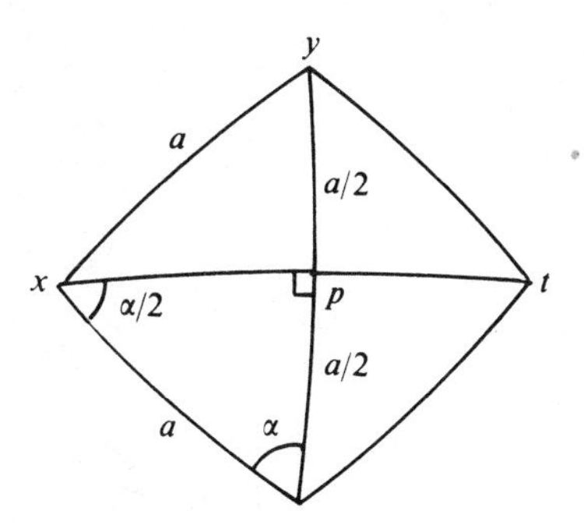

Figure 19.1.3.

Thus, from the considerations above, we have $2\sin(\alpha/2)\cos(a/2)=1$, and putting $b=\overline{xt}=2\overline{xp}$, we get

$$\frac{1}{\sin a}=\frac{\sin\alpha}{\sin\frac{b}{2}}.$$

The case $b=a$ leads to $\sin\alpha=\sin(\alpha/2)$, i.e. $\alpha=0$ or $2\pi/3$, both impossible for a triangle in case (i). There remains the possibility $b=\pi-a$, i.e. $\sin\alpha\sin a=\cos(a/2)$, which, together with

$$2\sin\frac{\alpha}{2}\cos\frac{a}{2}=1,$$

implies that $\cos a=1/\sqrt{5}$. □

An arduous calculation involving $\cos 5\alpha$ shows that $\alpha=2\pi/5$. To conclude the classification, we'd better start thinking about the regular icosahedron (1.F, 12.C or [B, 12.5.5]), even if we don't know of its existence yet. One way to introduce it is the following: We start from our set $\{x, y, z, t\}$, all of whose sides have length a except $\overline{xt}=\pi-a$, and such that the angles of $\{x, y, z\}$ are equal to $2\pi/5$. The essential observation is that the great half-circle defined by x and t and going from x to $-x$ (the antipodal point of x) is such that $\overline{t(-x)}=a$, since $\overline{xt}=\pi-a$. Applying rotations around the axis $\langle x, -x\rangle$ by multiples of $2\pi/5$, we obtain twelve points of S (see figure 19.1.4). They form a regular polyhedron with twelve vertices, by the properties of the set $\{x, y, z, t\}$ and the observation above.

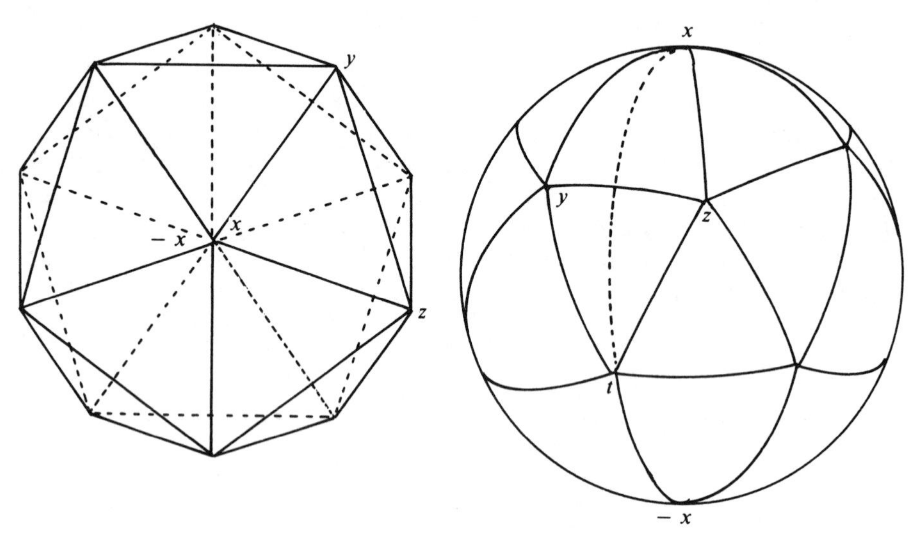

Figure 19.1.4.

We have thus shown the existence (and uniqueness) of the regular icosahedron. Now let's return to n-point equilateral sets in P (containing a triple of type (i)). They must necessarily be contained in the six-point set of projections of vertices of the icosahedron in S. Thus the only possible values are $n = 4,5,6$. An easy reasoning involving antipodal points shows that for each case $n = 4,5,6$, we only have one n-point equilateral set up to isometries. □

19.2 We draw the segment $[u, v]$ as in figure 19.2.1, calling α and β the angles at v and w, respectively, of the triangle $\{u, v, w\}$, and we put

$$d(v,w) = x.$$

Using the formulas of 19.C (or [B, 19.3.1, 19.3.5]), we obtain:

$$\operatorname{ch} x = \operatorname{ch} a \operatorname{ch} b, \quad \frac{\sin\alpha}{\operatorname{sh} a} = \frac{1}{\operatorname{sh} x}, \quad \frac{\sin\beta}{\operatorname{sh} b} = \frac{1}{\operatorname{sh} x},$$

$$\begin{aligned}\cos\gamma &= \sin\left(\frac{\pi}{2} - \alpha\right)\sin\left(\frac{\pi}{2} - \beta\right)\operatorname{ch} x - \cos\left(\frac{\pi}{2} - \alpha\right)\cos\left(\frac{\pi}{2} - \beta\right) \\ &= \cos\alpha\cos\beta\operatorname{ch} x - \sin\alpha\sin\beta.\end{aligned}$$

But

$$\cos^2\alpha = 1 - \sin^2\alpha = 1 - \frac{\operatorname{sh}^2 a}{\operatorname{sh}^2 x} = \frac{\operatorname{ch}^2 x - \operatorname{ch}^2 a}{\operatorname{sh}^2 x} = \frac{\operatorname{ch}^2 a \operatorname{sh}^2 b}{\operatorname{sh}^2 x},$$

so that $\cos\alpha = (\operatorname{ch} a \operatorname{sh} b/\operatorname{sh} x)$; and similarly $\cos\beta = (\operatorname{ch} b \operatorname{sh} a/\operatorname{sh} x)$. Thus

$$\begin{aligned}\cos\gamma &= \frac{\operatorname{ch}^2 a \operatorname{ch}^2 b \operatorname{sh} a \operatorname{sh} b}{\operatorname{sh}^2 x} - \frac{\operatorname{sh} a \operatorname{sh} b}{\operatorname{sh}^2 x} \\ &= \frac{\operatorname{sh} a \operatorname{sh} b(\operatorname{ch}^2 a \operatorname{ch}^2 b - 1)}{\operatorname{sh}^2 x} = \operatorname{sh} a \operatorname{sh} b.\end{aligned}$$ □

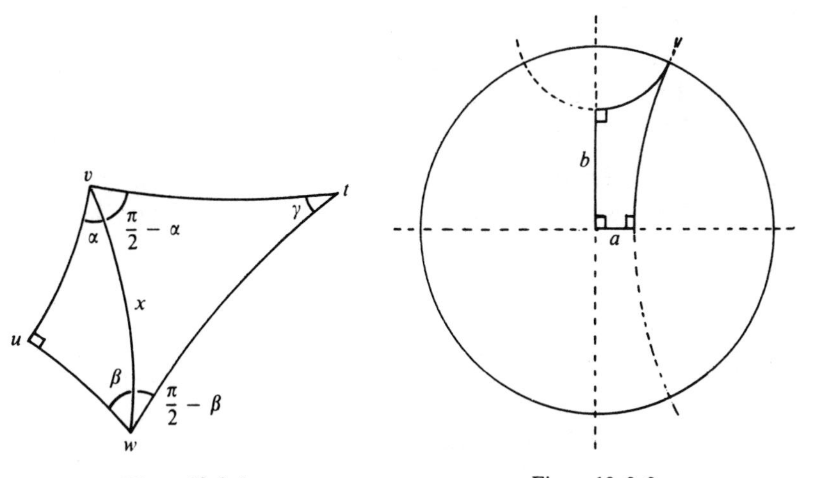

Figure 19.2.1. Figure 19.2.2.

We can add the following commentary to this formula: when a and b are very small, the angle γ is close to $\pi/2$, which is to be expected from the fact that the hyperbolic metric locally looks like the Euclidean metric (so the quadrilateral is close to a rectangle). On the other hand, a and b must be such that $\operatorname{sh} a \operatorname{sh} b < 1$, which is clear in the conformal model (cf. 19.D or [B, 19.6]). The case $\operatorname{sh} a \operatorname{sh} b = 1$ corresponds to the situation in which the fourth vertex is at infinity; the geodesic sides (represented by circles in the conformal model) are then tangent at their common point at infinity.

19.3 We shall use the projective model; for each z_i we take a representative ξ_i of z_i in $\mathbf{R}^{n+1}$ such that $q(\xi_i) = 1$. Thus we have by definition for all i, j: $\operatorname{ch}(d(z_i, z_j)) = P(\xi_i, \xi_j)$. Next we use the trick with the Gram determinant (cf. 8.J, [B, 8.11]) with an *ad hoc* modification: denoting by A the matrix of q in a higher-dimensional space $\mathbf{R}^{n+2}$, i.e.

$$A = \begin{pmatrix} -1 & & & 0 \\ & 1 & & \\ & & \ddots & \\ 0 & & & 1 \end{pmatrix},$$

we consider the $(n+2)\times(n+2)$ matrix $S = (\xi_1 \dots \xi_{n+2})$ formed by the column vectors ξ_i in $\mathbf{R}^{n+1}$, extended with a 0 as the $(n+2)$-th coordinate. We also consider its transpose ${}^tS$, formed with the corresponding row vectors. We thus have

$${}^tSAS = \begin{pmatrix} P(\xi_1, \xi_2) & \dots & P(\xi_1, \xi_{n+2}) \\ \vdots & & \vdots \\ P(\xi_{n+2}, \xi_1) & \dots & P(\xi_{n+2}, \xi_{n+2}) \end{pmatrix}.$$

The matrix S has determinant zero, since it contains a whole line of zeros; this shows $\det({}^tSAS) = 0$. □

Note. This proves that hyperbolic spaces share with Euclidean spaces (cf. 9.F or [B, 9.7.3]) and spheres (problem 18.4) the property that, in dimension n, the distances between $n+2$ arbitrary points satisfy a universal relation.

19.4 Uniqueness certainly holds. In fact, the formulas in 19.C or [B, 19.3.1, 19.3.5] show that, once we fix the side $d(x, y) = 2a$ (see figure 19.4.1), the

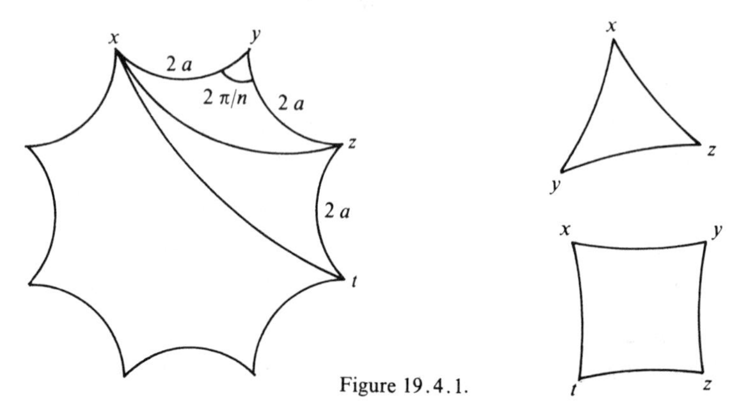

Figure 19.4.1.

distance $d(x, y)$ is well-determined. By induction, the distances from x to each of the other vertices are determined as well; uniqueness follows from 19.B (or [B, 19.4.4]). □

Existence. Observe first that the cases $n=3$ and 4 are impossible. The direct method below will show this fact, but we can also use the fundamental property that the sum of the angles of a triangle is always less than π (cf. 19.C or [B, 19.5.4]). Thus $3\times(2\pi/3)=2\pi$ clearly rules out $n=3$, and if we had a quadrilateral we could cut it into two triangles and we'd have $4\times(2\pi/4)=2\pi$ contradicting the fact that the sum of the angles of both triangles is less than 2π.

We will show that the polygons always exist for $n\geq 5$. The idea is to find the formulas the polygon would satisfy, were it to exist. Consider the center O of the prospective polygon, and the midpoint p of the side $[x, y]$; the angle at p of the triangle $\{x, O, p\}$ must be equal to $\pi/2$, and the angles at O and x must both equal $(1/2)\times(2\pi/n)=\pi/n$. But $d(x, p)=a$, so $d(O, p)$ also equals a by symmetry. Write $d(O, x)=b$ and apply the formulas:

$$\operatorname{ch} b=\operatorname{ch}^2 a, \qquad \frac{\sin\dfrac{\pi}{n}}{\operatorname{sh} a}=\frac{1}{\operatorname{sh} b}, \qquad 1+\frac{\operatorname{sh}^2 a}{\sin^2\dfrac{\pi}{n}}=\operatorname{ch}^4 a,$$

which implies

$$\operatorname{sh}^2 a=\frac{1}{\sin^2\dfrac{\pi}{n}}-2.$$

Since $[1/\sin^2(\pi/n)]>2$ for every $n>4$, there is always an isosceles triangles with base angles $2\pi/n$. We now label such a triangle $\{x, O, p\}$ and use reflections through the sides (first $\langle O, p\rangle$, then $\langle O, y\rangle$ and so on) to construct the rest of the desired polygon. This proves existence for $n\geq 5$.

Remark. Let P be a polygon as above; reflecting through the side $\langle x, y\rangle$ transforms it into an isometric polygon P' which is adjacent to P along

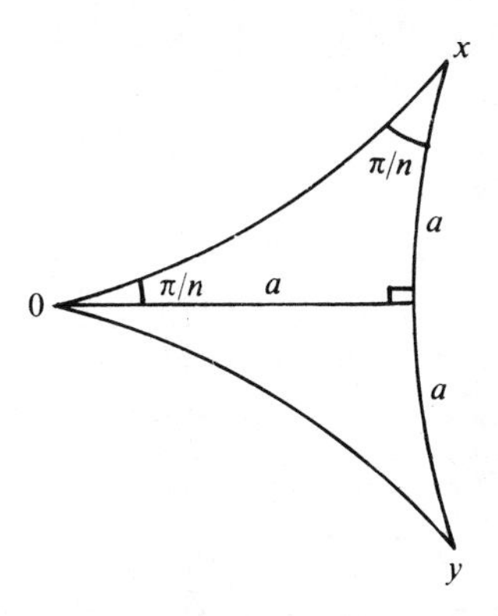

Figure 19.4.2.

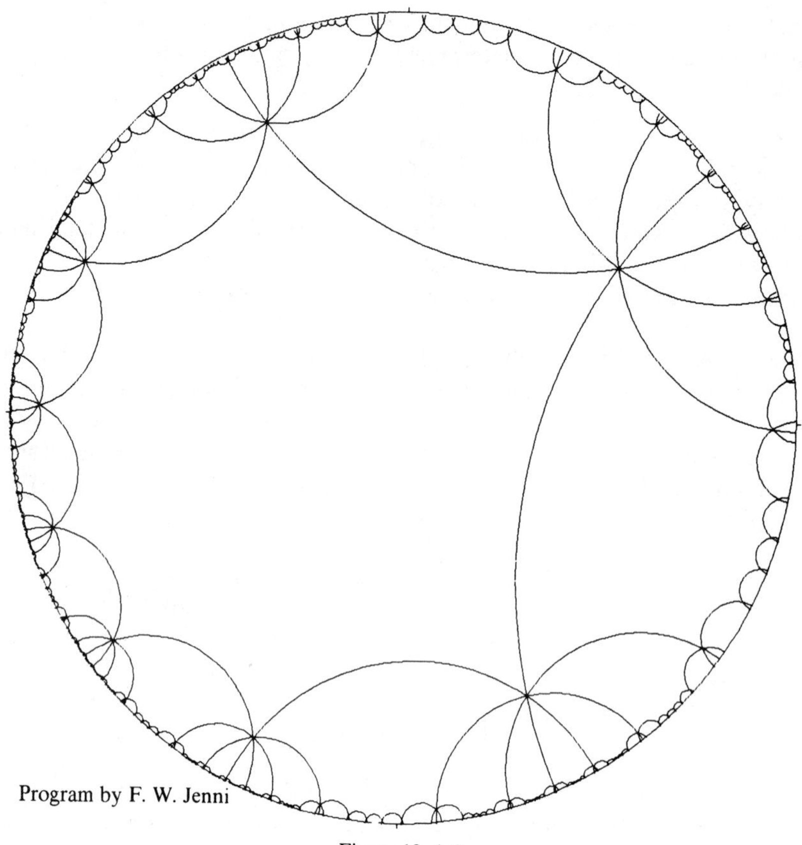

Figure 19.4.3.

$\langle x, y\rangle$. It is possible to show (though the rigorous proof is a bit delicate) that the images under all the reflections through the sides of P, P' and so on tile the hyperbolic plane. Thus for every $n \geq 5$ there are tilings of the plane by isometric n-sided regular polygons. This contrasts with the situation in the Euclidean plane, where 6 is the greatest possible number of sides (cf. [B, 1.9.13]). □

Chapter 20

20.1 Let p be a point in $\mathrm{im}(s)$ and $T_p s$ the tangent hyperplane to s at p (in the space E). The points q of $T_p s$ can be characterized by the fact that the sphere s' centered at q and passing through p is orthogonal to s. Thus the points s, s' of $S(E)$ corresponding to the two spheres must be conjugate with respect to ρ (or to $\hat{E}$).

Similarly, saying that s' contains p is the same as saying that s' and the punctual sphere p are conjugate with respect to $\hat{E}$. Thus the desired points q

are the centers of spheres s' which are conjugate both to p (an arbitrary point in $\mathrm{im}(s)$) and to s. The set of such spheres s' is exactly the intersection K of the polar hyperplanes of s and p, i.e. the intersection of the hyperplane H of $\mathrm{im}(s)$ with $p^{\perp}$, thus K is the tangent hyperplane to the quadric $\mathrm{im}(s)$ at p in the projective subspace H.

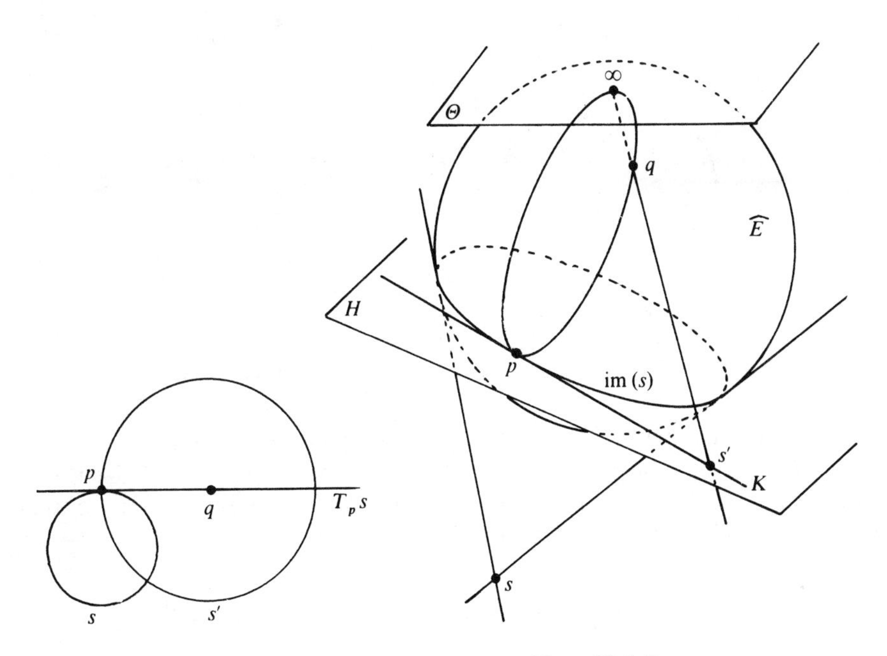

Figure 20.1.1. Figure 20.1.2.

As for the representation of $T_p s$, it will consist of the centers of these spheres s', i.e. (cf. 20.B or [B, 20.3.4]) the punctual spheres formed by the intersection points of E with lines $\langle \infty, s' \rangle$. Finally, this set is the intersection of $\hat{E}$ with the hyperplane spanned by K and the point at infinity. □

20.2 We have two things to do: first, finding the equation of a torus, in orthonormal coordinates (x, y, z) in $\mathbf{R}^3$ for instance. Then, check that this torus is indeed formed by punctual spheres which all belong to a projective quadric in $S(E)$, the space of spheres of $\mathbf{R}^3$. Thus we must be able to pass from ordinary coordinates in $\mathbf{R}^3$ to homogeneous coordinates (cf. 4.C) in the projective space $S(E)$. A point (a, b, c) in $\mathbf{R}^3$ is associated with the punctual sphere

$$(x-a)^2+(y-b)^2+(z-c)^2=0,$$

which, under the form $k\|\cdot\|^2+(\alpha|\cdot)+h=0$, corresponds, for example, to the values $k=1$, $\alpha=-2(a, b, c)$, $h=a^2+b^2+c^2$. But the numbers k, h, together with the coordinates of α, form a system of homogeneous coordinates for the sphere $k\|\cdot\|^2+(\alpha|\cdot)+h=0$ in $S(E)$. Thus the point (x, y, z) in $\mathbf{R}^3$ is associated with the point of $P^4(\mathbf{R})$ whose homogeneous coordinates are

$$(x_0, x_1, x_2, x_3, x_4)=(1, -2x, -2y, -2z, x^2+y^2+z^2).$$

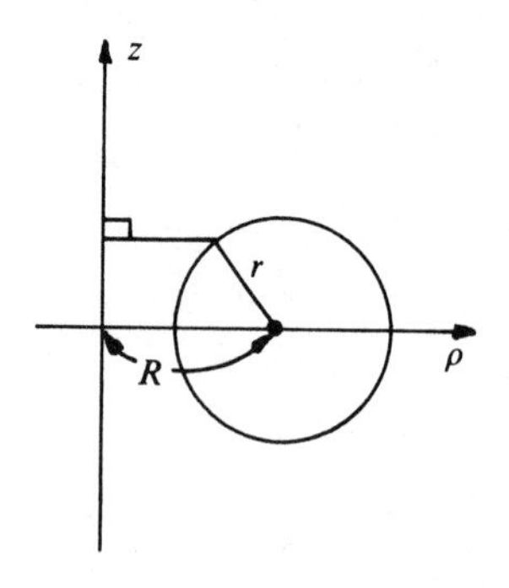

Figure 20.2.1.

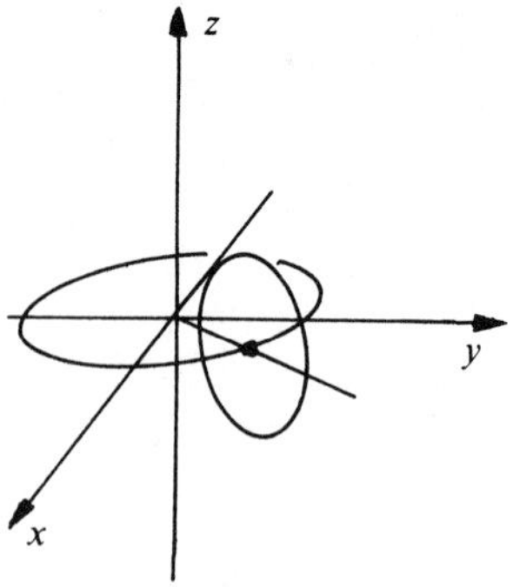

Figure 20.2.2.

We now find the equation of the torus obtained by rotating around the z-axis a circle of radius r whose center lies in the xy-plane at a distance R from the axis. Call ρ the distance from a point to the z-axis, so that $\rho^2=x^2+y^2$. The points in the torus satisfy the equation $(\rho-R)^2+z^2=r^2$. Thus

$$2\rho R=\rho^2+z^2+R^2-r^2=x^2+y^2+z^2+R^2-r^2,$$

or finally $4R^2(x^2+y^2)=(x^2+y^2+z^2+R^2-r^2)^2$. This corresponds, by virtue of the previous calculations, to the following equation in $P^4(\mathbf{R})$:

$$4R^2\left(x_1^2+x_2^2\right)=\left(x_4+(R^2-r^2)x_0\right)^2,$$

which clearly represents a quadric. □

20.3 We solve the problem analytically. Let u be the unit vector on the line D containing the centers x, x', x'' of the given spheres S, S', S'', let v be the unit vector on the variable line E containing the points y, y', y'' on S, S', S'', and w the unit vector on the line joining x to y. The conditions of the problem say that the distances $xy=r$, $x'y'=r'$, $x''y''=r''$ are fixed (and equal to the radii of S, S', S'', respectively). The distances $xx'=t'$, $xx''=t''$, yy' and yy'' are also fixed.

The clue to the solution lies in that the point y''' of E will always belong to a (variable) sphere whose center is the point x''' of D such that y''' is the image of x''' under the homography $D \to E$ that maps x to y, x' to y' and x'' to y''. All we have to show is that for two such points x''' and y''', the distance $x'''y'''$ is constant for any three unit vectors u, v, w satisfying the conditions above.

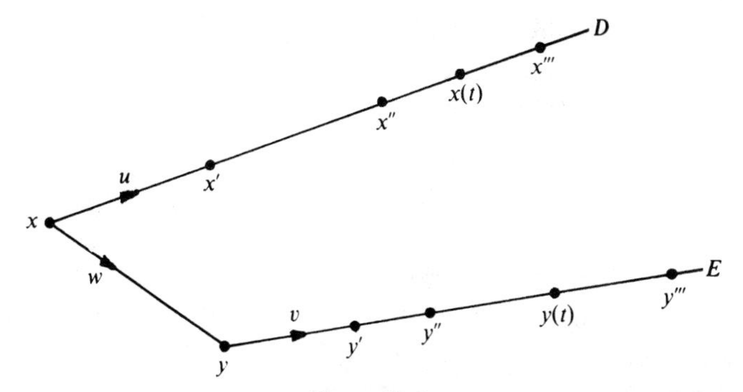

Figure 20.3.

Our homography, in coordinates t, $f(t)$ on D, E such that

$$t = d(x, x(t)), \quad f(t) = d(y, y(t)),$$

is necessarily of the form $f(t) = (at+b)/(ct+d)$ (cf. 6.D or [B, 5.2.3]). Since $f(x) = y$ for $t = 0$, we conclude that $b = 0$ (so we can normalize with $a = 1$), and also that $yy' = t'/(ct'+d)$, $yy'' = t''/(ct''+d)$. Since yy' and yy'' are fixed, the values c and d are fixed and well-determined.

We now calculate the distance $d(x(t), y(t))$. We have $x(t) = x + tu$,

$$y(t) = y + f(t)v = y + \frac{t}{ct+d}v;$$

thus

$$d^2(x(t), y(t)) = \left\| y + \frac{t}{ct+d}v - x - tu \right\|^2 = \left\| rw + \frac{t}{ct+d}v - tu \right\|^2$$

$$= r^2 + t^2 + \frac{t^2}{(ct+d)^2} + \frac{2t}{ct+d}\left[r(\alpha - d\beta) - t(rc\beta + \gamma) \right],$$

where we have defined $\alpha = (w|v)$, $\beta = (v|u)$, $\gamma = (u|v)$. Since $d(x(t'), y(t')) = r'$ and $d(x(t''), y(t'')) = r''$ are given, the quantities $\alpha - d\beta$ and $rc\beta + \gamma$ are also known. Thus, for an arbitrary fixed t''', the distance

$$d(x(t'''), y(t''')) = x'''y'''$$

is fixed.

The case when y''' belongs to a plane occurs for x''' at infinity, i.e. when the cross-ratios $[x, x', x'', \infty_D]$ and $[y, y', y'', y''']$, where y''' is uniquely defined by the condition

$$[y, y', y'', y'''] = \frac{\overrightarrow{x'x}}{\overrightarrow{x''x'}},$$

are the same (cf. 6.A or [B, 6.1.4, 6.2.5]).

We remark that, in fact, our proof is not valid in this case, since x''' is at infinity. To produce a rigorous proof, we can either pass to the limit or do it directly, and more simply, as follows: We have to prove that the orthogonal projection of y''' on D is a fixed point, i.e. that the scalar product $(u|rw + (1/c)v)$ (which is the formula for t''' at infinity) is a constant. But this scalar product is just

$$r\beta + \frac{1}{c}\gamma = \frac{1}{c}(rc\beta + \gamma),$$

which has just been proved to be constant. □

Note. In general the point y''' describes a subset only of the sphere S'''. This is clear, since S, S', S'' are bounded sets and yy''' is a constant (for example, the distance xy''' is bounded by $r + yy'''$). But when t''' tends towards infinity the radius of S''' also tends to infinity (think of the case of the plane above), unless ∞_D and ∞_E are preserved by the homography.

Index

A number followed by a letter represents a section in the text. A number followed by a number represents a problem; in some cases it is followed by an indication of solution only ("sol").

Problem Books in Mathematics

Series Editor: P.R. Halmos

Unsolved Problems in Intuitive Mathematics, Volume I: Unsolved Problems in Number Theory
by *Richard K. Guy*
1981. xviii, 161 pages. 17 illus.

Theorems and Problems in Functional Analysis
by *A.A. Kirillov* and *A.D. Gvishiani*
1982. ix, 347 pages. 6 illus.

Problems in Analysis
by *Bernard Gelbaum*
1982. vii, 228 pages. 9 illus.

A Problem Seminar
by *Donald J. Newman*
1982. viii, 113 pages.

Problem-Solving Through Problems
by *Loren C. Larson*
1983. xi, 344 pages. 104 illus.

Demography Through Problems
by *N. Keyfitz* and *J.A. Beekman*
1984. viii, 141 pages. 22 illus.

Problem Book for First Year Calculus
by *George W. Bluman*
1984. xvi, 384 pages. 384 illus.

Exercises in Integration
by *Claude George*
1984. x, 550 pages. 6 illus.

Exercises in Number Theory
by *D.P. Parent*
1984. x, 541 pages.

Problems in Geometry
by *Marcel Berger, Pierre Pansu, Jean-Pic Berry, and Xavier Saint-Raymond*
1984. viii, 266 pages. 244 illus.